Mary O. Haney
Harmony College
New Haven

ELEMENTS

OF

GEOMETRY:

CONTAINING

THE FIRST SIX BOOKS OF EUCLID,

WITH A

SUPPLEMENT

ON THE

QUADRATURE OF THE CIRCLE, AND THE GEOMETRY OF SOLIDS:

TO WHICH ARE ADDED

ELEMENTS OF PLANE AND SPHERICAL TRIGONOMETRY.

BY

JOHN PLAYFAIR, F.R.S. LOND. & EDIN.

PROFESSOR OF NATURAL PHILOSOPHY, FORMERLY OF MATHEMATICS, IN THE UNIVERSITY OF EDINBURGH.

FROM THE LAST LONDON EDITION, ENLARGED.

NEW-YORK;

PUBLISHED BY COLLINS & HANNAY, COLLINS & CO., O. A. ROORBACH, WHITE, GALLAHER & WHITE, AND G. & C. & H. CARVILL.

[illegible] PRINTER.

PREFACE.

It is a remarkable fact in the history of science, that the oldest book of Elementary Geometry is still considered as the best, and that the writings of EUCLID, at the distance of two thousand years, continue to form the most approved introduction to the mathematical sciences. This remarkable distinction the Greek Geometer owes not only to the elegance and correctness of his demonstrations, but to an arrangement most happily contrived for the purpose of instruction,—advantages which, when they reach a certain eminence, secure the works of an author against the injuries of time more effectually than even originality of invention. The elements of EUCLID, however, in passing through the hands of the ancient editors, during the decline of science, had suffered some diminution of their excellence, and much skill and learning have been employed by the modern mathematicians to deliver them from blemishes, which certainly did not enter into their original composition. Of these mathematicians, Dr. SIMson, as he may be accounted the last, has also been the most successful, and has left very little room for the ingenuity of future editors to be exercised in, either by amending the text of Euclid, or by improving the translations from it.

Such being the merits of Dr. Simson's edition, and the reception it has met with having been every way suitable, the work now offered to the public will perhaps appear unnecessary. And indeed, if the geometer just named had written with a view of accommodating the Elements of Euclid to the present state of the mathematical sciences, it is not likely that any thing new in Elementary Geometry would have been soon attempted. But his design was different: it was his object to restore the writings of Euclid to their original perfection, and to give them to Modern Europe as nearly as possible in the state wherein they made their first appearance in Ancient Greece. For this undertaking, nobody could be better qualified than Dr. Simson; who, to an accurate knowledge of the learned languages, and an indefatigable spirit of research, added a profound skill in the ancient Geometry, and an admiration of it almost enthusiastic. Accordingly, he not only restored the text of Euclid wherever it had been corrupted, but in some cases removed imperfections that probably belonged to the original work: though his extreme partiality for his author never permitted him to suppose, that such honour could fall to the share either of himself, or of any other of the moderns.

But, after all this was accomplished, something still remained to be done, since, notwithstanding the acknowledged excellence of EUCLID's Elements, it could not be doubted, that some alterations might be made, that would accomodate them better to a state of the mathematical sciences, so much more improved and extended than at the period when they were written. Accordingly, the object of the edition now offered to the public, is not so much to give to the writings of EUCLID the form which they originally had, as that which may at present render them most useful.

One of the alterations made with this view, respects the Doctrine of Proportion, the method of treating which, as it is laid down in the fifth of EUCLID, has great advantages, accompanied with considerable defects; of which, however, it must be observed, that the advantages are essential, and the defects only accidental. To explain the nature of the former, requires a more minute examination than is suited to this place, and must, therefore, be reserved for the Notes; but, in the mean time, it may be remarked, that no definition, except that of EUCLID, has ever been given, from which the properties of proportionals can be deduced by reasonings, which, at the same time that they are perfectly rigorous, are also simple and direct. As to the defects, the prolixness and obscurity, that have so often been complained of in the fifth Book, they seem to arise chiefly from the nature of the language employed, which being no other than that of ordinary discourse, cannot express, without much tediousness and circumlocution, the relations of mathematical quantities, when taken in their utmost generality, and when no assistance can be received from diagrams. As it is plain, that the concise language of Algebra is directly calculated to remedy this inconvenience, I have endeavoured to introduce it here, in a very simple form however, and without changing the nature of the reasoning, or departing in any thing from the rigour of geometrical demonstration. By this means, the steps of the reasoning which were before far separated, are brought near to one another, and the force of the whole is so clearly and directly perceived, that I am persuaded no more difficulty will be found in understanding the propositions of the fifth Book, than those of any other of the Elements.

In the second Book, also, some algebraic signs have been introduced, for the sake of representing more readily the addition and subtraction of the rectangles on which the demonstrations depend. The use of such symbolical writing, in translating from an original, where no symbols are used, cannot, I think, be regarded as an unwarrantable liberty: for, if by that means the translation is not made into English, it is made into that universal language so much sought after in all the sciences, but destined, it would seem, to be enjoyed only by the mathematical.

The alterations above mentioned are the most material that have been attempted on the books of EUCLID. There are, however, a few others, which, though less considerable, it is hoped may in some degree facilitate the study of the Elements. Such are those made on the definitions in the first Book, and particularly on that of a straight line. A new axiom is also introduced in the room of the 12th, for the purpose of demonstrating more easily some of the properties of parallel lines. In the third Book, the remarks concerning the angles made by a straight line, and the circumference of a circle, are left out, as tending to perplex one who has advanced no farther than the elements of the science. The

27th, 28th, and 29th of the sixth are changed for easier and more simple propositions, which do not materially differ from them, and which answer exactly the same purpose. Some propositions also have been added; but, for a fuller detail concerning these changes, I must refer to the Notes, in which several of the more difficult, or more interesting subjects of Elementary Geometry are treated at considerable length.

The SUPPLEMENT now added to the Six Books of EUCLID is arranged differently from what it was in the first edition of these Elements.

The First of the three Books, into which it is divided, treats of the rectification and quadrature of the circle,—subjects that are often omitted altogether in works of this kind. They are omitted, however, as I conceive, without any good reason, because, to measure the length of the simplest of all the curves which Geometry treats of, and the space contained within it, are problems that certainly belong to the elements of the science, especially as they are not more difficult than other propositions which are usually admitted into them. When I speak of the rectification of the circle, or of measuring the length of the circumference, I must not be supposed to mean, that a straight line is to be made equal to the circumference *exactly*—a problem which, as is well known, Geometry has never been able to resolve. All that is proposed is, to determine two straight lines that shall differ very little from one another, not more, for instance, than the four hundred and ninety-seventh part of the diameter of the circle, and of which the one shall be greater than the circumference of that circle, and the other less. In the same manner, the quadrature of the circle is performed only by approximation, or by finding two rectangles nearly equal to one another, one of them greater, and another less, than the space contained within the circle.

In the second Book of the Supplement, which treats of the intersection of Planes, I have departed as little as possible from Euclid's method of considering the same subject in his eleventh Book. The demonstration of the fourth proposition is from Legendre's Elements of Geometry; that of the sixth is new, as far as I know; as is also the solution of the problem in the ninteenth proposition,—a problem which, though in itself extemely simple, has been omitted by Euclid, and hardly ever treated of, in an elementary form, by any geometer.

With respect to the Geometry of Solids, in the

third Book, I have departed from EUCLID altogether, with a view of rendering it both shorter and more comprehensive. This, however, is not attempted by introducing a mode of reasoning less rigorous than that of the Greek Geometer; for this would be to pay too dear even for the time that might thereby be saved; but it is done chiefly by laying aside a certain rule, which, though it be not essential to the accuracy of demonstration, EUCLID has thought it proper, as much as possible, to observe.

The rule referred to, is one which influences the arrangement of his propositions through the whole of the Elements, viz. That in the demonstration of a theorem, he never supposes any thing to be done, as any line to be drawn, or any figure to be constructed, the manner of doing which he has not previously explained. Now, the only use of this rule is to prevent the admission of impossible or contradictory suppositions, which, no doubt, might lead into error; and it is a rule well calculated to answer that end, as it does not allow the existence of any thing to be supposed, unless the thing itself be actually exhibited. But it is not always necessary to make use of this defence; for the existence of many things is obviously possible, and very far from implying a contradiction, where the method of actually exhibiting them may be altogether unknown. Thus, it is plain, that on any given figure as a base, a solid may be constituted, or conceived to exist, equal in solid contents to any given solid, (because a solid, whatever be its base, as its height may be indefinitely varied, is capable of all degrees of magnitude, from nothing upwards), and yet it may in many cases be a problem of extreme difficulty to assign the height of such a solid, and actually to exhibit it. Now, this very supposition, that on a given base a solid of a given magnitude may be constituted, is one of those, by the introduction of which, the Geometry of solids is much shortened, while all the real accuracy of the demonstrations is preserved; and therefore, to follow, as Euclid has done, the rule that excludes this, and such like hypotheses, is to create artificial difficulties, and to embarrass geometrical investigation with more obstacles than the nature of things has thrown in its way. It is a rule, too, which cannot always be followed, and from which even Euclid himself has been forced to depart, in more than one instance.

In the Book, therefore, on the Properties of Solids, which I now offer to the public, I have not sought to subject the demonstrations to the law just mentioned, and have never hesitated to admit the existence of such solids, or such lines as are evidently possible, though the manner of actually describing them may not have been explained. In this way, I have been enabled to offer that very refined artifice in geometrical reasoning, to which we give the name of the Method of Exhaustions, under a much simpler form than it appears in the 12th of EUCLID; and the spirit of the method may, I think, be best learned when it is thus disengaged from every thing not essential. That it may be the better understood, and because the demonstrations which require exhaustions are, no doubt, the most difficult in the Elements, they are all conducted as nearly as possible in the same way, in the cases of the different solids, from the pyramid to the sphere. The comparison of this last solid with the cylinder, concludes the last Book of the Supplement, and is a proposition that may not improperly be considered as terminating the elementary part of Geometry.

The Book of the Data has been annexed to several editions of EUCLID's Elements, and particularly to Dr. SIMSON's, but in this it is omitted altogether. It is omitted, however, not from any opinion of its being in itself useless, but because it does not belong to this place, and is not often read by beginners. It contains the rudiments of what is properly called the Geometrical Analysis, and has itself an analytical form; and for these reasons, I would willingly reserve it, or rather a compend of it, for a separate work, intended as an introduction to the study of that analysis.

In explaining the elements of Plane and Spherical Trigonometry, there is not much new that can be attempted, or that will be expected by the intelligent reader. Except, perhaps, some new demonstrations, and some changes in the arrangement, these two treatises have, accordingly, no novelty to boast of. The Plane Trigonometry is so divided, that the part of it that is barely sufficient for the resolution of Triangles may be easily taught by itself. The method of constructing the Trigonometrical Tables is explained, and a demonstration is added of those properties of the sines and cosines of arches, which are the foundation of those applications of Trigonometry lately introduced, with so much advantage, into the higher Geometry.

In the Spherical Trigonometry, the rules for preventing the ambiguity of the solutions, wherever it can be prevented, have been particularly attended to; and I have availed myself as much as possible of that excellent abstract of the rules of this science which Dr. MASKELYNE has prefixed to TAYLOR'S Tables of Logarithms.

An explanation of Napier's very ingenious and useful rule of the *Circular Parts* is here added as an Appendix to Spherical Trigonometry.

It has been objected to many of the writers on Elementary Geometry, and particularly to Euclid, that they have been at great pains to prove the truth of many simple propositions, which every body is ready to admit, without any demonstration, and that thus they take up the time, and fatigue the attention of the student, to no purpose. To this objection, if there be any force in it, the present treatise is certainly as much exposed as any other; for no attempt is here made to abridge the Elements, by considering as self-evident any thing that admits of being proved. Indeed, those who make the objection just stated, do not seem to have reflected sufficiently on the end of Mathematical Demonstration, which is not only to prove the truth of a certain proposition, but to shew its necessary connection with other propositions, and its dependance on them. The truths of Geometry are all necessarily connected with one another, and the system of such truths can never be rightly explained, unless that connection be accurately traced, wherever it exists. It is upon this that the beauty and peculiar excellence of the mathematical sciences depend: it is this, which, by preventing any one truth from being single and insulated, connects the different parts so firmly, that they must all stand, or all fall together. The demonstration, therefore, even of an obvious proposition, answers the purpose of connecting that proposition with others, and ascertaining its place in the general system of mathematical truth. If, for example, it be alleged, that it is needless to demonstrate that any two sides of a triangle are greater than the third; it may be replied, that this is no doubt a truth which, without proof, most men will be inclined to admit; but are we for that reason to account it of no consequence to know what the propositions are, which must cease to be true if this proposition were supposed to be false? Is it not useful to know, that unless it be true, that any two sides of a triangle are greater than the third, neither could it be true, that the greater side of every triangle is opposite to the greater angle, nor that the equal sides are opposite to equal angles, nor lastly, that things equal to the same thing are equal to one another? By a scientific mind this information will not be thought lightly of; and it is exactly that which we receive from Euclid's demonstration.

To all this it may be added, that the mind, especially when beginning to study the art of reasoning, cannot be employed to greater advantage than in analysing those judgments, which, though they appear simple, are in reality complex, and capable of being distinguished into parts. No progress in ascending higher can be expected, till a regular habit of demonstration is thus acquired; it is much to be feared, that he who has declined the trouble of tracing the connection between the proposition already quoted, and those that are more simple, will not be very expert in tracing its connection with those that are more complex; and that, as he has not been careful in laying the foundation, he will never be successful in raising the superstructure.

College of Edinburgh,

Dec. 1, 1813.

ELEMENTS

OF

GEOMETRY.

BOOK I.

DEFINITIONS.

I.

" A Point is that which has position, but not magnitude*." (See Notes.)

II.

A line is length without breadth.

" Corollary. The extremities of a line are points; and the inter-
" sections of one line with another are also points."

III.

" If two lines are such that they cannot coincide in any two points,
" without coinciding altgether, each of them is called a straight
" line."

" Cor. Hence two straight lines cannot inclose a space. Neither can
" two straight lines have a common segment; that is, they cannot
" coincide in part, without coinciding altogether."

IV.

A superficies is that which has only length and breadth.

" Cor. The extremities of a superficies are lines; and the intersec-
" tions of one superficies with another are also lines."

V.

A plane superficies is that in which any two points being taken, the straight line between them lies wholly in that superficies.

VI.

A plane rectilineal angle is the inclination of two straight lines to one another, which meet together, but are not in the same straight line.

* The definitions marked with inverted commas are different from those of Euclid.

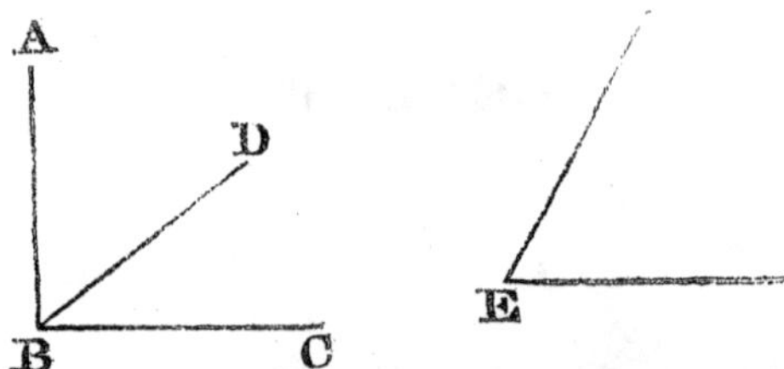

N. B. ' When several angles are at one point B, any one of them ' is expressed by three letters, of which the letter that is at the ver- ' tex of the angle, that is, at the point in which the straight lines that ' contain the angle meet one another, is put between the other two ' letters, and one of these two is somewhere upon one of those straight ' lines, and the other upon the other line: Thus the angle which ' is contained by the straight lines, AB, CB, is named the angle ' ABC, or CBA; that which is contained by AB, BD, is named the ' angle ABD, or DBA; and that which is contained by BD, CB, is ' called the angle DBC, or CBD; but, if there be only one angle at ' a point, it may be expressed by a letter placed at that point; as the ' angle at E.'

VII.

When* a straight line standing on another straight line makes the adjacent angles equal to one another, each of the angles is called a right angle; and the straight line which stands on the other is called a perpendicular to it.

VIII.

An obtuse angle is that which is greater than a right angle.

IX.

An acute angle is that which is less than a right angle.

X.

A figure is that which is inclosed by one or more boundaries.—*The word area denotes the quantity of space contained in a figure, without any reference to the nature of the line or lines which bound it.*

XI.

A circle is a plane figure contained by one line, which is called the circumference, and is such that all straight lines drawn from a certain point within the figure to the circumference, are equal to one another.

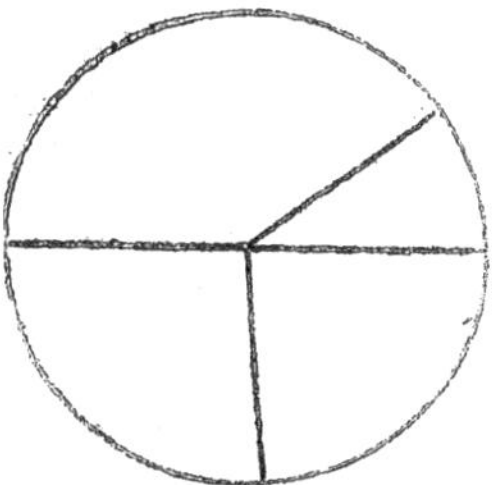

XII.

And this point is called the centre of the circle.

XIII.

A diameter of a circle is a straight line drawn through the centre, and terminated both ways by the circumference.

XIV.

A semicircle is the figure contained by a diameter and the part of the circumference cut off by the diameter.

XV.

Rectilineal figures are those which are contained by straight lines.

XVI.

Trilateral figures, or triangles, by three straight lines

XVII.

Quadrilateral, by four straight lines.

XVIII.

Multilateral figures, or polygons, by more than four straight lines.

XIX.

Of three sided figures, an equilateral triangle is that which has three equal sides.

XX.

An isosceles triangle is that which has only two sides equal.

XXI.

A scalene triangle, is that which has three unequal sides.

XXII.

A right angled triangle is that which has a right angle.

XXIII.

An obtuse angled triangle, is that which has an obtuse angle.

XXIV.

An acute angled triangle, is that which has three acute angles.

XXV.

Of four sided figures, a square is that which has all its sides equal, and all its angles right angles.

XXVI.

An oblong, is that which has all its angles right angles, but has not all its sides equal.

XXVII.

A rhombus, is that which has all its sides equal, but its angles are not right angles.

XXVIII.

A rhomboid, is that which has its opposite sides equal to one another, but all its sides are not equal, nor its angles right angles.

XXIX.

All other four sided figures besides these, are called Trapeziums.

XXX.

Parallel straight lines, are such as are in the same plane, and which, being produced ever so far both ways, do not meet.

POSTULATES.

I.

Let it be granted that a straight line may be drawn from any one point to any other point.

II.

That a terminated straight line may be produced to any length in a straight line.

III.

And that a circle may be described from any centre, at any distance from that centre.

AXIOMS.

I.

Things which are equal to the same thing are equal to one another.

II.

If equals be added to equals, the wholes are equal.

III.

If equals be taken from equals, the remainders are equal.

IV.

If equals be added to unequals, the wholes are unequal.

V.

If equals be taken from unequals, the remainders are unequal.

VI.

Things which are doubles of the same thing, are equal to one another.

VII.

Thihgs which are halves of the same thing, are equal to one another.

VIII.

Magnitudes which coincide with one another, that is, which exactly fill the same space, are equal to one another.

IX.

The whole is greater than its part.

X.

All right angles are equal to one another.

XI.

"Two straight lines which intersect one another, cannot be both parallel to the same straight line."

PROPOSITION I. PROBLEM.

To describe an equilateral triangle upon a given finite straight line.

Let AB be the given straight line; it is required to describe an equilateral triangle upon it.

From the centre A, at the distance AB, describe (3. Postulate) the circle BCD, and from the centre B, at the distance BA, describe the circle ACE; and from the point C, in which the circles cut one another, draw the straight lines (1. Post.) CA, CB to the points A, B; ABC is an equilateral triangle.

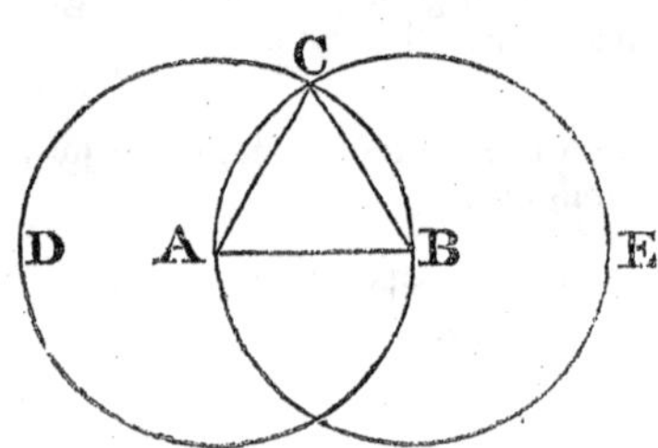

Because the point A is the centre of the circle BCD, AC is equal (11. Definition) to AB; and because the point B is the centre of the circle ACE, BC is equal to AB: But it has been proved that CA is equal to AB; therefore CA, CB are each of them equal to AB; now things which are equal to the same are equal to one another, (1. Axiom); therefore CA is equal to CB; wherefore CA, AB, CB are equal to one another; and the triangle ABC is therefore equilateral, and it is described upon the given straight line AB. Which was required to be done.

PROP. II. PROB.

From a given point to draw a straight line equal to a given straight line.

Let A be the given point, and BC the given straight line; it is required to draw, from the point A, a straight line equal to BC.

From the point A to B draw (1. Post.) the straight line AB; and upon it describe (1. 1.) the equilateral triangle DAB, and produce (2. Post.) the straight lines DA, BD, to E and F; from the centre B, at the distance BC, describe (3. Post.) the circle CGH, and from the centre D, at the distance DG, describe the circle GKL. AL is equal to BC.

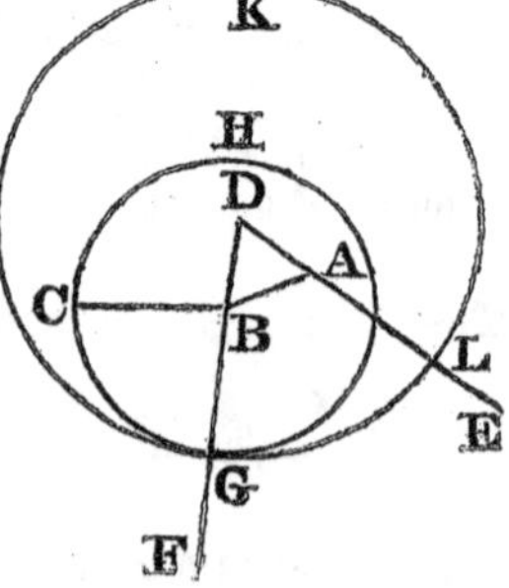

Because the point B is the centre of the circle CGH, BC is equal (11. Def.) to BG; and because D is the centre of the circle GKL, DL is equal to DG, and DA, DB, parts of them, are equal; therefore the remainder AL is equal to the remainder (3. Ax.) BG: But it has been shewn that BC is equal to BG; wherefore AL and BC are each of them equal to BG; and things that are equal to the same are equal to one another; therefore the straight line AL is equal to BC. Wherefore, from the given point A, a straight line AL has been drawn equal to the given straight line BC. Which was to be done.

PROP. III. PROB.

From the greater of two given straight lines to cut off a part equal to the less.

Let AB and C be the two given straight lines, whereof AB is the greater. It is required to cut off from AB, the greater, a part equal to C, the less.

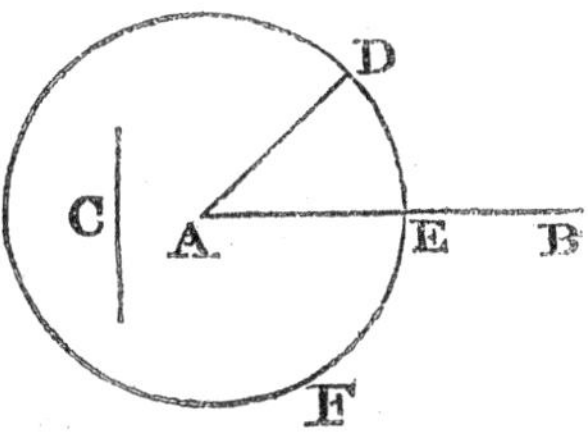

From the point A draw (2. 1.) the straight line AD equal to C; and from the centre A, and at the distance AD, describe (3. Post.) the circle DEF; and because A is the centre of the circle DEF, AE is equal to AD; but the straight line C is likewise equal to AD; whence AE and C are each of them equal to AD; wherefore the straight line AE is equal to (1. Ax.) C, and from AB the greater of two straight lines, a part AE has been cut off equal to C the less. Which was to be done.

PROP. IV. THEOREM.

If two triangles have two sides of the one equal to two sides of the other, each to each; and have likewise the angles contained by those sides equal to one another, their bases, or third sides, *shall be equal; and the areas of the triangles shall be equal; and their other angles shall be equal, each to each, viz. those to which the equal sides are opposite.**

Let ABC, DEF be two triangles which have the two sides AB, AC equal to the two sides DE, DF, each to each, viz. AB to DE, and AC to DF; and let the angle BAC be also equal to the angle EDF: then shall the base BC be equal to the base EF; and the triangle ABC to the triangle DEF; and the other angles, to which the equal sides are op-

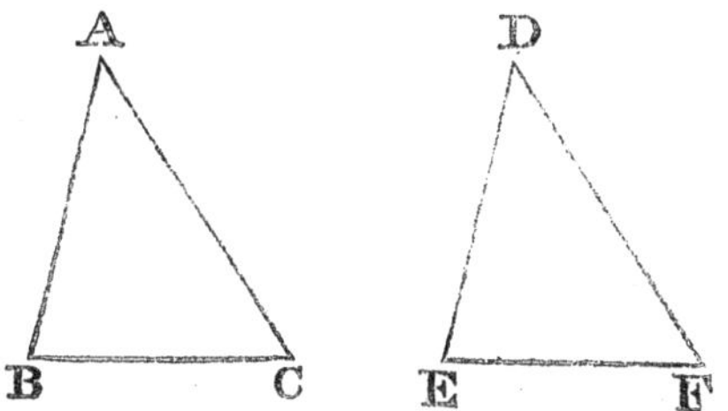

* The three conclusions in this enunciation are more briefly expressed by saying, that the triangles are every way equal.

posite, shall be equal, each to each, viz. the angle ABC to the angle DEF, and the angle ACB to DFE.

For, if the triangle ABC be applied to the triangle DEF, so that the point A may be on D, and the straight line AB upon DE; the point B shall coincide with the point E, because AB is equal to DE; and AB coinciding with DE, AC shall coincide with DF, because the angle BAC is equal to the angle EDF; wherefore also the point C shall coincide with the point F, because AC is equal to DF: But the point B coincides with the point E; wherefore the base BC shall coincide with the base EF (cor def. 3.), and shall be equal to it. Therefore also the whole triangle ABC shall coincide with the whole triangle DEF, so that the spaces which they contain or their areas are equal; and the remaining angles of the one shall coincide with the remaining angles of the other, and be equal to them, viz. the angle ABC to the angle DEF, and the angle ACB to the angle DFE. Therefore, if two triangles have two sides of the one equal to two sides of the other, each to each, and have likewise the angles contained by those sides equal to one another; their bases shall be equal, and their areas shall be equal, and their other angles, to which the equal sides are opposite, shall be equal, each to each. Which was to be demonstrated.

PROP. V. THEOR.

The angles at the base of an Isosceles triangle are equal to one another; and if the equal sides be produced, the angles upon the other side of the base shall also be equal.

Let ABC be an isosceles triangle, of which the side AB is equal to AC, and let the straight lines AB, AC be produced to D and E, the angle ABC shall be equal to the angle ACB, and the angle CBD to the angle BCE.

In BD take any point F, and from AE the greater cut off AG equal (3. 1.) to AF, the less, and join FC, GB.

Because AF is equal to AG, and to AC, the two sides FA, AC are equal to the two GA, AB, each to each; and they contain the angle FAG common to the two triangles, AFC, AGB; therefore the base FC is equal (4. 1.) to the base GB, and the triangle AFC to the triangle AGB; and the remaining angles of the one are equal (4. 1.) to the remaining angles of the other, each to each, to which the equal sides are opposite, viz. the angle ACF to the angle ABG, and the angle AFC to the angle AGB: And because the whole AF is equal to the whole AG, and the part AB to the part AC; the remainder BF shall be equal (3. Ax.) to the

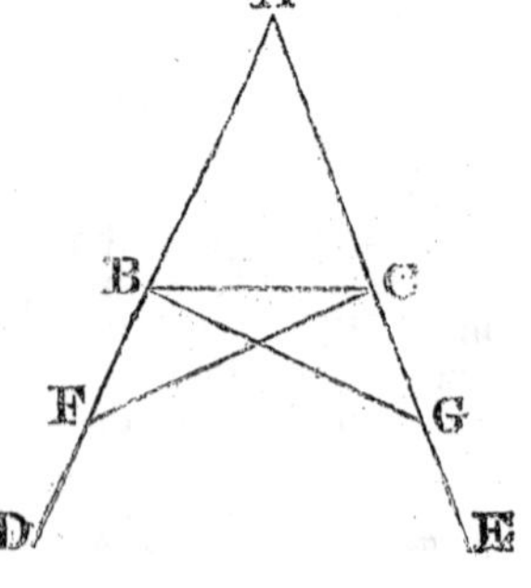

remainder CG ; and FC was proved to be equal to GB, therefore the two sides BF, FC are equal to the two CG, GB, each to each ; but the angle BFC is equal to the angle CGB ; wherefore the triangles BFC, CGB are equal (3. .), and their remaining angles are equal, to which the equal sides are opposite; therefore the angle FBC is equal to the angle GCB, and the angle BCF to the angle CBG. Now, since it has been demonstrated, that the whole angle ABG is equal to the whole ACF, and the part CBG to the part BCF, the remaining angle ABC is therefore equal to the remaining angle ACB, which are the angles at the base of the triangle ABC : And it has also been proved that the angle FBC is equal to the angle GCB, which are the angles upon the other side of the base. Therefore, the angles at the base, &c. Q. E. D.

COROLLARY. Hence every equilateral triangle is also equiangular.

PROP. VI. THEOR.

If two angles of a triangle be equal to one another, the sides which subtend, or are opposite to them, are also equal to one another.

Let ABC be a triangle having the angle ABC equal to the angle ACB ; the side AB is also equal to the side AC.

For, if AB be not equal to AC, one of them is greater than the other : Let AB be the greater, and from it cut (3. 1.) off DB equal to AC the less, and join DC ; therefore, because in the triangles DBC, ACB, DB is equal to AC, and BC common to both, the two sides DB, BC are equal to the two AC, CB, each to each ; but the angle DBC is also equal to the angle ACB ; therefore the base DC is equal to the base AB, and the area of the triangle DBC is equal to that of the triangle (4. 1.) ACB, the less to the greater ; which is absurd. Therefore, AB is not unequal to AC, that is, it is equal to it. Wherefore, if two angles, &c. Q. E. D.

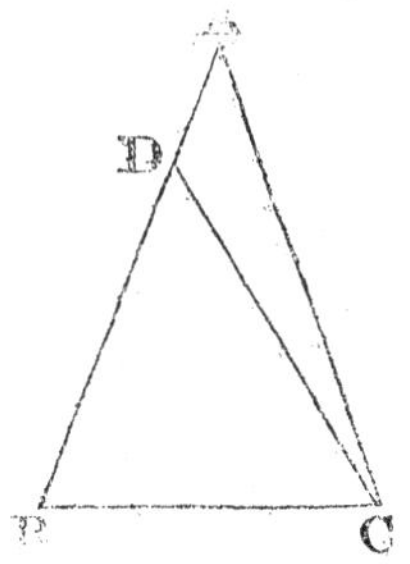

COR. Hence every equiangular triangle is also equilateral.

PROP. VII. THEOR.

Upon the same base, and on the same side of it, there cannot be two triangles, that have their sides which are terminated in one extremity of the base equal to one another, and likewise those which are terminated in the other extremity, equal to one another.

Let there be two triangles ACB, ADB, upon the same base AB, and upon the same side of it, which have their sides CA, DA, terminated in A equal to one another : then their sides, CB, DB, terminated in

B, cannot be equal to one another.

Join CD, and if possible let CB be equal to DB; then, in the case in which the vertex of each of the triangles is without the other triangle, because AC is equal to AD, the angle ACD is equal (5. 1.) to the angle ADC: But the angle ACD is greater than the angle BCD; therefore the angle ADC is greater also than BCD; much more then is the angle BDC greater than the angle BCD. Again, because CB is equal to DB, the angle BDC is equal (5. 1.) to the angle BCD; but it has been demonstrated to be greater than it; which is impossible.

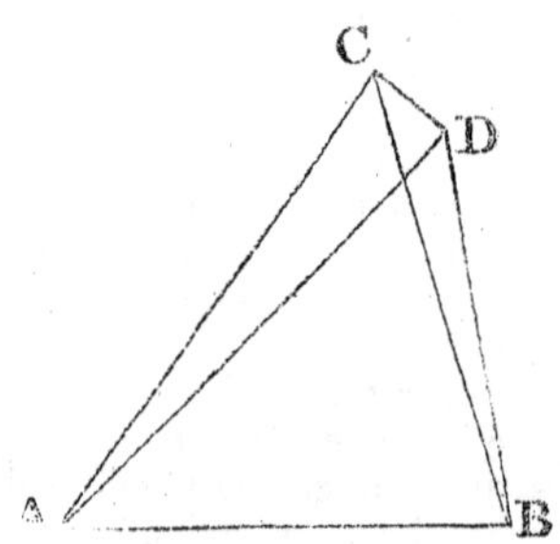

But if one of the vertices, as D, be within the other triangle ACB; produce AC, AD to E, F; therefore, because AC is equal to AD in the triangle ACD, the angles ECD, FDC upon the other side of the base CD are equal (5. 1.) to one another, but the angle ECD is greater than the angle BCD; wherefore the angle FDC is likewise greater than BCD; much more then is the angle BDC greater than the angle BCD. Again, because CB is equal to DB, the angle BDC is equal (5. 1.) to the angle BCD; but BDC has been proved to be greater than the same BCD; which is impossible. The case in which the vertex of one triangle is upon a side of the other, needs no demonstration.

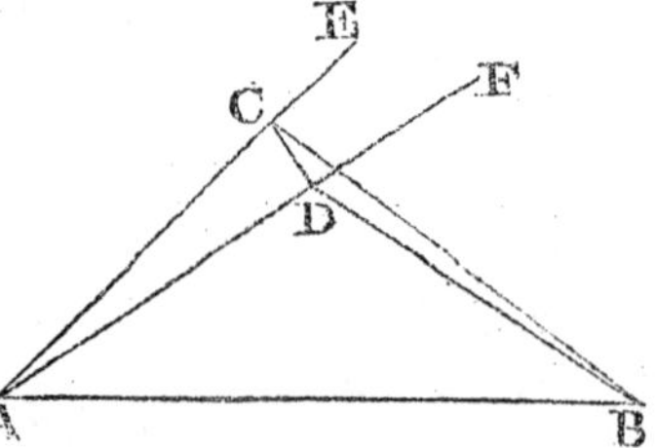

Therefore, upon the same base, and on the same side of it, there cannot be two triangles that have their sides which are terminated in one extremity of the base equal to one another, and likewise those which are terminated in the other extremity equal to one another. Q. E. D.

PROP. VIII. THEOR.

If two triangles have two sides of the one equal to two sides of the other, each to each, and have likewise their bases equal; the angle which is contained by the two sides of the one shall be equal to the angle contained by the two sides of the other.

Let ABC, DEF be two triangles having the two sides AB, AC, equal to the two sides DE, DF, each to each, viz. AB to DE, and AC

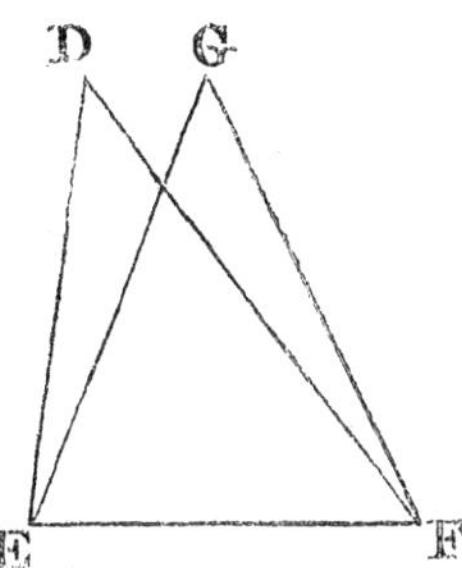

to DF; and also the base BC equal to the base EF. The angle BAC is equal to the angle EDF.

For, if the triangle ABC be applied to the triangle DEF, so that the point B be on E, and the straight line BC upon EF; the point C shall also coincide with the point F, because BC is equal to EF: therefore BC coinciding with EF, BA and AC shall coincide with ED, and DF; for, if BA, and CA do not coincide with ED, and FD, but have a different situation as EG and FG; then, upon the same base EF, and upon the same side of it, there can be two triangles EDF, EGF, that have their sides which are terminated in one extremity of the base equal to one another, and likewise their sides terminated in the other extremity; but this is impossible (7. 1.); therefore, if the base BC coincides with the base EF, the sides BA, AC cannot but coincide with the sides ED, DF; wherefore likewise the angle BAC coincides with the angle EDF, and is equal (8. Ax.) to it. Therefore if two triangles, &c. Q. E. D.

PROP. IX. PROB.

To bisect a given rectilineal angle, that is, to divide it into two equal angles.

Let BAC be the given rectilineal angle, it is required to bisect it.

Take any point D in AB, and from AC cut (3. 1.) off AE equal to AD; join DE, and upon it describe (1. 1.) an equilateral triangle DEF; then join AF; the straight line AF bisects the angle BAC.

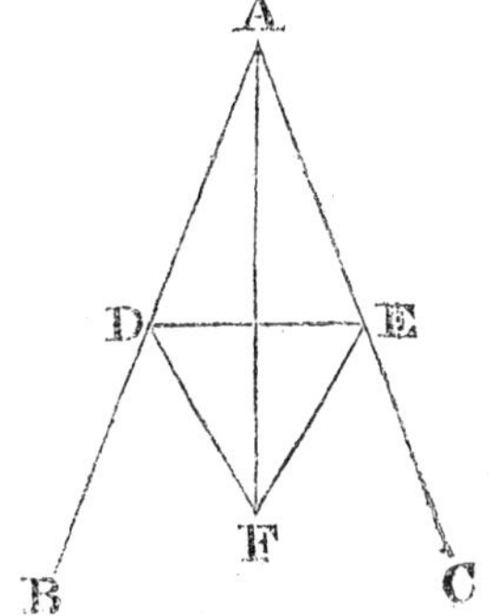

Because AD is equal to AE, and AF is common to the two triangles DAF, EAF; the two sides DA, AF, are equal to the two sides EA, AF, each to each; but the base DF is also equal to the base EF; therefore the angle DAF is equal (8. 1.) to the angle EAF: wherefore the given rectilineal angle BAC is bisected by the straight line AF. Which was to be done.

PROP. X. PROB.

To bisect a given finite straight line, that is, to divide it into two equal parts.

Let AB be the given straight line; it is required to divide it into two equal parts.

Describe (1. 1.) upon it an equilateral triangle ABC, and bisect (9. 1.) the angle ACB by the straight line CD. AB is cut into two equal parts in the point D.

Because AC is equal to CB, and CD common to the two triangles ACD, BCD: the two sides AC, CD, are equal to the two BC, CD, each to each; but the angle ACD is also equal to the angle BCD; therefore the base AD is equal to the base (4. 1.) DB, and the straight line AB is divided into two equal parts in the point D. Which was to be done.

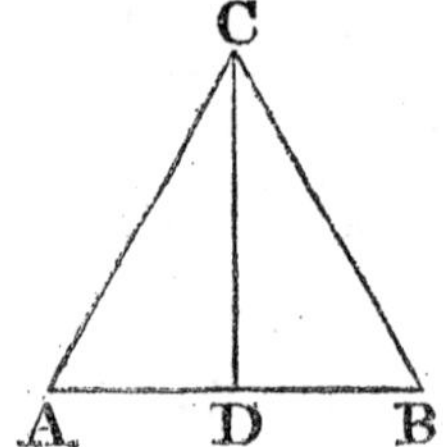

PROP. XI. PROB.

To draw a straight line at right angles to a given straight line, from a given point in that line.

Let AB be a given straight line, and C a point given in it; it is required to draw a straight line from the point C at right angles to AB.

Take any point D in AC, and (3. 1.) make CE equal to CD, and upon DE describe (1. 1.) the equilateral triangle DFE, and join FC; the straight line FC, drawn from the given point C, is at right angles to the given straight line AB.

Because DC is equal to CE, and FC common to the two triangles DCF, ECF, the two sides DC, CF are equal to the two EC, CF, each to each; but the base DF is also equal to the base EF; therefore the angle DCF is equal (8. 1.) to the angle ECF; and they are adjacent angles. But, when the adjacent angles which one straight line makes with another straight line are equal to one another, each of them is called a right (7. def.) angle; therefore each of the angles DCF, ECF, is a right angle. Wherefore, from the given point C, in the given straight line AB, FC has been drawn at right angles to AB. Which was to be done.

PROP. XII. PROB.

To draw a straight line perpendicular to a given straight line, of an unlimited length, from a given point without it.

Let AB be a given straight line, which may be produced to any length both ways, and let C be a point without it. It is required to draw a straight line perpendicular to AB from the point C.

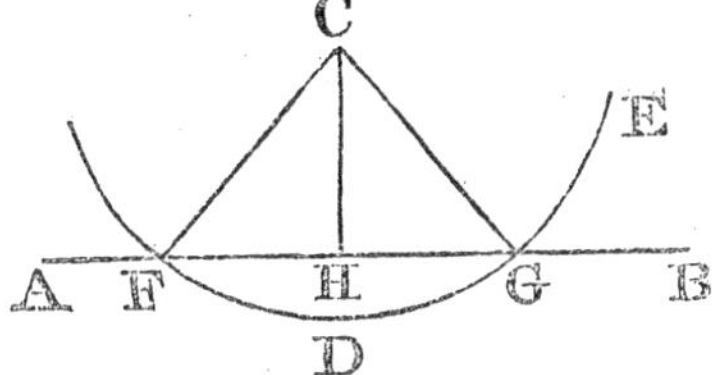

Take any point D upon the other side of AB, and from the centre C, at the distance CD, describe (3. Post.) the circle EGF meeting AB in F, G: and bisect (10. 1.) FG in H, and join CF, CH, CG; the straight line CH, drawn from the given point C, is perpendicular to the given straight line AB.

Because FH is equal to HG, and HC common to the two triangles FHC, GHC, the two sides FH, HC are equal to the two GH, HC, each to each; but the base CF is also equal (11. Def. 1.) to the base CG; therefore the angle CHF is equal (8. 1.) to the angle CHG; and they are adjacent angles; now when a straight line standing on a straight line makes the adjacent angles equal to one another, each of them is a right angle, and the straight line which stands upon the other is called a perpendicular to it; therefore from the given point C a perpendicular CH has been drawn to the given straight line AB. Which was to be done.

PROP. XIII. THEOR.

The angles which one straight line makes with another upon one side of it, are either two right angles, or are together equal to two right angles.

Let the straight line AB make with CD, upon one side of it the angles CBA, ABD; these are either two right angles, or are together equal to two right angles.

For, if the angle CBA be equal to ABD, each of them is a right angle (Def. 7.); but, if not, from the point B draw BE at right an-

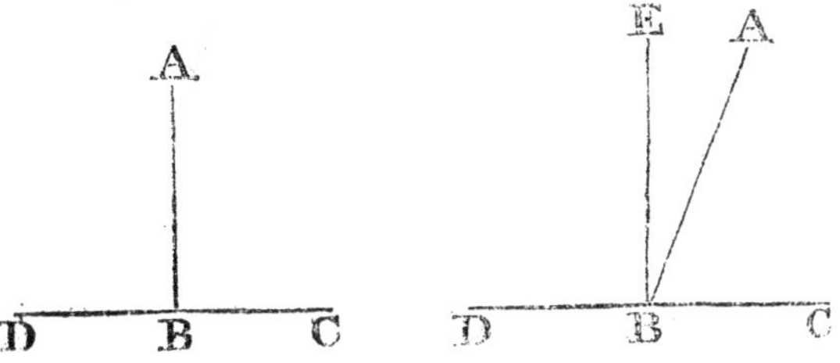

gles (11. 1.) to CD; therefore the angles CBE, EBD are two right angles. Now, the angle CBE is equal to the two angles CBA, ABE together; add the angle EBD to each of these equals, and the two angles CBE, EBD will be equal (2. Ax.) to the three CBA, ABE, EBD. Again, the angle DBA is equal to the two angles DBE, EBA; add to each of these equals the angle ABC; then will the two angles DBA, ABC be equal to the three angles DBE, EBA, ABC; but the angles CBE, EBD have been demonstrated to be equal to the same three angles; and things that are equal to the same are equal (1. Ax.) to one another; therefore the angles CBE, EBD are equal to the angles DBA, ABC; but CBE, EBD, are two right angles; therefore DBA, ABC are together equal to two right angles. Wherefore when a straight line, &c. Q. E. D.

PROP. XIV. THEOR.

If, at a point in a straight line, two other straight lines, upon the opposite sides of it, make the adjacent angles together equal to two right angles, these two straight lines are in one and the same straight line.

At the point B in the straight line AB, let the two straight lines BC, BD upon the opposite sides of AB, make the adjacent angles ABC, ABD equal together to two right angles. BD is in the same straight line with CB.

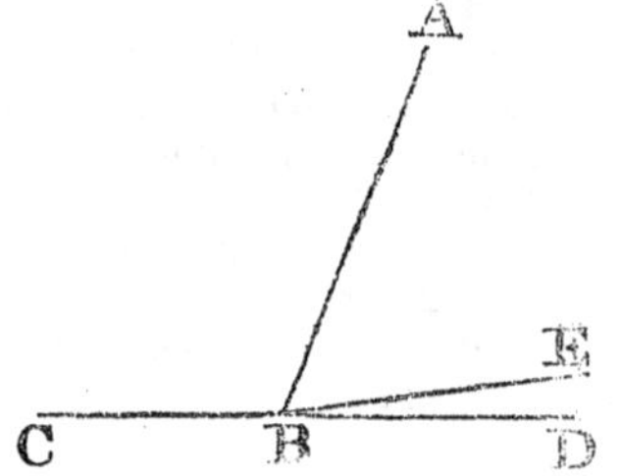

For if BD be not in the same straight line with CB, let BE be in the same straight line with it; therefore, because the straight line AB makes angles with the straight line CBE, upon one side of it, the angles ABC, ABE are together equal (13. 1.) to two right angles; but the angles ABC, ABD are likewise together equal to two right angles: therefore the angles CBA, ABE are equal to the angles CBA, ABD: Take away the common angle ABC, and the remaining angle ABE is equal (3. Ax.) to the remaining angle ABD, the less to the greater, which is impossible; therefore BE is not in the same straight line with BC. And in like manner, it may be demonstrated, that no other can be in the same straight line with it but BD, which therefore is in the same straight line with CB. Wherefore, if at a point, &c. Q. E. D.

PROP. XV. THEOR.

If two straight lines cut one another, the vertical, or opposite angles are equal.

Let the two straight lines AB, CD cut one another in the point E: the angle AEC shall be equal to the angle DEB, and CEB to AED.

For the angles CEA, AED, which the straight line AE makes with the straight line CD, are together equal (13. 1.) to two right angles; and the angles AED, DEB, which the straight line DE makes with the straight line AB, are also together equal (13. 1.) to two right angles; therefore the two angles CEA, AED are equal to the two AED, DEB. Take away the common angle AED, and the remaining angle CEA is equal (3. Ax.) to the remaining angle DEB. In the same manner it may be demonstrated that the angles CEB, AED are equal. Therefore, if two straight lines, &c. Q. E. D.

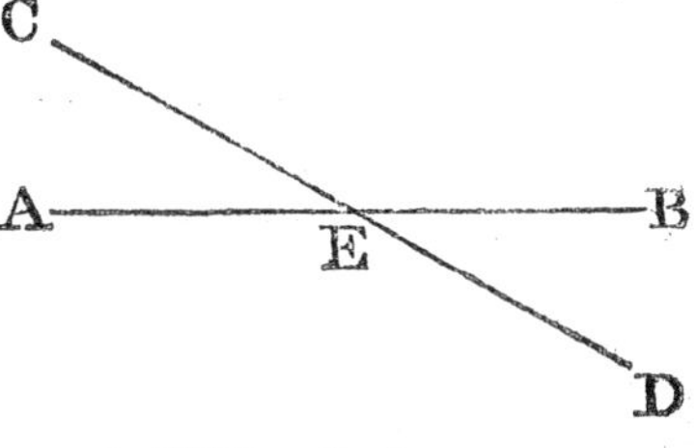

COR. 1. From this it is manifest, that if two straight lines cut one another, the angles which they make at the point of their intersection, are together equal to four right angles.

COR. 2. And hence, all the angles made by any number of straight lines meeting in one point, are together equal to four right angles.

PROP. XVI. THEOR.

If one side of a triangle be produced, the exterior angle is greater than either of the interior, and opposite angles.

Let ABC be a triangle, and let its side BC be produced to D, the exterior angle ACD is greater than either of the interior opposite angles CBA, BAC.

Bisect (10. 1.) AC in E, join BE and produce it to F, and make EF equal to BE; join also FC, and produce AC to G.

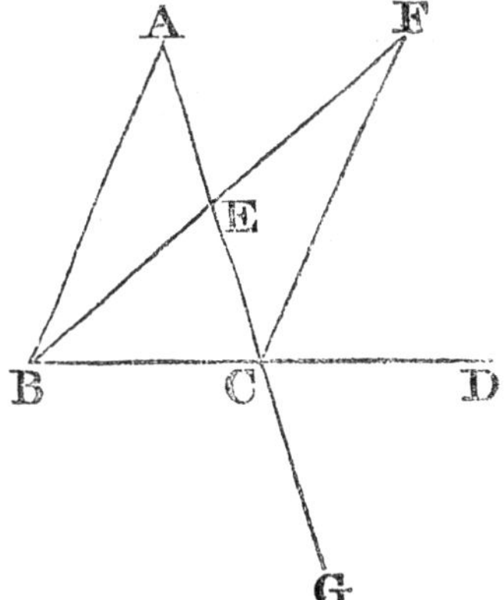

Because AE is equal to EC, and BE to EF; AE, EB are equal to CE, EF, each to each; and the angle AEB is equal (15. 1.) to the angle CEF, because they are vertical angles; therefore the base AB is equal (4. 1.) to the base CF, and the triangle AEB to the triangle CEF, and the remaining angles to the remaining angles each to each, to which the equal sides are opposite; wherefore the angle BAE is equal to the angle ECF; but the angle ECD is greater than the angle ECF; therefore the angle ECD, that is ACD, is greater than BAE: In the same manner, if the side

BC be bisected, it may be demonstrated that the angle BCG, that is (15. 1.), the angle ACD, is greater than the angle ABC. Therefore if one side, &c. Q. E. D.

PROP. XVII. THEOR.

Any two angles of a triangle are together less than two right angles.

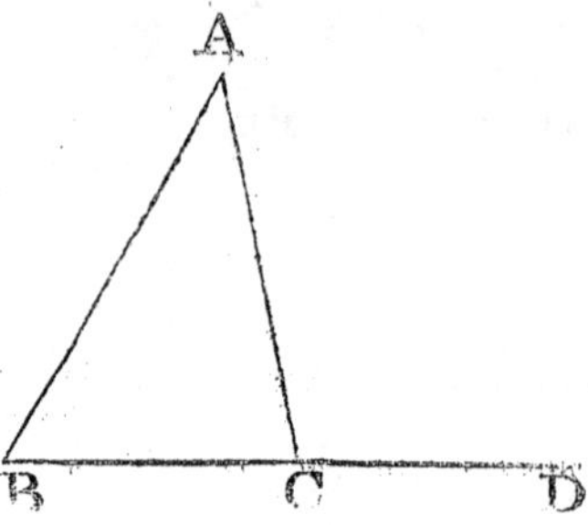

Let ABC be any triangle; any two of its angles together are less than two right angles.

Produce BC to D; and because ACD is the exterior angle of the triangle ABC, ACD is greater (16. 1.) than the interior and opposite angle ABC; to each of these add the angle ACB; therefore the angles ACD, ACB are greater than the angles ABC, ACB; but ACD, ACB are together equal (13. 1.) to two right angles: therefore the angles ABC, BCA are less than two right angles. In like manner, it may be demonstrated, that BAC, ACB, as also, CAB, ABC, are less than two right angles. Therefore, any two angles, &c. Q. E. D.

PROP. XVIII. THEOR.

The greater side of every triangle has the greater angle opposite to it.

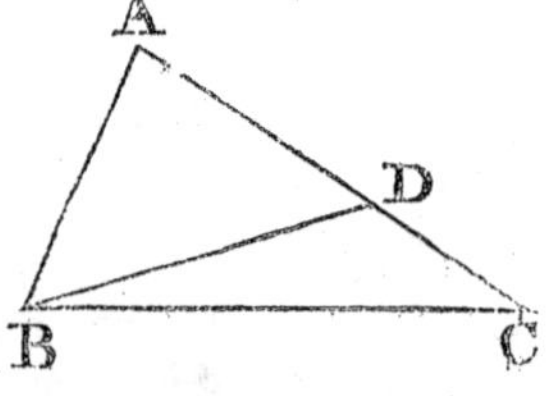

Let ABC be a triangle of which the side AC is greater than the side AB; the angle ABC is also greater than the angle BCA.

From AC, which is greater than AB, cut off (3. 1.) AD equal to AB, and join BD: and because ADB is the exterior angle of the triangle BDC, it is greater (16. 1.) than the interior and opposite angle DCB; but ADB is equal (5. 1.) to ABD, because the side AB is equal to the side AD; therefore the angle ABD is likewise greater than the angle ACB; wherefore much more is the angle ABC greater than ACB. Therefore the greater side, &c. Q. E. D.

PROP. XIX. THEOR.

The greater angle of every triangle is subtended by the greater side, or has the greater side opposite to it.

Let ABC be a triangle, of which the angle ABC is greater than the angle BCA; the side AC is likewise greater than the side AB.

For, if it be not greater, AC must either be equal to AB, or less than it; it is not equal, because then the angle ABC would be equal (5. 1.) to the angle ACB; but it is not; therefore AC is not equal to AB; neither is it less; because then the angle ABC would be less (18. 1.) than the angle ACB; but it is not; therefore the side AC is not less than AB; and it has been shewn that it is not equal to AB; therefore AC is greater than AB. Wherefore the greater angle, &c. Q. E. D.

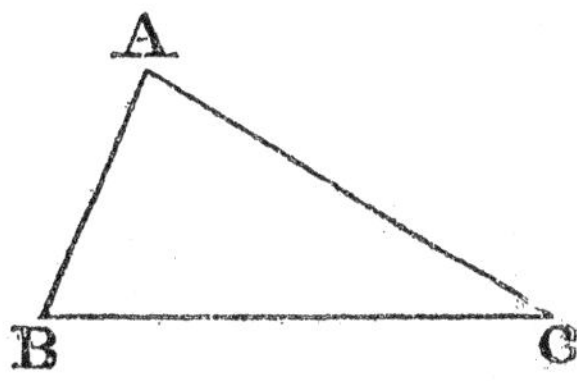

PROP. XX. THEOR.

Any two sides of a triangle are together greater than the third side.

Let ABC be a triangle; any two sides of it together are greater than the third side, viz. the sides BA, AC greater than the side BC; and AB, BC greater than AC; and BC, CA greater than AB.

Produce BA to the point D, and make (3. 1.) AD equal to AC; and join DC.

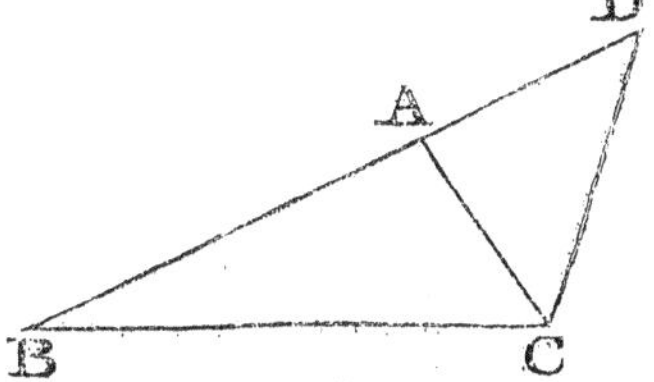

Because DA is equal to AC, the angle ADC is likewise equal (5. 1.) to ACD; but the angle BCD is greater than the angle ACD; therefore the angle BCD is greater than the angle ADC; and because the angle BCD of the triangle DCB is greater than its angle BDC, and that the greater (19. 1.) side is opposite to the greater angle: therefore the side DB is greater than the side BC; but DB is equal to BA and AC together; therefore BA and AC together are greater than BC. In the same manner it may be demonstrated, that the sides AB, BC are greater than CA, and BC, CA greater than AB. Therefore any two sides, &c. Q. E. D.

PROP. XXI. THEOR.

If from the ends of one side of a triangle, there be drawn two straight lines to a point within the triangle, these two lines shall be less than the other two sides of the triangle, but shall contain a greater angle.

Let the two straight lines BD, CD be drawn from B, C, the ends of the side BC of the triangle ABC, to the point D within it; BD and DC are less than the other two sides BA, AC of the triangle, but contain an angle BDC greater than the angle BAC.

Produce BD to E; and because two sides of a triangle (20. 1.) are greater than the third side, the two sides BA, AE of the triangle ABE are greater than BE. To each of these add EC; therefore the sides BA, AC are greater than BE, EC: Again, because the two sides CE, ED, of the triangle CED are greater than CD, if DB be added to each, the sides CE, EB, will be greater than CD, DB; but it has been shewn that BA, AC are greater that BE, EC; much more then are BA, AC greater than BD, DC.

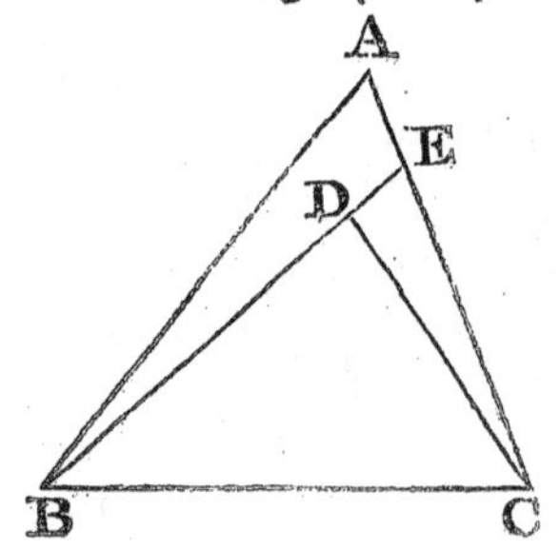

Again, because the exterior angle of a triangle (16. 1.) is greater than the interior and opposite angle, the exterior angle BDC of the triangle CDE is greater than CED; for the same reason, the exterior angle CEB of the triangle ABE is greater than BAC; and it has been demonstrated that the angle BDC is greater than the angle CEB; much more then is the angle BDC greater than the angle BAC. Therefore, if from the ends of, &c. Q. E. D.

PROP. XXII. PROB.

To construct a triangle of which the sides shall be equal to three given straight lines; but any two whatever of these lines must be greater than the third (20. 1.)

Let A, B, C be the three given straight lines, of which any two whatever are greater than the third, viz. A and B greater than C; A and C greater than B; and B and C than A. It is required to make a triangle of which the sides shall be equal to A, B, C, each to each.

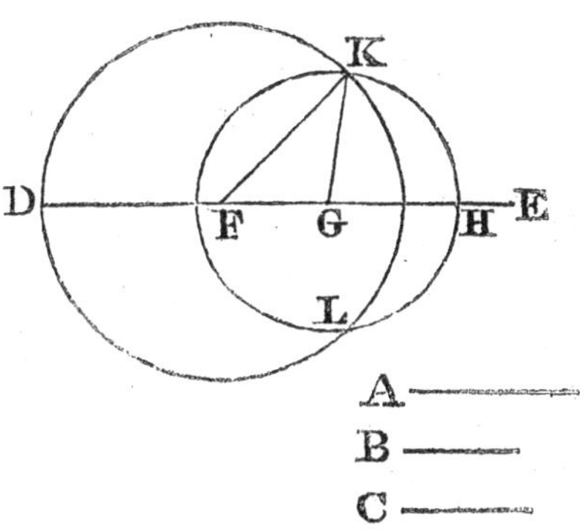

Take a straight line DE, terminated at the point D, but unlimited towards E, and make (3. 1.) DF equal to A, FG to B, and GH equal to C; and from the centre F, at the distance FD, describe (3. Post.) the circle DKL; and from the centre G, at the distance GH, describe (3. Post.) another circle HLK; and join KF, KG; the triangle KFG has its sides equal to the three straight lines, A, B, C.

Because the point F is the centre of the circle DKL, FD is equal (11. Def.) to FK; but FD is equal to the straight line A; therefore FK is equal to A: Again, because G is the centre of the circle LKH, GH is equal (11. Def.) to GK; but GH is equal to C; therefore, also GK is equal to C; and FG is equal to B; therefore the three straight lines KF, FG, GK, are equal to the three A, B, C: And therefore the triangle KFG has its three sides KF, FG, GK equal to the three given straight lines, A, B, C. Which was to be done.

PROP. XXIII. PROB.

At a given point in a given straight line, to make a rectilineal angle equal to a given rectilineal angle.

Let AB be the given straight line, and A the given point in it, and DCE the given rectilineal angle; it is required to make an angle at the given point A in the given straight line AB, that shall be equal to the given rectilineal angle DCE.

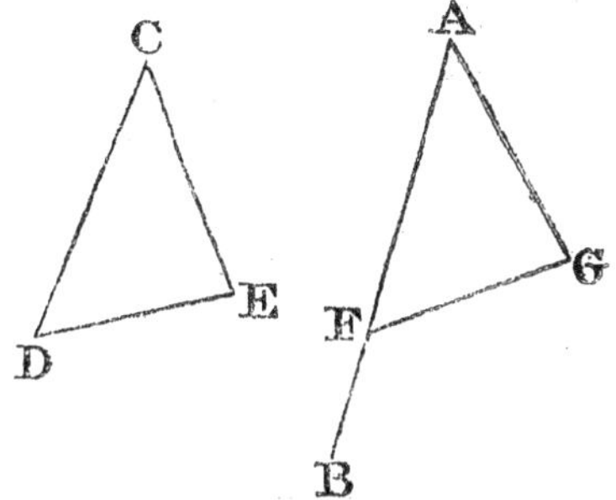

Take in CD, CE any points D, E, and join DE; and make (22. 1.) the triangle AFG, the sides of which shall be equal to the three straight lines, CD, DE, CE, so that CD be equal to AF, CE to AG, and DE to FG; and because DC, CE are equal to FA, AG, each to each, and the base DE to the base FG: the angle DCE is equal (8. 1.) to the angle

FAG. Therefore, at the given point A in the given straight line AB, the angle FAG is made equal to the given rectilineal angle DCE. Which was to be done.

PROP. XXIV. THEOR.

If two triangles have two sides of the one equal to two sides of the other, each to each, but the angle contained by the two sides of the one greater than the angle contained by the two sides of the other; the base of that which has the greater angle shall be greater than the base of the other.

Let ABC, DEF be two triangles which have the two sides AB, AC equal to the two DE, DF each to each, viz. AB equal to DE, and AC to DF; but the angle BAC greater than the angle EDF; the base BC is also greater than the base EF.

Of the two sides DE, DF, let DE be the side which is not greater than the other, and at the point D, in the straight line DE make (23. 1.) the angle EDG equal to the angle BAC: and make DG equal (3. 1.) to AC or DF, and join EG, GF.

Because AB is equal to DE, and AC to DG, the two sides BA, AC are equal to the two ED, DG, each to each, and the angle BAC is equal to the angle EDG, therefore the base BC is equal (4. 1.) to the base EG; and because DG is equal to DF, the angle DFG is equal (5. 1.) to the angle DGF; but the angle DGF is greater than the angle EGF; therefore the angle DFG is greater than EGF; and much more is the angle EFG greater than the angle EGF; and because the angle EFG of the triangle EFG is greater than its angle EGF, and because the greater (9. 1.) side is opposite to the greater angle, the side EG is greater than the side EF; but EG is equal to BC; and therefore also BC is greater than EF. Therefore, if two triangles, &c. Q. E. D.

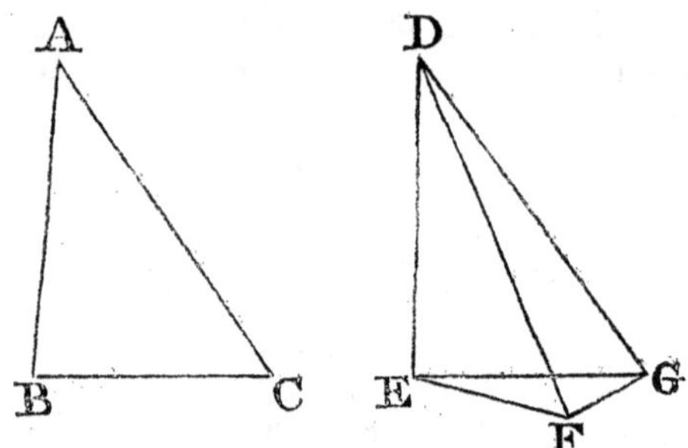

PROP. XXV. THEOR.

If two triangles have two sides of the one equal to two sides of the other, each to each, but the base of the one greater than the base of the other; the angle contained by the sides of that which has the greater base, shall be greater than the angle contained by the sides of the other.

Let ABC, DEF be two triangles which have the two sides AB, AC, equal to the two sides DE, DF, each to each, viz. AB equal to DE, and AC to DF: but let the base CB be greater than the base EF, the angle BAC is likewise greater than the angle EDF.

For, if it be not greater, it must either be equal to it, or less; but the angle BAC is not equal to the angle EDF, because then the base BC would be equal (4. 1.) to EF; but it is not; therefore the angle BAC is not equal to the angle EDF; neither is it less; because then the base BC would be less (24. 1.) than the base EF; but it is not; therefore the angle BAC is not less than the angle EDF: and it was shewn that it is not equal to it: therefore the angle BAC is greater than the angle EDF. Wherefore, if two triangles, &c. Q. E. D.

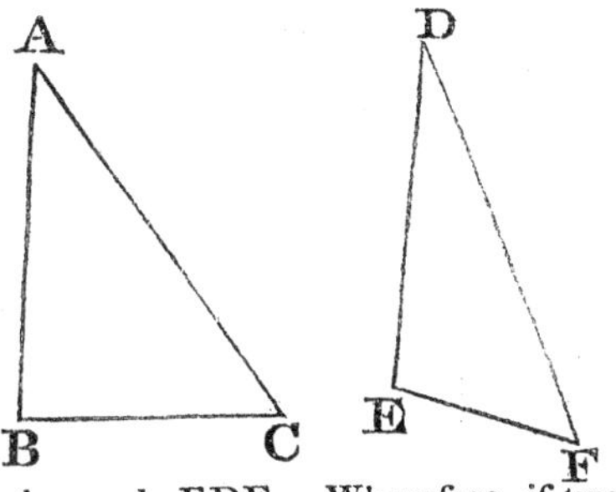

PROP. XXVI. THEOR.

If two triangles have two angles of the one equal to two angles of the other, each to each; and one side equal to one side, viz. either the side adjacent to the equal angles, or the sides opposite to the equal angles in each; then shall the other sides be equal, each to each; and also the third angle of the one to the third angle of the other.

Let ABC, DEF be two triangles which have the angles ABC, BCA equal to the angles DEF, EFD, viz. ABC to DEF, and BCA to EFD; also one side equal to one side; and first, let those sides be equal which are adjacent to the angles that are equal in the two triangles, viz. BC to EF; the other sides shall be equal, each to each, viz. AB to DE, and AC to DF; and the third angle BAC to the third angle EDF.

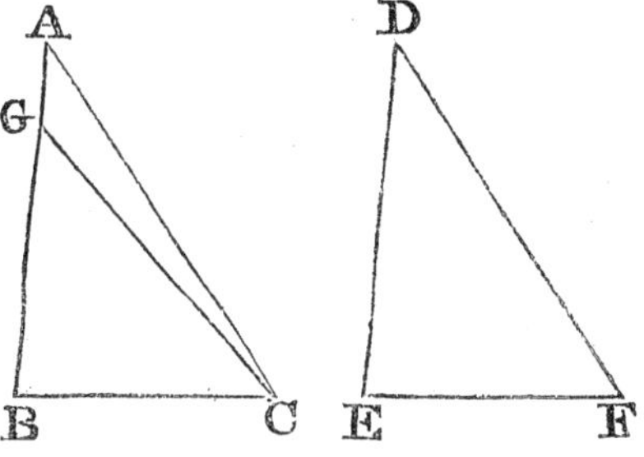

For, if AB be not equal to DE, one of them must be the greater. Let AB be the greater of the two, and make BG equal to DE, and join GC; therefore, because BG is equal to DE, and BC to EF, the two sides GB, BC are equal to the two DE, EF, each to each; and the angle GBC is equal to the angle DEF; therefore the base GC is equal (4. 1.) to the base DF, and the triangle GBC to the triangle

DEF, and the other angles to the other angles, each to each, to which the equal sides are opposite; therefore the angle GCB is equal to the angle DFE, but DFE is, by the hypothesis, equal to the angle BCA; wherefore also the angle BCG is equal to the angle BCA, the less to the greater, which is impossible; therefore AB is not unequal to DE, that is, it is equal to it; and BC is equal to EF; therefore the two AB, BC are equal to the two DE, EF, each to each; and the angle ABC is equal to the angle DEF; therefore the base AC is equal (4. 1.) to the base DF, and the angle BAC to the angle EDF.

Next, let the sides which are opposite to equal angles in each triangle be equal to one another, viz. AB to DE; likewise in this case, the other sides shall be equal, AC to DF, and BC to EF; and also the third angle BAC to the third EDF.

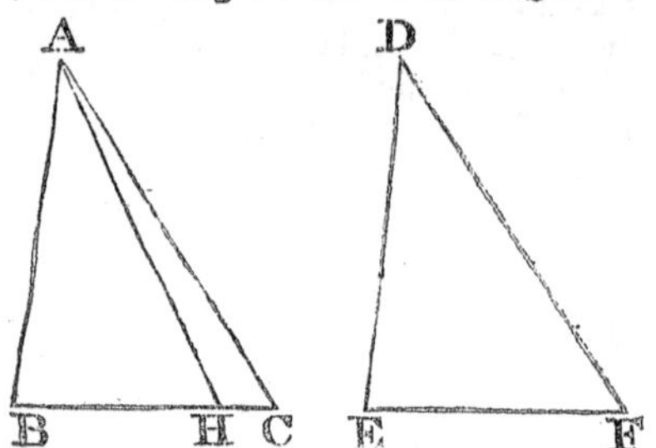

For, if BC be not equal to EF, let BC be the greater of them, and make BH equal to EF, and join AH; and because BH is equal to EF, and AB to DE; the two AB, BH are equal to the two DE, EF, each to each; and they contain equal angles; therefore (4. 1.) the base AH is equal to the base DF, and the triangle ABH to the triangle DEF, and the other angles are equal, each to each, to which the equal sides are opposite; therefore the angle BHA is equal to the angle EFD; but EFD is equal to the angle BCA; therefore also the angle BHA is equal to the angle BCA, that is, the exterior angle BHA of the triangle AHC is equal to its interior and opposite angle BCA, which is impossible (16. 1.); wherefore BC is not unequal to EF, that is, it is equal to it; and AB is equal to DE; therefore the two AB, BC are equal to the two DE, EF, each to each; and they contain equal angles; wherefore the base AC is equal to the base DF, and the third angle BAC to the third angle EDF. Therefore, if two triangles, &c. Q. E. D.

PROP. XXVII. THEOR.

If a straight line falling upon two other straight lines makes the alternate angles equal to one another, these two straight lines are parallel.

Let the straight line EF, which falls upon the two straight lines AB, CD make the alternate angles AEF, EFD equal to one another; AB is parallel to CD.

For, if it be not parallel, AB and CD being produced shall meet either towards B, D, or towards A, C; let them be produced and meet towards B, D in the point G; therefore GEF is a triangle, and its exterior angle AEF is greater (16. 1.) than the interior and opposite angle EFG; but it is also equal to it, which is impossible: therefore, AB and CD being produced, do not meet towards B, D. In like manner it may be demonstrated that they do not meet towards A, C; but those straight lines which meet neither way, though produced ever so far, are parallel (30. Def.) to one another. AB therefore is parallel to CD. Wherefore, if a straight line, &c. Q. E. D.

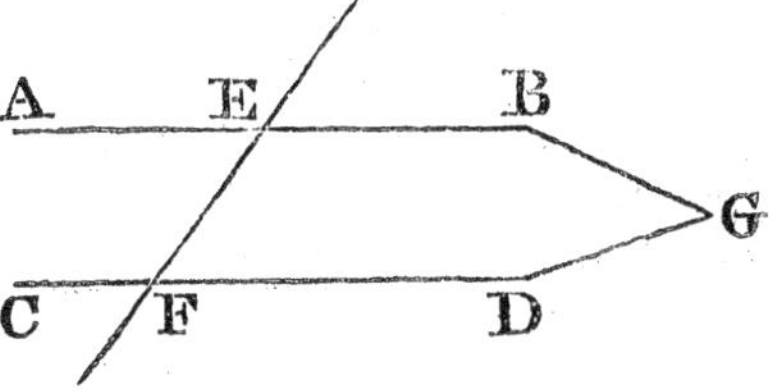

PROP. XXVIII. THEOR.

If a straight line falling upon two other straight lines makes the exterior angle equal to the interior and opposite upon the same side of the line; or makes the interior angles upon the same side together equal to two right angles; the two straight lines are parallel to one another.

Let the straight line EF, which falls upon the two straight lines AB, CD, make the exterior angle EGB equal to GHD, the interior and opposite angle upon the same side; or let it make the interior angles on the same side BGH, GHD together equal to two right angles; AB is parallel to CD.

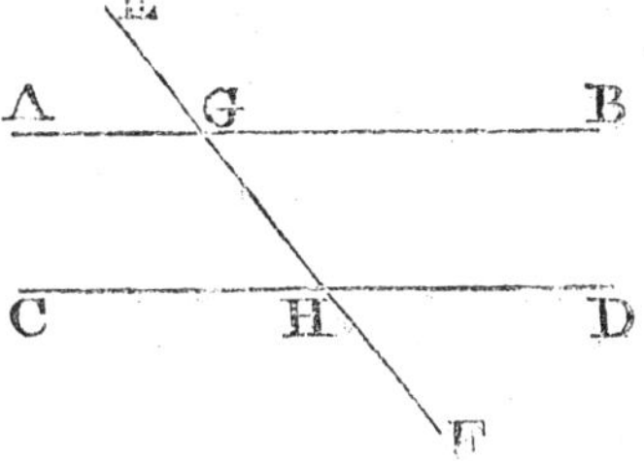

Because the angle EGB is equal to the angle GHD, and also (15. 1.) to the angle AGH, the angle AGH is equal to the angle GHD; and they are the alternate angles; therefore AB is parallel (27. 1.) to CD. Again, because the angles BGH, GHD are equal (By Hyp) to right angles, and AGH, BGH, are also equal (13. 1.) to two right angles, the angles AGH, BGH are equal to the angles BGH, GHD: Take away the common angle BGH; therefore the remaining angle AGH is equal to the remaining angle GHD; and they are alternate angles; therefore AB is parallel to CD. Wherefore, if a straight line, &c. Q. E. D.

PROP. XXIX. THEOR.

If a straight line fall upon two parallel straight lines, it makes the alternate angles equal to one another; and the exterior angle equal to the interior and opposite upon the same side; and likewise the two interior angles upon the same side together equal to two right angles.

Let the straight line EF fall upon the parallel straight lines AB, CD; the alternate angles AGH, GHD are equal to one another; and the exterior angle EGB is equal to the interior and opposite, upon the same side, GHD ; and the two interior angles BGH, GHD upon the same side are together equal to two right angles.

For if AGH be not equal to GHD, let KG be drawn making the angle KGH equal to GHD, and produce KG to L ; then KL will be parallel to CD (27. 1.) ; but AB is also parallel to CD ; therefore two straight lines are drawn through the same point G, parallel to CD, and yet not coinciding with one another, which is impossible (11. Ax.) The angles AGH, GHD therefore are not unequal, that is, they are equal to one another. Now, the angle EGB is equal to AGH (15. 1.) ; and AGH is proved to be equal to GHD ; therefore EGB is likewise equal to GHD ; add to each of these the angle BGH ; therefore the angles EGB, BGH are equal to the angles BGH, GHD ; but EGB, BGH are equal (13. 1.) to two right angles ; therefore also BGH, GHD are equal to two right angles. Wherefore, if a straight line, &c. Q. E. D.

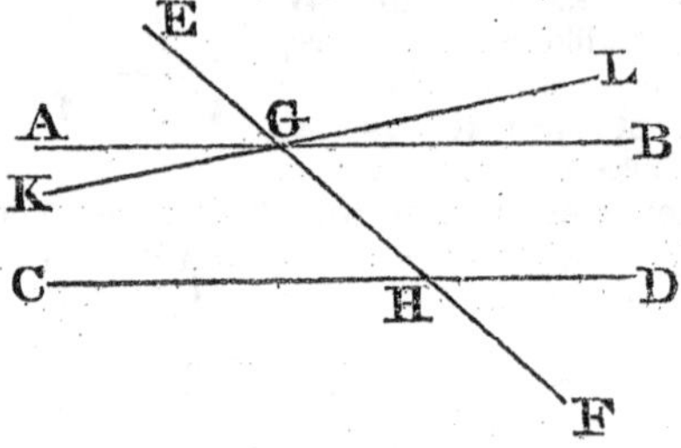

COR. If two lines KL and CD make, with EF, the two angles KGH, GHC together less than two right angles, KG and CH will meet on the side of EF on which the two angles are that are less than two right angles.

For, if not, KL and CD are either parallel, or they meet on the other side of EF ; but they are not parallel ; for the angles KGH, GHC would then be equal to two right angles. Neither do they meet on the other side of EF ; for the angles LGH, GHD would then be two angles of a triangle, and less than two right angles ; but this is impossible ; for the four angles KGH, HGL, CHG, GHD are together equal to four right angles (13. 1.) of which the two, KGH, CHG are by supposition less than two right angles ; therefore the other two, HGL, GHD are greater than two right angles. Therefore since KL and CD are not parallel, and since they do not meet towards L and D, they must meet if produced towards K and C.

PROP. XXX. THEOR.

Straight lines which are parallel to the same straight line are parallel to one another.

Let AB, CD, be each of them parallel to EF ; AB is also parallel to CD.

Let the straight line GHK cut AB, EF, CD ; and because GHK cuts the parallel straight lines AB, EF, the angle AGH is equal

(29. 1.) to the angle GHF. Again, because the straight line GK cuts the parallel straight lines EF, CD, the angle GHF is equal (29. 1.) to the angle GKD: and it was shewn that the angle AGK is equal to the angle GHF; therefore also AGK is equal to GKD; and they are alternate angles; therefore AB is parallel (27. 1.) to CD. Wherefore straight lines, &c. Q. E. D.

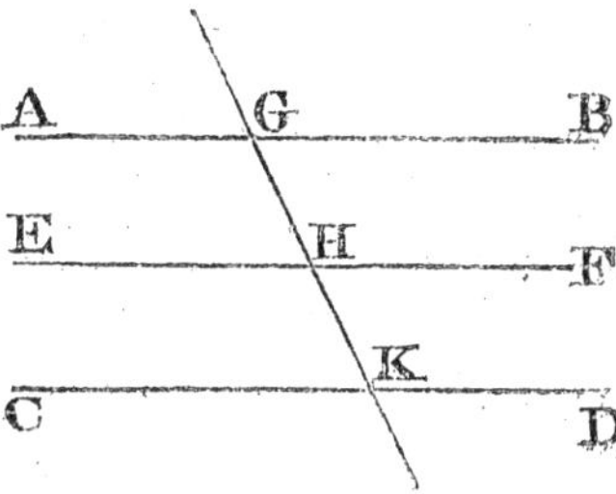

PROP. XXXI. PROB.

To draw a straight line through a given point parallel to a given straight line.

Let A be the given point, and BC the given straight line, it is required to draw a straight line through the point A, parallel to the straight line BC.

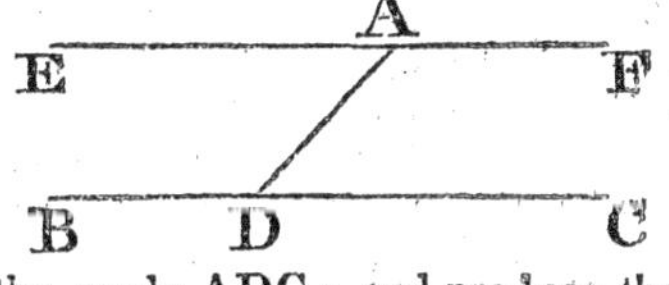

In BC take any point D, and join AD; and at the point A, in the straight line AD, make (23. 1.) the angle DAE equal to the angle ADC; and produce the straight line EA to F.

Because the straight line AD, which meets the two straight lines BC, EF, makes the alternate angles EAD, ADC equal to one another, EF is parallel (27. 1.) to BC. Therefore the straight line EAF is drawn through the given point A parallel to the given straight line BC. Which was to be done.

PROP. XXXII. THEOR.

If a side of any triangle be produced, the exterior angle is equal to the two interior and opposite angles; and the three interior angles of every triangle are equal to two right angles.

Let ABC be a triangle, and let one of its sides BC be produced to D; the exterior angle ACD is equal to the two interior and opposite angles CAB, ABC; and the three interior angles of the triangle, viz. ABC, BCA, CAB, are together equal to two right angles.

Through the point C draw CE parallel (31. 1.) to the straight line AB; and because AB is parallel to CE, and AC meets them, the alternate angles BAC, ACE are equal (29. 1.) Again, because AB is

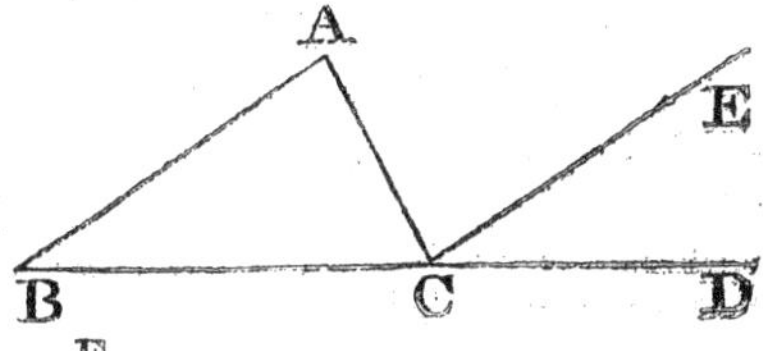

parallel to CE, and BD falls upon them, the exterior angle ECD is equal to the interior and opposite angle ABC, but the angle ACE was shewn to be equal to the angle BAC; therefore the whole exterior angle ACD is equal to the two interior and opposite angles CAB, ABC; to these angles add the angle ACB, and the angles ACD, ACB are equal to the three angles CBA, BAC, ACB; but the angles ACD ACB are equal (13. 1.) to two right angles; therefore also the angles CBA, BAC, ACB are equal to two right angles. Wherefore, if a side of a triangle, &c. Q. E. D.

Cor. 1. All the interior angles of any rectilineal figure are equal to twice as many right angles as the figure has sides, wanting four right angles.

For any rectilineal figure ABCDE can be divided into as many triangles as the figure has sides, by drawing straight lines from a point F within the figure to each of its angles. And, by the preceding proposition, all the angles of these triangles are equal to twice as many right angles as there are triangles, that is, as there are sides of the figure; and the same angles are equal to the angles of the figure, together with the angles at the point F, which is the common vertex of the triangles; that is, (2 Cor. 15. 1.) together with four right angles. Therefore, twice as many right angles as the figure has sides, are equal to all the angles of the figure, together with four right angles, that is, the angles of the figure are equal to twice as many right angles as the figure has sides, wanting four.

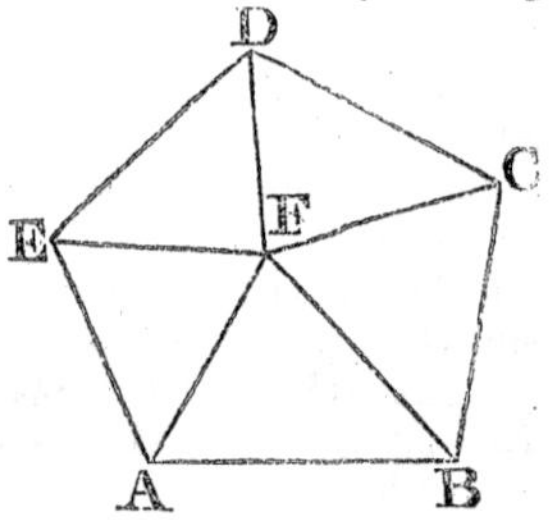

Cor. 2. All the exterior angles of any rectilineal figure are together equal to four right angles.

Because every interior angle ABC, with its adjacent exterior ABD, is equal (13. 1.) to two right angles; therefore all the interior, together with all the exterior angles of the figure, are equal to twice as many right angles as there are sides of the figure; that is, by the foregoing corollary, they are equal to all the interior angles of the figure, together with four right angles; therefore all the exterior angles are equal to four right angles.

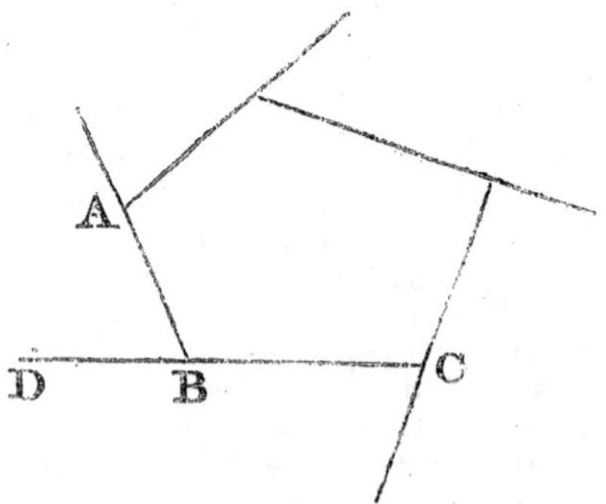

PROP. XXXIII. THEOR.

The straight lines which join the extremities of two equal and parallel straight lines, towards the same parts, are also themselves equal and parallel.

Let AB, CD, be equal and parallel straight lines, and joined towards the same parts by the straight lines AC, BD; AC, BD are also equal and parallel.

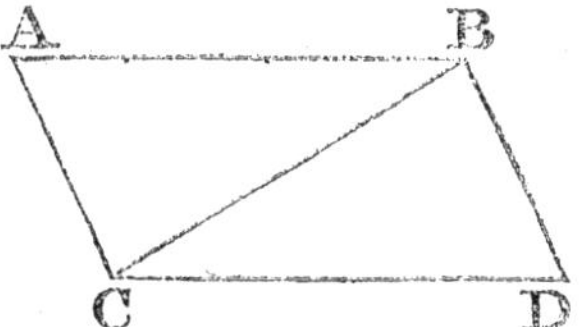

Join BC; and because AB is parallel to CD, and BC meets them, the alternate angles ABC, BCD are equal (29. 1.); and because AB is equal to CD, and BC common to the two triangles ABC, DCB, the two sides AB, BC are equal to the two DC, CB; and the angle ABC is equal to the angle BCD; therefore the base AC is equal (4. 1.) to the base BD, and the triangle ABC to the triangle BCD, and the other angles to the other angles (4 1.) each to each, to which the equal sides are opposite; therefore the angle ACB is equal to the angle CBD; and because the straight line BC meets the two straight lines AC, BD, and makes the alternate angles, ACB, CBD equal to one another, AC is parallel (27. 1.) to BD; and it was shewn to be equal to it. Therefore straight lines, &c. Q. E. D.

PROP. XXXIV. THEOR.

The opposite sides and angles of a parallelogram are equal to one another, and the diameter bisects it; that is, divides it into two equal parts.

N. B. A Parallelogram is a four-sided figure, of which the opposite sides are parallel; and the diameter is the straight line joining two of its opposite angles.

Let ACDB be a parallelogram, of which BC is a diameter; the opposite sides and angles of the figure are equal to one another: and the diameter BC bisects it.

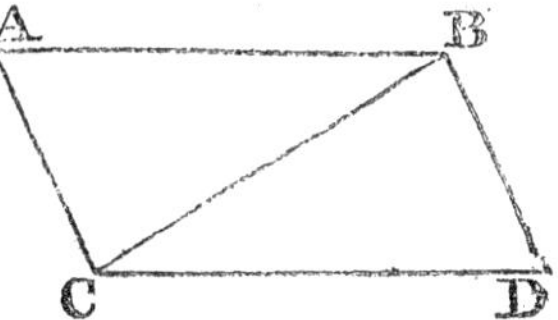

Because AB is parallel to CD, and BC meets them, the alternate angles ABC, BCD are equal (29. 1.) to one another; and because AC is parallel to BD, and BC meets them, the alternate angles ACB, CBD are equal (29. 1.) to one another; wherefore the two triangles ABC, CBD have two angles ABC, BCA in one, equal to two angles BCD, CBD in the other, each to each, and the side BC, which is adjacent to these equal angles, common to the two triangles; therefore their other sides are equal, each to each, and the third angle of the one to the third angle of the other (26. 1.) *viz.* the side AB to the side CD, and AC to BD, and the angle BAC equal to the angle BDC. And because the angle ABC is equal to the angle BCD, and the angle CBD to the angle ACB, the whole angle ABD is equal to the whole angle ACD: And the angle BAC has been shewn to be equal to the angle BDC; therefore the opposite sides and angles of a parallelogram are equal to one another; also, its diameter bisects it: for AB being equal to CD, and BC common, the two AB, BC are equal to the two DC, CB, each to each; now the angle ABC is equal to the angle BCD; therefore the triangle ABC is equal (4. 1.) to the triangle BCD, and the diameter BC divides the parallelogram ACDB into two equal parts. Therefore, &c. Q. E. D.

PROP. XXXV. THEOR.

Parallelograms upon the same base and between the same parallels, are equal to one another.

(SEE THE 2d AND 3d FIGURES.)

Let the parallelograms ABCD, EBCF be upon the same base BC, and between the same parallels AF, BC; the parallelogram ABCD is equal to the parallelogram EBCF.

If the sides AD, DF of the parallelograms ABCD, DBCF opposite to the base BC be terminated in the same point D; it is plain that each of the parallelograms is double (34. 1.) of the triangle BDC; and they are therefore equal to one another.

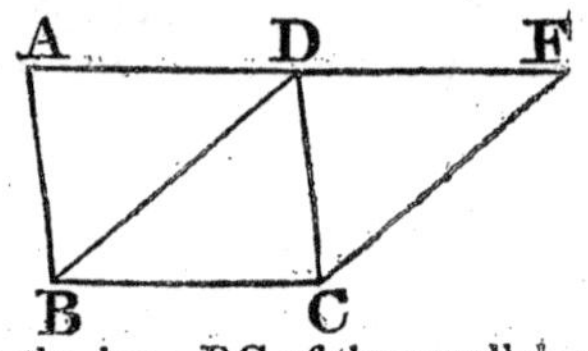

But, if the sides AD, EF, opposite to the base BC of the parallelograms ABCD, EBCF, be not terminated in the same point; then, because ABCD is a parallelogram, AD is equal (34. 1.) to BC; for the same reason EF is equal to BC; wherefore AD is equal (1. Ax.) to EF; and DE is common; therefore the whole, or the remainder, AE is equal (2. or 3. Ax.) to the whole, or the remainder DF; now AB is also equal to DC; therefore the two EA, AB are equal to the two

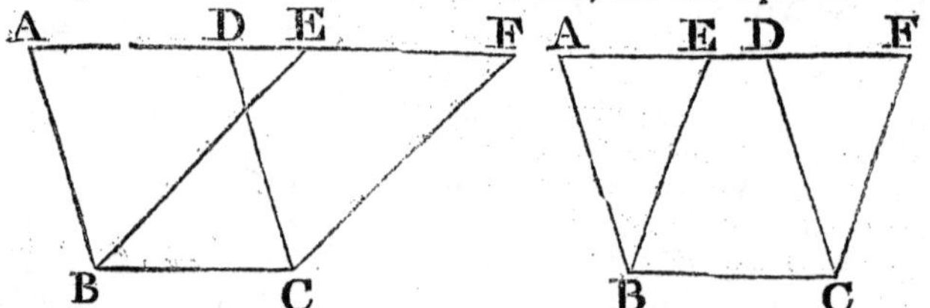

FD, DC, each to each; but the exterior angle FDC is equal (29. 1.) to the interior EAB, wherefore the base EB is equal to the base FC, and the triangle EAB (4. 1.) to the triangle FDC. Take the triangle FDC from the trapezium ABCF, and from the same trapezium take the triangle EAB; the remainders will then be equal (3. Ax.), that is, the parallelogram ABCD is equal to the parallelogram EBCF. Therefore, parallelograms upon the same base, &c. Q. E. D.

PROP. XXXVI. THEOR.

Parallelograms upon equal bases, and between the same parallels, are equal to one another.

Let ABCD, EFGH be parallelograms upon equal bases BC, FG and between the same parallels AH, BG; the parallelogram ABCD is equal to EFGH.

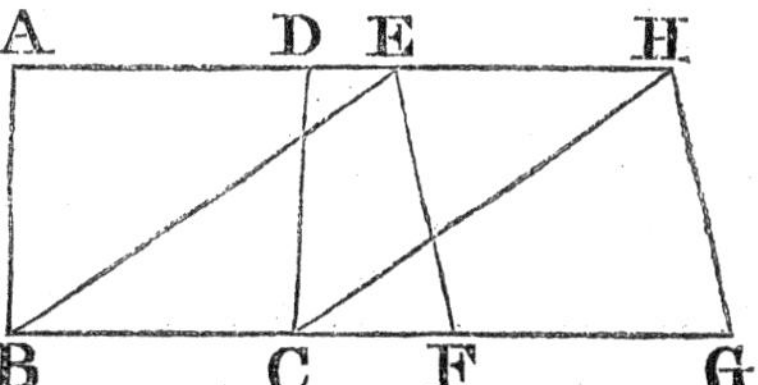

Join BE, CH; and because BC is equal to FG, and FG to (34. 1.) EH, BC is equal to EH; and they are parallels, and joined towards the same parts by the straight lines BE, CH: But straight lines which join equal and parallel straight lines towards the same parts, are themselves equal and parallel (33. 1.); therefore EB, CH are both equal and parallel, and EBCH is a parallelogram; and it is equal (35. 1.) to ABCD, because it is upon the same base BC, and between the same parallels BC, AH; For the like reason, the parallelogram EFGH is equal to the same EBCH: Therefore also the parallelogram ABCD is equal to EFGH. Wherefore, parallelograms, &c. Q. E. D.

PROP. XXXVII. THEOR.

Triangles upon the same base, and between the same parallels, are equal to one another.

Let the triangles ABC, DBC be upon the same base BC, and between the same parallels, AD, BC: The triangle ABC, is equal to the triangle DBC.

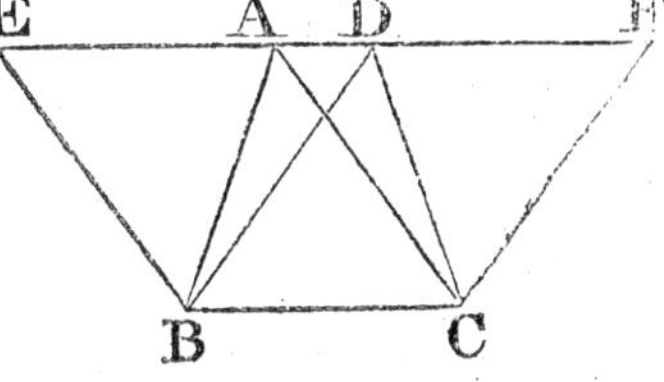

Produce AD both ways to the points E, F, and through B draw (31. 1.) BE parallel to CA; and through C draw CF parallel to BD: Therefore, each of the figures EBCA, DBCF is a parallelogram; and DBCA is equal (35. 1.) to DBCF, because they are upon the same base BC, and between the same parallels BC, EF:

but the triangle ABC is the half of the parallelogram EBCA, because the diameter AB bisects (34. 1.) it; and the triangle DBC is the half of the parallelogram DBCF, because the diameter DC bisects it; and the halves of equal things are equal. (7. Ax.); therefore the triangle ABC is equal to the triangle DBC. Wherefore triangles, &c. Q. E. D.

PROP. XXXVIII. THEOR.

Triangles upon equal bases, and between the same parallels, are equal to one another.

Let the triangles ABC, DEF be upon equal bases BC, EF, and between the same parallels BF, AD: The triangle ABC is equal to the triangle DEF.

Produce AD both ways to the points G, H, and through B draw BG parallel (31. 1.) to CA, and through F draw FH parallel to ED: Then each of the figures GBCA, DEFH is a parallelogram; and they are equal to (36. 1.) one another, because they are upon equal bases BC, EF, and between the same parallels BF, GH; and the triangle ABC is the half (34. 1.) of the parallelogram GBCA, because the diameter AB bisects it; and the triangle DEF is the half (34. 1.) of the parallelogram DEFH, because the diameter DF bisects it; But the halves of equal things are equal (7. Ax.); therefore the triangle ABC is equal to the triangle DEF. Wherefore triangles, &c. Q. E. D.

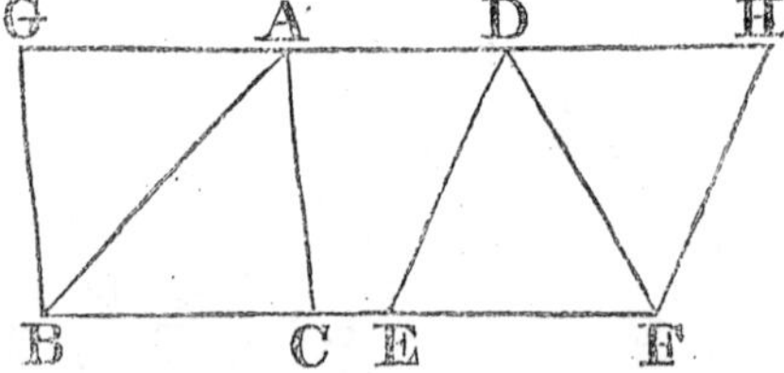

PROP. XXXIX. THEOR.

Equal triangles upon the same base, and upon the same side of it, are between the same parallels.

Let the equal triangles ABC, DBC be upon the same base BC, and upon the same side of it; they are between the same parallels.

Join AD: AD is parallel to BC; for, if it is not, through the point A draw (31. 1.) AE parallel to BC, and join EC; The triangle ABC, is equal (37. 1.) to the triangle EBC, because it is upon the same base BC, and between the same parallels BC, AE; But the triangle ABC is equal to the triangle BDC; therefore also the triangle BDC is equal to the triangle EBC, the greater to the less, which is impossible: Therefore AE is not parallel to BC. In the same manner, it may be demonstrated that no other line but AD is parallel to BC; AD is therefore parallel to it. Wherefore equal triangles upon, &c. Q. E. D.

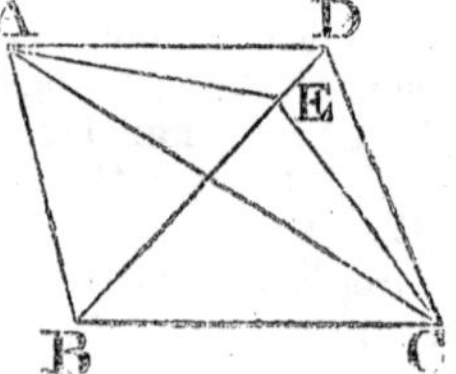

PROP. XL. THEOR.

Equal triangles on the same side of bases, which are equal and in the same straight line, are between the same parallels.

Let the equal triangles ABC, DEF be upon equal bases BC, EF, in the same straight line BF, and towards the same parts; they are between the same parallels.

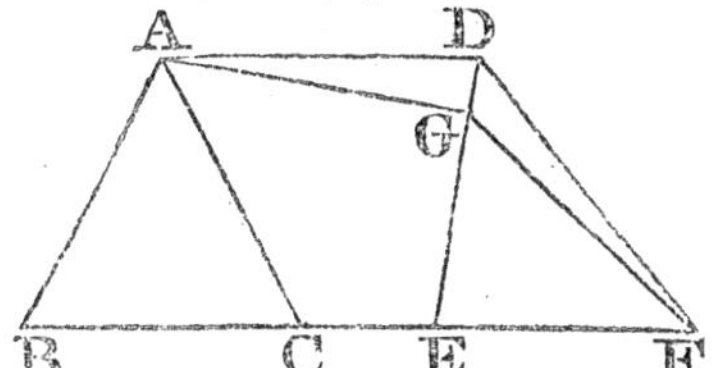

Join AD; AD is parallel to BC: For, if it is not, through A draw (31. 1.) AG parallel to BF, and join GF. The triangle ABC is equal (38. 1.) to the triangle GEF, because they are upon equal bases BC, EF, and between the same parallels BF, AG: But the triangle ABC is equal to the triangle DEF; therefore also the triangle DEF is equal to the triangle GEF, the greater to the less, which is impossible; Therefore AG is not parallel to BF; and in the same manner it may be demonstrated that there is no other parallel to it but AD; AD is therefore parallel to BF. Wherefore equal triangles, &c. Q. E. D.

PROP. XLI. THEOR.

If a parallelogram and a triangle be upon the same base, and between the same parallel; the parallelogram is double of the triangle.

Let the parallelogram ABCD and the triangle EBC be upon the same base BC and between the same parallels BC, AE; the parallelogram ABCD is double of the triangle EBC.

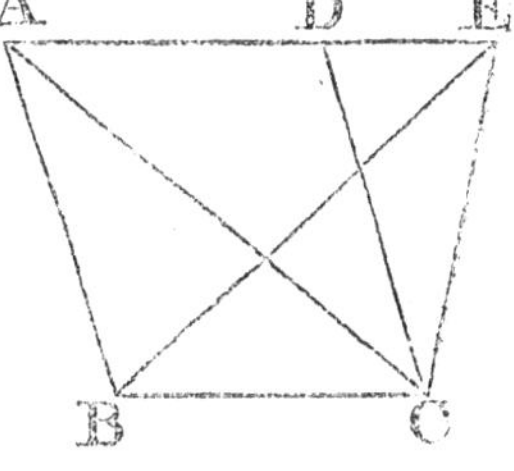

Join AC; then the triangle ABC is equal (37. 1.) to the triangle EBC, because they are upon the same base BC, and between the same parallels BC, AE. But the parallelogram ABCD is double (34. 1.) of the triangle ABC, because the diameter AC divides it into two equal parts; wherefore ABCD is also double of the triangle EBC. Therefore, if a parallelogram, &c. Q. E. D.

PROP. XLII. PROB.

To describe a parallelogram that shall be equal to a given triangle, and have one of its angles equal to a given rectilineal angle.

Let ABC be the given triangle, and D the given rectilineal angle.

It is required to describe a parallelogram that shall be equal to the given triangle ABC, and have one of its angles equal to D.

Bisect (10. 1.) BC in E, join AE, and at the point E in the straight line EC make (23. 1.) the angle CEF equal to D; and through A draw (31. 1.) AG parallel to BC, and through C draw CG (31. 1.) parallel to EF: Therefore F CG is a parallelogram: And because BE is equal to EC, the triangle ABE is likewise equal (38. 1.) to the triangle AEC, since they are upon equal bases BE, EC, and between the same parallels BC AG; therefore the triangle ABC is double of the triangle AEC. And the parallelogram FECG is likewise double (41. 1.) of the triangle AEC, because it is upon the same base, and between the same parallels: Therefore the parallelogram FECG is equal to the triangle ABC, and it has one of its angles CEF equal to the given angle D; Wherefore there has been described a parallelogram FECG equal to a given triangle ABC, having one of its angles CEF equal to the given angle D. Which was to be done.

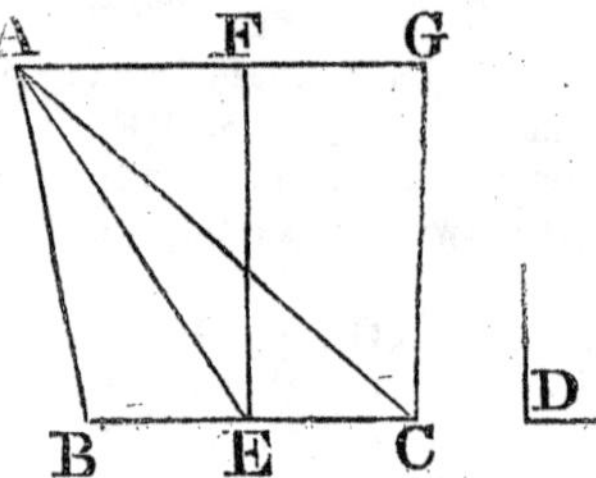

PROP. XLIII. THEOR.

The complements of the parallelograms which are about the diameter of any parallelogram, are equal to one another.

Let ABCD be a parallelogram of which the diameter is AC; let EH, FG be the parallelograms about AC, that is, through which AC passes, and let BK, KD be the other parallelograms, which make up the whole figure ABCD, and are therefore called the complements: The complement BK is equal to the complement KD.

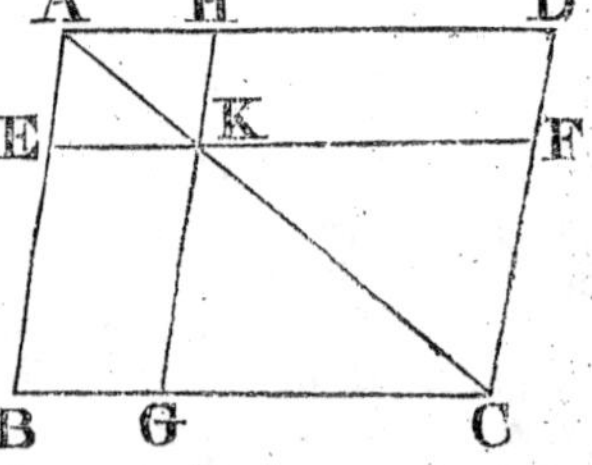

Because ABCD is a parallelogram and AC its diameter, the triangle ABC is equal (34. 1.) to the triangle ADC: And because EKHA is a parallelogram, and AK its diameter, the triangle AEK is equal to the triangle AHK: For the same reason, the triangle KGC is equal to the triangle KFC. Then, because the triangle AEK is equal to the triangle AHK, and the triangle KGC to the triangle KFC; the triangle AEK, together with the triangle KGC, is equal to the triangle AHK, together with the triangle KFC: But the whole triangle ABC is equal to the whole ADC; therefore the remaining complement BK is equal to the remaining complement KD. Wherefore, the complements, &c. Q. E. D.

PROP. XLIV. PROB.

To a given straight line to apply a parallelogram, which shall be equal to a given triangle, and have one of its angles equal to a given rectilineal angle.

Let AB be the given straight line, and C the given triangle, and D the given rectilineal angle. It is required to apply to the straight line AB a parallelogram equal to the triangle C, and having an angle equal to D. Make (42. 1.) the parallelogram BEFG equal to the triangle

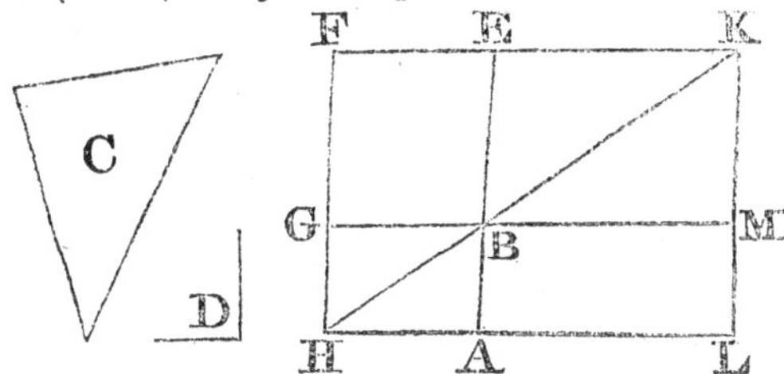

C, having the angle EBG equal to the angle D, and the side BE in the same straight line with AB : produce FG to H, and through A draw (31. 1.) AH parallel to BG or EF, and join HB. Then because the straight line HF falls upon the parallels AH, EF, the angles AHF, HFE, are together equal (29. 1.) to two right angles ; wherefore the angles BHF, HFE are less than two right angles : But straight lines which with another straight line make the interior angles, upon the same side less than two right angles, do meet if produced (Cor. 29. 1.): Therefore HB, FE will meet, if produced ; let them meet in K, and through K draw KL parallel to EA or FH, and produce HA, GB to the points L, M : Then HLKF is a parallelogram, of which the diameter is HK, and AG, ME are the parallelograms about HK ; and LB, BF are the complements ; therefore LB is equal (43. 1.) to BF : but BF is equal to the triangle C ; wherefore LB is equal to the triangle C ; and because the angle GBE is equal (15. 1.) to the angle ABM, and likewise to the angle D ; the angle ABM is equal to the angle D : Therefore the parallelogram LB, which is applied to the straight line AB, is equal to the triangle C, and has the angle ABM equal to the angle D : Which was to be done.

PROP. XLV. PROB.

To describe a parallelogram equal to a given rectilineal figure, and having an angle equal to a given rectilineal angle.

Let ABCD be the given rectilineal figure, and E the given rectilineal angle. It is required to describe a parallelogram equal to ABCD, and having an angle equal to E.

Join DB, and describe (42. 1.) the parallelogram FH equal to the triangle ADB, and having the angle HKF equal to the angle E; and to the straight line GH (44. 1.) apply the parallelogram GM equal to the triangle DBC, having the angle GHM equal to the angle E. And because the angle E is equal to each of the angles FKH, GHM, the angle FKH is equal to GHM; add to each of these the angle KHG; therefore the angles FKH, KHG are equal to the angles KHG, GHM; but FKH, KHG are equal (29. 1.) to two right angles; therefore also KHG, GHM are equal to two right angles: and because at the point

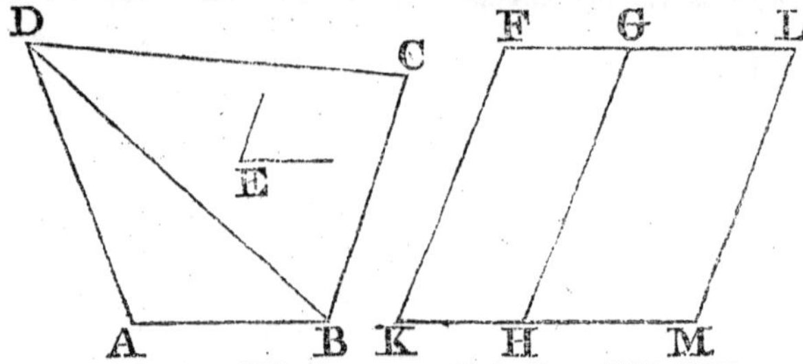

H in the straight line GH, the two straight lines KH, HM, upon the opposite sides of GH, make the adjacent angles equal to two right angles, KH, is in the same straight line (14. 1.) with HM. And because the straight line HG meets the parallels KM, FG, the alternate angles MHG, HGF are equal (29. 1.); add to each of these the angle HGL; therefore the angles MHG, HGL, are equal to the angles HGF, HGL: But the angles MHG, HGL, are equal (29. 1.) to two right angles; wherefore also the angles HGF, HGL, are equal to two right angles, and FG is therefore in the same straight line with GL. And because KF is parallel to HG, and HG to ML, KF is parallel (30. 1.) to ML; but KM, FL are parallels: wherefore KFLM is a parallelogram. And because the triangle ABD is equal to the parallelogram H F, and the triangle DBC to the parallelogram GM, the whole rectilineal figure ABCD is equal to the whole parallelogram KFLM; therefore the parallelogram KFLM has been described equal to the given rectilineal figure ABCD, having the angle FKM equal to the given angle E. Which was to be done.

Cor. From this it is manifest how to a given straight line to apply a parallelogram, which shall have an angle equal to a given rectilineal angle, and shall be equal to a given rectilineal figure, *viz.* by applying (44. 1.) to the given straight line a parallelogram equal to the first triangle ABD, and having an angle equal to the given angle.

PROP. XLVI. PROB.

To describe a square upon a given straight line.

Let AB be the given straight line: it is required to describe a square upon AB.

From the point A draw (11. 1.) AC at right angles to AB; and make (3. 1.) AD equal to AB, and through the point D draw DE parallel (31. 1.) to AB, and through B draw BE parallel to AD; therefore ADEB is a parallelogram; whence AB is equal (34. 1.) to DE, and AD to BE; but BA is equal to AD: therefore the four straight lines BA, AD, DE, EB are equal to one another, and the parallelogram ADEB is equilateral; it is likewise rectangular; for the straight line AD meeting the parallels, AB, DE, makes the angles BAD, ADE equal (29. 1.) to two right angles; but BAD is a right angle; therefore also ADE is a right angle now the opposite angles of parallelograms are equal (34. 1.); therefore each of the opposite angles ABE, BED is a right angle; wherefore the figure ADEB is rectangular, and it has been demonstrated that it is equilateral; it is therefore a square, and it is described upon the given straight line AB; Which was to be done.

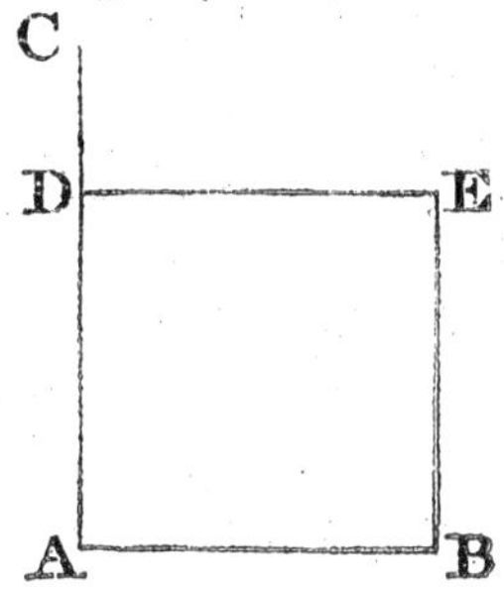

Cor. Hence every parallelogram that has one right angle has all its angles right angles.

PROP. XLVII. THEOR.

In any right angled triangle, the square which is described upon the side subtending the right angle, is equal to the squares described upon the sides which contain the right angle.

Let ABC be a right angled triangle having the right angle BAC; the square described upon the side BC is equal to the squares described upon BA, AC.

On BC describe (46. 1.) the square BDEC, and on BA, AC the squares GB, HG; and through A draw (31. 1.) AL parallel to BD or CE, and join AD, FC; then, because each of the angles BAC, BAG is a right angle (25. def.), the two straight lines AC, AG upon the opposite sides of AB, make with it at the point A the adjacent angles equal to two right angles; therefore CA is in the same straight line (14. 1) with AG; for the same reason, AB and AH are in the same straight line. Now because the angle DBC is equal to the angle FBA, each of them being a right angle, adding to each the angle ABC, the whole angle DBA will be equal (2. Ax.) to the whole FBC; and because the two sides AB, BD, are equal to the two FB, BC each to each, and the angle DBA equal to the

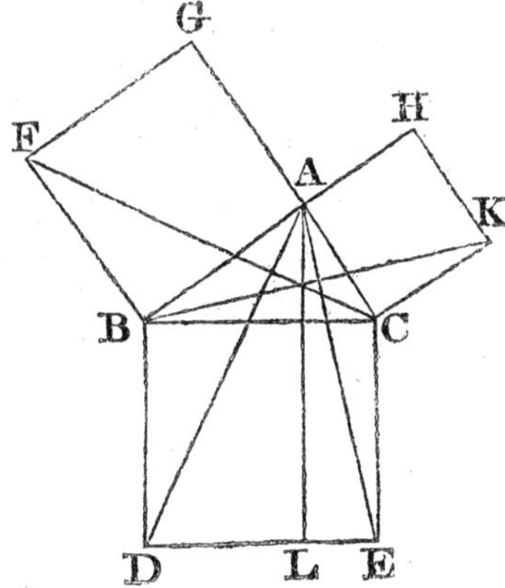

angle FBC, therefore the base AD is equal (4. 1.) to the base FC, and the triangle ABD to the triangle FBC. But the parallelogram BL is double (41. 1.) of the triangle ABD, because they are upon the same base, BD, and between the same parallels, BD, AL; and the square GB is double of the triangle BFC, because these also are upon the same base FB, and between the same parallels FB, GC. Now the doubles of equals are equal (6. Ax.) to one another; therefore the parallelogram BL is equal to the square GB: And in the same manner, by joining AE, BK, it is demonstrated that the parallelogram CL is equal to the square HC. Therefore, the whole square BDEC is equal to the two squares GB, HC; and the square BDEC is described upon the straight line BC, and the squares GB, HC upon BA, AC: wherefore the square upon the side BC is equal to the squares upon the sides BA, AC. Therefore, in any right angled triangle, &c. Q. E. D.

PROP. XLVIII. THEOR.

If the square described upon one of the sides of a triangle, be equal to the squares described upon the other two sides of it; the angle contained by these two sides is a right angle.

If the square described upon BC, one of the sides of the triangle ABC, be equal to the squares upon the other sides BA, AC, the angle BAC is a right angle.

From the point A draw (11. 1.) AD at right angles to AC, and make AD equal to BA, and join DC. Then because DA is equal to AB, the square of DA is equal to the square of AB; To each of these add the square of AC; therefore the squares of DA, AC are equal to the squares of BA, AC. But the square of DC is equal (47. 1.) to the squares of DA, AC, because DAC is a right angle; and the square of BC, by hypothesis, is equal to the squares of BA, AC; therefore, the square of DC is equal to the square of BC; and therefore also the side DC is equal to the side BC. And because the side DA is equal to AB, and AC common to the two triangles DAC, BAC, and the base DC likewise equal to the base BC, the angle DAC is equal (8. 1.) to the angle BAC; But DAC is a right angle; therefore also BAC is a right angle. Therefore, if the square, &c. Q. E. D.

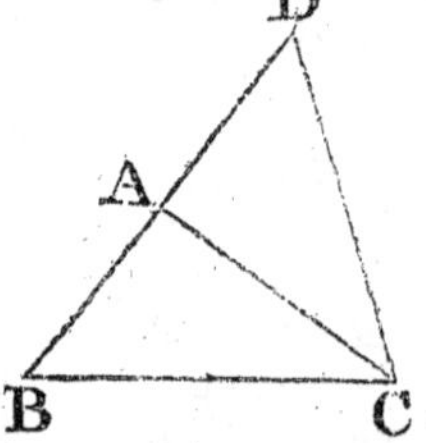

ELEMENTS

OF

GEOMETRY.

BOOK II.

DEFINITIONS.

I.

Every right angled parallelogram, or *rectangle*, is said to be contained by any two of the straight lines which are about one of the right angles.

"Thus the right angled parallelogram AC is called the rectangle "contained by AD and DC, or by AD and AB, &c. For the sake of "brevity, instead of the *rectangle contained* by AD and DC, we shall "simply say the rectangle AD.DC, placing a point between the two "sides of the rectangle. Also, instead of the square of a line, for "instance, of AD, we shall frequently in what follows write AD^2."

"The sign + placed between the names of two magnitudes, signi"fies that those magnitudes are to be added together, and the sign "− placed between them, signifies that the latter is to be taken away "from the former."

"The sign = signifies, that the things between which it is placed "are equal to one another."

II.

In every parallelogram, any of the parallelograms about a diameter, together with the two complements, is called a Gnomon. "Thus the parallelo"gram HG, together with the "complements AF, FC, is the "gnomon of the parallelogram "AC. This gnomon may also, "for the sake of brevity, be "called the gnomon AGK or "EHC."

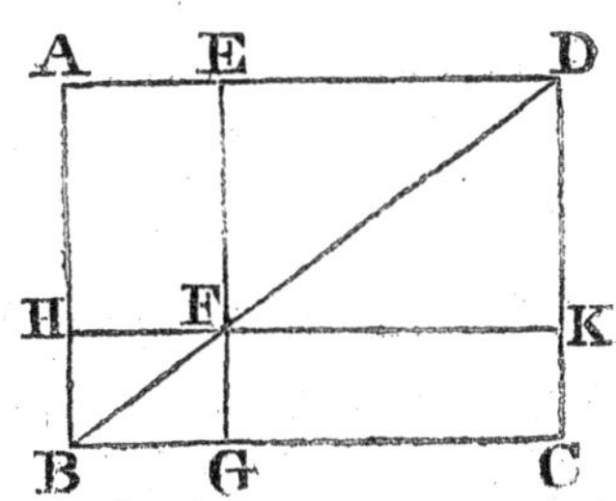

PROP. I. THEOR.

If there be two straight lines, one of which is divided into any number of parts; the rectangle contained by the two straight lines is equal to the rectangles contained by the undivided line, and the several parts of the divided line.

Let A and BC be two straight lines; and let BC be divided into any parts in the points D, E; the rectangle A.BC is equal to the several rectangles A.BD, A.DE, A.EC.

From the point B draw (11. 1.) BF at right angles to BC, and make BG equal (3. 1.) to A; and through G draw (31. 1.) GH parallel to BC; and through D, E, C, draw (31. 1.) DK, EL, CH parallel to BG; then BH, BK, DL, and EH are rectangles, and BH = BK + DL + EH.

But BH=BG.BC= A.BC, because BG = A: Also BK = BG. BD = A.BD, because BG = A; and DL=DK.DE=A.DE, because (34. 1.) DK=BG=A. In like manner, EH = A.EC. Therefore A.BC = A.BD+A.DE+A.EC; that is, the rectangle A.BC is equal to the several rectangles A.BD, A.DE, A.EC. Therefore, if there be two straight lines, &c. Q. E. D.

PROP. II. THEOR.

If a straight line be divided into any two parts, the rectangles contained by the whole and each of the parts, are together equal to the square of the whole line.

Let the straight line AB be divided into any two parts in the point C; the rectangle AB.BC, together with the rectangle AB.AC, is equal to the square of AB; or AB.AC + AB.BC = AB².

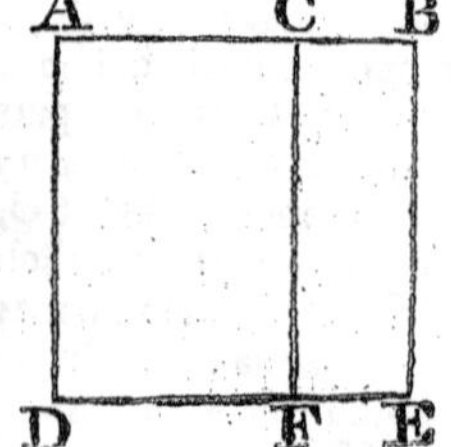

On AB describe (46. 1.) the square ADEB, and through C draw CF (31. 1.) parallel to AD or BE; then AF + CE = AE. But AF = AD.AC = AB.AC, because AD = AB; CE = BE.BC = AB.BC; and AE = AB². Therefore AB.AC + AB.BC = AB². Therefore, if a straight line, &c. Q. E. D.

PROP. III. THEOR.

If a straight line be divided into any two parts, the rectangle contained by the whole and one of the parts, is equal to the rectangle contained by the two parts, together with the square of the aforesaid part.

Let the straight line AB be divided into two parts in the point C; the rectangle AB.BC is equal to the rectangle AC.BC, together with BC^2.

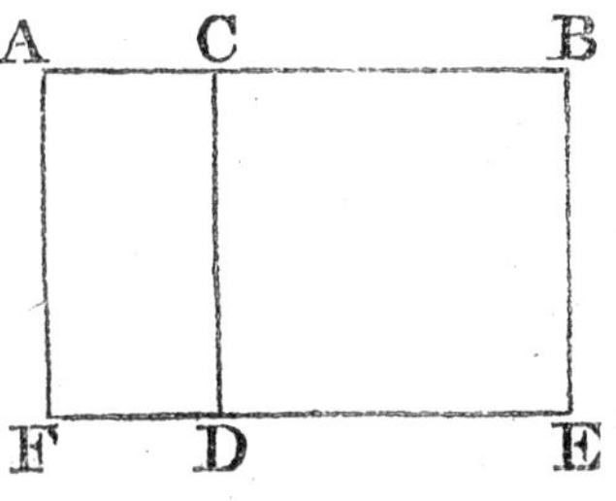

Upon BC describe (46. 1.) the square CDEB, and produce ED to F, and through A draw (31. 1.) AF parallel to CD or BE; then AE = AD + CE.

But AE = AB.BE = AB.BC, because BE = BC. So also AD = AC.CD = AC.CB; and CE = BC^2; therefore AB.BC = AC.CB + BC^2. Therefore, if a straight line, &c. Q. E. D.

PROP. IV. THEOR.

If a straight line be divided into any two parts, the square of the whole line is equal to the squares of the two parts, together with twice the rectangle contained by the parts.

Let the straight line AB be divided into any two parts in C; the square of AB is equal to the squares of AC, CB, and to twice the rectangle contained by AC, CB, that is, $AB^2 = AC^2 + CB^2 + 2AC.CB$.

Upon AB describe (46. 1.) the square ADEB, and join BD, and through C draw (31. 1.) CGF parallel to AD or BE, and through G draw HK parallel to AB or DE. And because CF is parallel to AD, and BD falls upon them, the exterior angle BGC is equal (29. 1.) to the interior and opposite angle ADB; but ADB is equal (5. 1.) to the angle ABD, because BA is equal to AD, being sides of a square; wherefore the angle CGB is equal to the angle GBC; and therefore the side BC is equal (6. 1.) to the side CG: but CB is equal (34. 1.) also to GK and CG to BK: wherefore the figure CGKB is equilateral. It is likewise rectangular; for the angle CBK being a right angle, the other angles of the parallelogram CGKB are also right angles (Cor. 46. 1.) Wherefore CGKB is a square, and it is upon the side CB. For the same reason HF also is a square, and

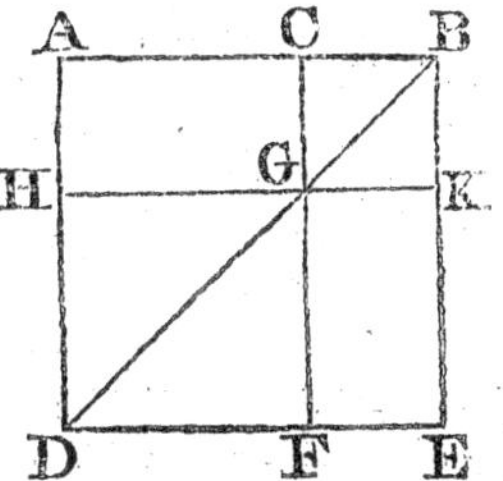

it is upon the side HG, which is equal to AC: therefore HF, CK are the squares of AC, CB. And because the complement AG is equal (43. 1.) to the complement GE; and because AG=AC.CG=AC.CB, therefore also GE=AC.CB, and AG+GE=2AC.CB. Now, HF= AC^2 and CK=CB^2; therefore, HF+CK+AG+GE=AC^2+CB^2+ 2AC.CB.

But HF+CK+AG+GE=the figure AE, or AB^2; therefore AB^2 =AC^2+CB^2+2AC.CB. Wherefore, if a straight line be divided &c. Q. E. D.

COR. From the demonstration, it is manifest that the parallelograms about the diameter of a square are likewise squares.

PROP. V. THEOR.

If a straight line be divided into two equal parts, and also into two unequal parts; the rectangle contained by the unequal parts, together with the square of the line between the points of section, is equal to the square of half the line.

Let the straight line AB be divided into two equal parts in the point C, and into two unequal parts in the point D; the rectangle AD.DB, together with the square of CD, is equal to the square of CB, or AD. DB+CD^2=CB^2.

Upon CB describe (46. 1.) the square CEFB, join BE, and through D draw (31. 1.) DHG parallel to CE or BF; and through H draw KLM parallel to CB or EF; and also through A draw AK parallel to CL or BM: And because CH = HF, if DM be added to both, CM=DF. But AL=(36. 1.) CM, therefore AL = DF, and adding CH to both, AH=gnomon CMG. But AH= AD.DH = AD.DB, because DH=DB (Cor. 4. 2.); therefore gnomon CMG=AD.DB. To each add LG=CD^2, then, gnomon CMG + LG=AD.DB+CD^2. But CMG+ LG = BC^2; therefore AD.DB+CD^2=BC^2. Wherefore, if a straight line, &c. Q. E. D.

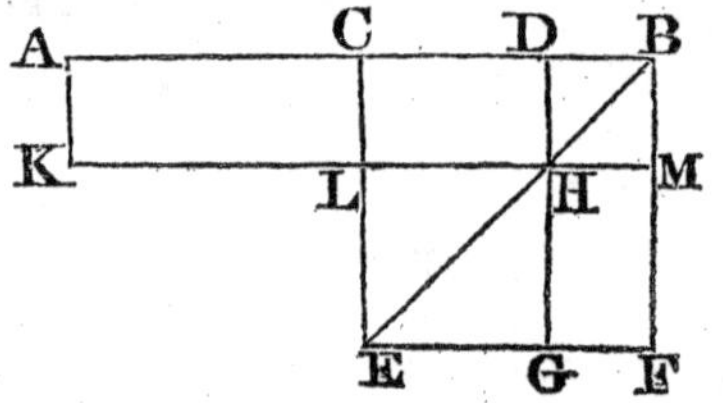

" COR. From this proposition it is manifest, that the difference of " the squares of two unequal lines, AC, CD, is equal to the rectangle " contained by their sum and difference, or that AC^2—CD^2=(AC+ " CD) (AC − CD."

PROP. VI. THEOR.

If a straight line be bisected, and produced to any point; the rectangle contained by the whole line thus produced, and the part of it produced, together with the square of half the line bisected, is equal to the square of the straight line which is made up of the half and the part produced.

Let the straight line AB be bisected in C, and produced to the point

D; the rectangle AD.DB, together with the square of CB, is equal to the square of CD.

Upon CD describe (46. 1.) the square CEFD, join DE, and through B draw (31. 1.) BHG parallel to CE or DF, and through H draw KLM parallel to AD or EF, and also through A draw AK parallel to CL or DM. And because AC is equal to CB, the rectangle AL is equal (36. 1.) to CH; but CH is equal (43. 1.) to HF; therefore also AL is equal to HF: To each of these add CM; therefore the whole AM is equal to the gnomon CMG. Now AM=AD.DM=AD.DB, because DM = DB. Therefore gnomon CMG=AD.DB, and CMG+LG = AD.DB+CB². But CMG + LG = CF = CD², therefore AD.DB + CB² = CD². Therefore, if a straight line, &c. Q. E. D.

PROP. VII. THEOR.

If a straight line be divided into any two parts, the squares of the whole line, and of one of the parts, are equal to twice the rectangle contained by the whole and that part, together with the square of the other part.

Let the straight line AB be divided into any two parts in the point C; the squares of AB, BC, are equal to twice the rectangle AB.BC, together with the square of AC, or $AB^2+BC^2=2AB.BC+AC^2$.

Upon AB describe (46. 1.) the square ADEB, and construct the figure as in the preceding propositions: Because AG=GE, AG+CK =GE+CK, that is, AK=CE, and therefore AK+CE=2AK. But AK+CE = gnomon AKF+CK; and therefore, AKF + CK = 2AK=2AB.BK=2AB.BC, because BK = (Cor. 4. 2.) BC. Since then, AKF+CK=2AB.BC, AKF+CK+HF =2AB.BC+HF; and because AKF+HF=AE=AB², $AB^2+CK=2AB.BC+HF$, that is, (since $CK = CB^2$, and $HF = AC^2$,) $AB^2 + CB^2 = 2AB.BC + AC^2$. Wherefore, if a straight line, &c. Q. E. D.

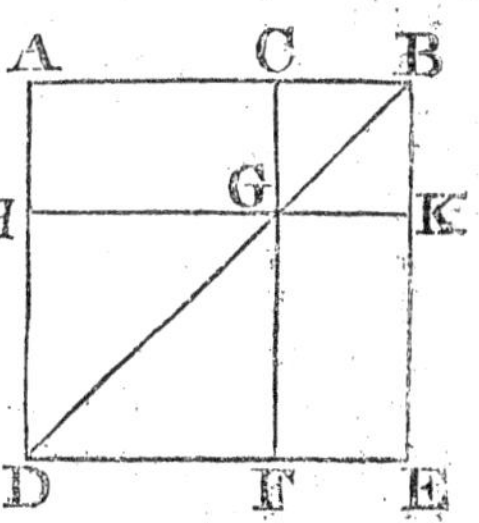

Otherwise,

"Because $AB^2 = AC^2 + BC^2 + 2AC.BC$ (4. 2.), adding BC^2 to "both, $AB^2+BC^2 = AC^2+2BC^2+2AC.BC$.
"But $BC^2+AC.BC=AB.BC$ (3. 2.); and "therefore, $2BC^2+2AC.BC=2AB.BC$; and "therefore $AB^2+BC^2=AC^2+2AB.BC$."

A C B

"Cor. Hence, the sum of the squares of any two lines is equal "to twice the rectangle contained by the lines together with the "square of the difference of the lines."

PROP. VIII. THEOR.

If a straight line be divided into any two parts, four times the rectangle contained by the whole line, and one of the parts, together with the square of the other part, is equal to the square of the straight line which is made up of the whole and the first-mentioned part.

Let the straight line AB be divided into any two parts in the point C; four times the rectangle AB.BC, together with the square of AC, is equal to the square of the straight line made up of AB and BC together.

Produce AB to D, so that BD be equal to CB, and upon AD describe the square AEFD; and construct two figures such as in the preceding. Because GK is equal (34. 1.) to CB, and CB to BD, and BD to KN, GK is equal to KN. For the same reason, PR is equal to RO; and because CB is equal to BD, and GK to KN, the rectangles CK and BN are equal, as also the rectangles GR and RN: But CK is equal (43. 1.) to RN, because they are the complements of the parallelogram CO: therefore also BN is equal to GR; and the four rectangles BN, CK, GR, RN are therefore equal to one another, and so CK+BN+GR+RN=4CK. Again, because CB is equal to BD, and BD equal (Cor. 4. 2.) to BK, that is, to CG; and CB equal to GK, that (Cor. 21. 2.) is, to GP; therefore CG is equal to GP; and because CG is equal to GP, and PR to RO, the rectangle AG is equal to MP, and PL to RF: But MP is equal (43. 1.) to PL, because they are the complements of the parallelogram ML; wherefore AG is equal also to RF. Therefore the four rectangles AG, MP, PL, RF, are equal to one another, and so AG+MP+PL+RF=4AG. And it was demonstrated, that CK+BN+GR+RN=4CK; wherefore, adding equals to equals, the whole gnomon AOH=4AK. Now AK=AB.BK =AB.BC, and 4AK=4AB.BC; therefore, gnomon AOH=4AB.BC; and adding XH, or (Cor. 4. 2.) AC^2, to both, gnomon AOH+XH= $4AB.BC+AC^2$. But $AOH+XH=AF=AD^2$; therefore $AD^2= 4AB.BC+AC^2$. Now AD is the line that is made up of AB and BC, added together into one line: Wherefore, if a straight line, &c. Q. E. D.

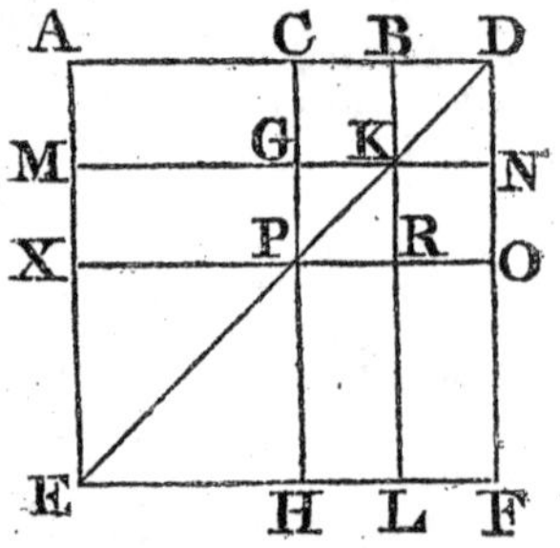

"Cor. 1. Hence, because AD is the sum, and AC the difference of "the lines AB and BC, four times the rectangle contained by any two "lines, together with the square of their difference, is equal to the "square of the sum of the lines."

"Cor. 2. From the demonstration it is manifest, that since the "square of CD is quadruple of the square of CB, the square of any "line is quadruple of the square of half that line."

Otherwise;

"Because AD is divided any how in C (4. 2.). $AD^2 = AC^2 + CD^2$ "$+ 2CD.AC$. But $CD = 2CB$; and therefore $CD = CB^2 + BD^2 +$ "$2CB.BD$ (4. 2.) $= 4CB^2$, and also $2CD.AC = 4CB.AC$; therefore, "$AD^2 = AC^2 + 4BC^2 + 4BC.AC$. Now $BC^2 + BC.AC = AB.BC$ "(3. 2.) and therefore $AD^2 = AC + 4AB.BC$. Q. E. D."

PROP. IX. THEOR.

If a straight line be divided into two equal, and also into two unequal parts; the squares of the two unequal parts are together double of the square of half the line, and of the square of the line between the points of section.

Let the straight line AB be divided at the point C into two equal, and at D into two unequal parts; The squares of AD, DB are together double of the squares of AC, CD.

From the point C draw (11. 1.) CE at right angles to AB, and make it equal to AC or CB, and join EA, EB; through D draw (31. 1. DF parallel to CE, and through F draw FG parallel to AB; and join AF; Then, because AC is equal to CE, the angle EAC is equal (5. 1.) to the angle AEC; and because the angle ACE is a right angle, the two others AEC, EAC together make one right angle (32. 1.); and they are equal to one another; each of them therefore is half of a right angle. For the same reason each of the angles CEB, EBC is half a right angle; and therefore the whole AEB is a right angle; And because the angle GEF is half a right angle, and EGF a right angle, for it is equal (29. 1.) to the interior and opposite angle ECB, the remaining angle EFG is half a right angle; therefore the angle GEF is equal to the angle EFG, and the side EG equal (6. 1.) to the side GF; Again, because the angle at B is half a right angle, and FDB a right angle, for it is equal (29. 1.) to the interior and opposite angle ECB, the remaining angle BFD is half a right angle; therefore the angle at B is equal to the angle BFD, and the side DF to (6. 1.) the side DB. Now, because $AC = CE$, $AC^2 = CE^2$, and $AC^2 + CE^2 = 2AC^2$. But (47. 1.) $AE^2 = AC^2 + CE^2$; therefore $AE^2 = 2AC^2$. Again, because $EG = GF$, $EG^2 = GF^2$, and $EG^2 + GF^2 = 2GF^2$. But $EF^2 = EG^2 + GF^2$; therefore, $EF^2 = 2GF^2 = 2CD^2$, because (34. 1.) $CD = GF$. And it was shown that $AE^2 = 2AC^2$; therefore $AE^2 + EF^2 = 2AC^2 + 2CD^2$. But (47. 1.) $AF^2 = AE^2 +$

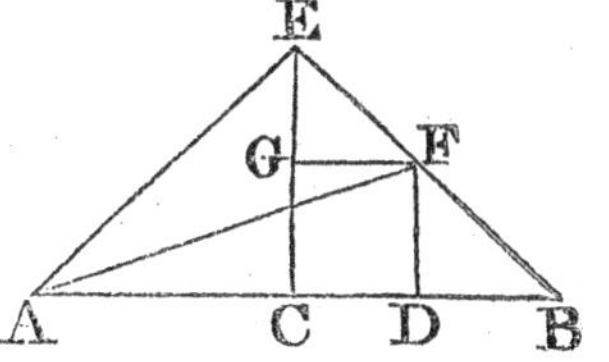

EF^2, and $AD^2+DF^2=AF^2$, or $AD^2+DB^2=AF^2$; therefore, also, $AD^2+DB^2=2AC^2+2CD^2$. Therefore, if a straight line, &c. Q. E. D.

Otherwise;

" Because $AD^2=$ (4. 2.) $AC^2+CD^2+2AC.CD$, and $DB^2+2BC.$
" $CD=$ (7. 2.) $BC^2+CD^2=AC^2+CD^2$, by adding equals to equals,
" $AD^2+BD^2+2BC.CD=2AC^2+2CD^2+2AC.CD$; and therefore
" taking away the equal rectangles 2BC.CD and 2AC.CD, there re-
" mains $AD^2+DB^2=2AC^2+2CD^2$."

PROP. X. THEOR.

If a straight line be bisected, and produced to any point, the square of the whole line thus produced, and the square of the part of it produced, are together double of the square of half the line bisected, and of the square of the line made up of the half and the part produced.

Let the straight line AB be bisected in C, and produced to the point D; the squares of AD, DB are double of the squares of AC, CD.

From the point C draw (11. 1.) CE at right angles to AB, and make it equal to AC or CB; join AE, EB; through E draw (31. 1.) EF parallel to AB, and through D draw DF parallel to CE. And because the straight line EF meets the parallels EC, FD, the angles CEF, EFD are equal (29. 1.) to two right angles; and therefore the angles BEF, EFD are less than two right angles; But straight lines, which with another straight line make the interior angles upon the same side less than two right angles, do meet, (Cor 29. 1.) if produced far enough; therefore EB, FD will meet, if produced, towards B, D: let them meet in G, and join AG. Then because AC is equal to CE, the angle CEA is equal (5. 1.) to the angle EAC; and the angle ACE is a right angle; therefore each of the angles CEA, EAC is half a right angle (32. 1.); For the same reason, each of the angles CEB, EBC is half a right angle; therefore AEB is a right angle; And because EBC is half a right angle, DBG is also (15. 1.) half a right angle, for they are vertically opposite: but BDG is a right angle, because it is equal (29. 1.) to the alternate angle DCE; therefore the remaining angle DGB is half a right angle, and is therefore equal to the angle DBG; wherefore also the side DB is equal (6. 1.) to the side DG. Again, because EGF is half a right angle, and the angle at F a right angle, being equal (34. 1.) to the opposite angle ECD, the remaining angle FEG is half a right angle, and equal to the angle EGF; wherefore also the side GF is equal (6. 1.) to the side FE. And because EC = CA, $EC^2+CA^2=2CA^2$. Now $AE^2=$(47. 1.) AC^2+CE; therefore, $AE^2=2AC^2$. Again,

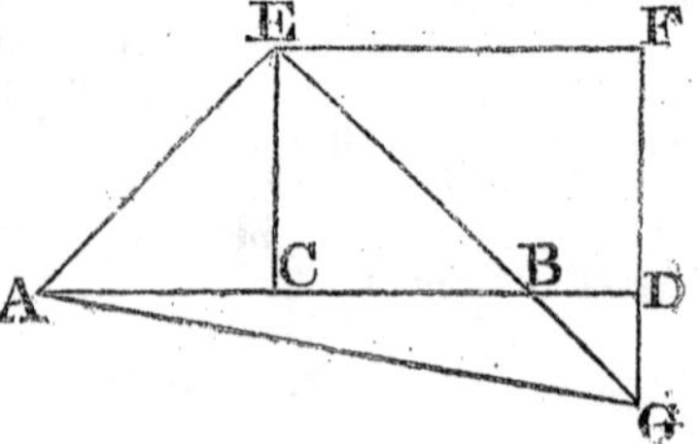

because $EF=FG$, $EF^2=FG^2$, and $EF^2+FG^2=2EF^2$. But $EG^2=$ (47. 1.) EF^2+FG^2; therefore $EG^2=2EF^2$; and since $EF=CD$, $EG^2=2CD^2$. And it was demonstrated, that $AE^2=2AC^2$; therefore, $AE^2+EG^2=2AC^2+2CD^2$. Now, $AG^2=AE^2+EG^2$, wherefore $AG^2=2AC^2+2CD^2$. But AG^2 (47. 1.) $=AD^2+DG^2=AD^2+DB^2$, because $DG=DB$: Therefore, $AD^2+DB^2=2AC^2+2CD^2$. Wherefore, if a straight line, &c. Q. E. D.

PROP. XI. PROB.

To divide a given straight line into two parts, so that the rectangle contained by the whole, and one of the parts, may be equal to the square of the other part.

Let AB be the given straight line; it is required to divide it into two parts, so that the rectangle contained by the whole, and one of the parts, shall be equal to the square of the other part.

Upon AB describe (46. 1.) the square ABDC; bisect (10. 1.) AC in E, and join BE; produce CA to F, and make (3. 1.) EF equal to EB, and upon AF describe (46. 1.) the square FGHA, AB is divided in H, so that the rectangle AB, BH is equal to the square of AH.

Produce GH to K: Because the straight line AC is bisected in E, and produced to the point F, the rectangle CF.FA, together with the square of AE, is equal (6. 2.) to the square of EF: But EF is equal to EB; therefore the rectangle CF.FA, together with the square of AE, is equal to the square of EB: And the squares of BA, AE are equal (47. 1.) to the square of EB, because the angle EAB is a right angle; therefore the rectangle CF.FA, together with the square of AE, is equal to the squares of BA, AE: take away the square of AE, which is common to both, therefore the remaining rectangle CF.FA is equal to the square of AB. Now the figure FK is the rectangle CF.FA, for AF is equal to FG; and AD is the square of AB; therefore FK is equal to AD: take away the common part AK, and the remainder FH is equal to the remainder HD. But HD is the rectangle AB.BH, for AB is equal to BD; and FH is the square of AH; therefore the rectangle AB.BH is equal to the square of AH: Wherefore the straight line AB is divided in H, so that the rectangle AB.BH is equal to the square of AH. Which was to be done.

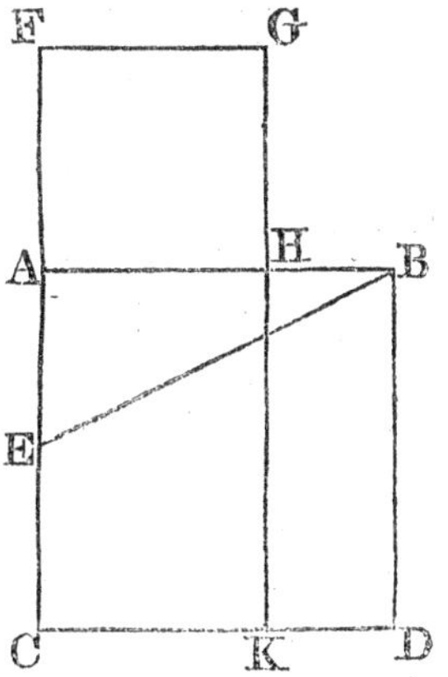

PROP. XII. THEOR.

In obtuse angled triangles, if a perpendicular be drawn from any of the acute angles to the opposite side produced, the square of the side subtending the obtuse angle is greater than the squares of the sides containing the obtuse angle, by twice the rectangle contained by the side upon which, when produced, the perpendicular falls, and the straight line intercepted between the perpendicular and the obtuse angle.

Let ABC be an obtuse angled triangle, having the obtuse angle ACB, and from the point A let AD be drawn (12. 1.) perpendicular to BC produced: The square of AB is greater than the squares of AC, CB, by twice the rectangle BC.CD.

Because the straight line BD is divided into two parts in the point C, $BD^2 =$ (4. 2.) $BC^2+CD^2+2BC.CD$; add AD^2 to both: Then $BD^2+AD^2=BC^2+CD^2+AD^2+2BC.CD$. But $AB^2=BD^2+AD^2$ (47. 1.), and $AC^2=CD^2+AD^2$ (47. 1.); therefore, $AB^2=BC^2+AC^2+2BC.CD$; that is, AB^2 is greater than BC^2+AC^2 by 2BC.CD. Therefore, in obtuse angled triangles, &c. Q.E.D.

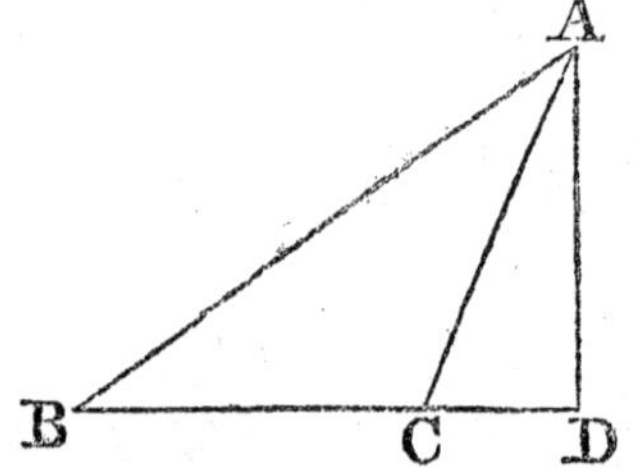

PROP. XIII. THEOR.

In every triangle the square of the side subtending any of the acute angles, is less than the squares of the sides containing that angle, by twice the rectangle contained by either of these sides, and the straight line intercepted between the perpendicular, let fall upon it from the opposite angle, and the acute angle.

Let ABC be any triangle, and the angle at B one of its acute angles, and upon BC, one of the sides containing it, let fall the perpendicular (12. 1.) AD from the opposite angle: The square of AC, opposite to the angle B, is less than the squares of CB, BA by twice the rectangle CB.BD.

First, let AD fall within the triangle ABC; and because the straight line CB is divided into two parts in the point D (7. 2.), $BC^2+BD^2=2BC.BD+CD^2$. Add to each AD^2; then $BC+BD+AD^2=2BC.BD+CD^2+AD^2$. But $BD^2+AD^2=AB^2$, and $CD^2+DA^2=AC^2$ (47. 1.); therefore $BC^2+AB^2=2BC.BD+AC^2$; that is, AC^2 is less than BC^2+AB^2 by 2BC.BD.

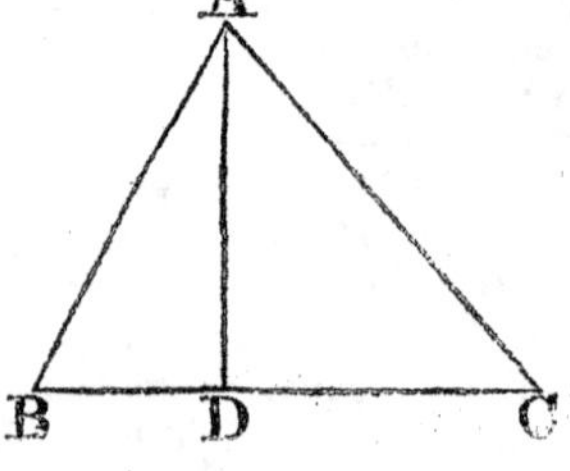

Secondly, Let AD fall without the triangle ABC:* Then because the angle at D is a right angle, the angle ACB is greater (16. 1.) than a right angle, and $AB^2=$(12. 2.) $AC^2+BC^2+2BC.CD$. Add BC^2 to each; then $AB^2+BC^2=AC^2+2BC^2+2BC.CD$. But because BD is divided into two parts in C, $BC^2+BC.CD=$(3. 2.) $BC.BD$, and $2BC^2+2BC.CD=2BC.BD$: therefore $AB^2+BC^2=AC^2+2BC.BD$; and AC^2 is less than AB^2+BC^2, by $2BD.BC$.

Lastly, Let the side AC be perpendicular to BC; then is BC the straight line between the perpendicular and the acute angle at B; and it is manifest that (47. 1.) $AB^2+BC^2=AC^2+2BC^2=AC^2+2BC.BC$. Therefore in every triangle, &c. Q. E. D.

PROP. XIV. PROB.

To describe a square that shall be equal to a given retilineal figure.

Let A be the given rectilineal figure; it is required to describe a square that shall be equal to A.

Describe (45. 1.) the rectangular parallelogram BCDE equal to the rectilineal figure A. If then the sides of it, BE, ED are equal to one another, it is a square, and what was required is done; but if they are not equal, produce one of them, BE to F, and make EF equal to ED, and bisect BF in G: and from the centre G, at the distance GB, or GF, describe the semicircle BHF, and produce DE to H, and join GH. Therefore, because the straight line BF is divided into two equal parts in the point G, and into two unequal in the point E, the rectangle BE.EF, together with the square of EG, is equal (5. 2.) to the square of GF: but GF is equal to GH; therefore the rectangle BE.EF, together with the square of EG, is equal to the square of GH: But the squares of HE and EG are equal (47. 1.) to the square of GH: Therefore also the rectangle BE.EF, together with the square of EG, is equal to the squares of HE and EG. Take away the square of EG, which is common to both, and the remaining rectangle BE.EF is equal to the square of EH: But BD is the rectangle con-

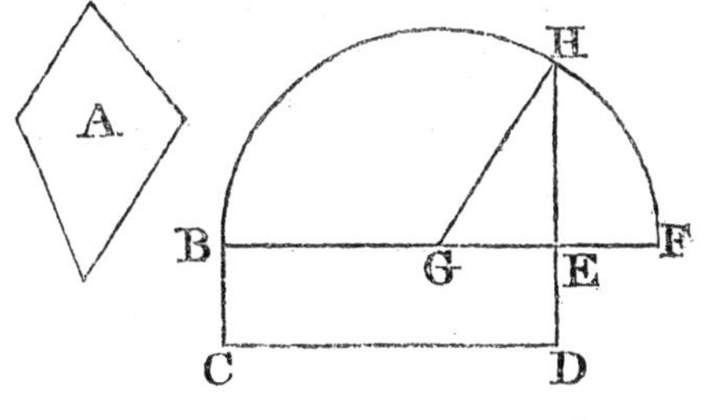

* See figure of the last Proposition.

tained by BE and EF, because EF is equal to ED; therefore BD is equal to the square of EH; and BD is also equal to the rectilineal figure A; therefore the rectilineal figure A is equal to the square of EH: Wherefore a square has been made equal to the given rectilineal figure A, viz. the square described upon EH. Which was to be done.

PROP. A. THEOR.

If one side of a triangle be bisected, the sum of the squares of the other two sides is double of the square of half the side bisected, and of the square of the line drawn from the point of bisection to the opposite angle of the triangle.

Let ABC be a triangle, of which the side BC is bisected in D, and DA drawn to the opposite angle; the squares of BA and AC are together double of the squares of BD and DA.

From A draw AE perpendicular to BC, and because BEA is a right angle, $AB^2=$(47. 1.) BE^2+AE^2 and $AC^2=CE^2+AE^2$; wherefore $AB^2+AC^2=BE^2+CE^2+2AE^2$. But because the line BC is cut equally in D, and unequally in E, $BE^2+CE^2=$ (9. 2.) $2BD^2+2DE^2$; therefore $AB^2+AC^2=2BD^2+2DE^2.2AE^2$.

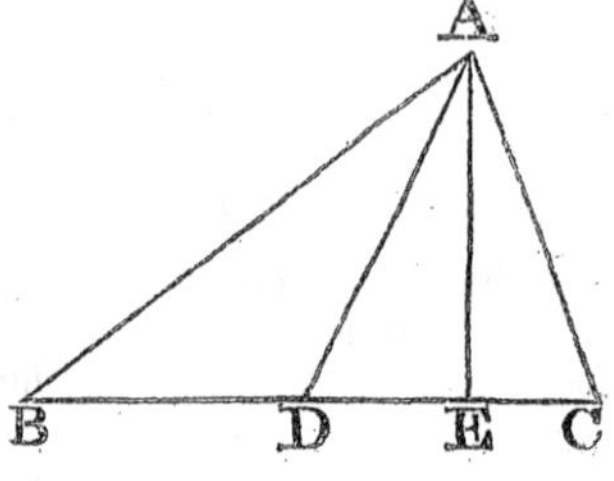

Now $DE^2+AE^2=$(47. 1.) AD^2, and $2DE^2+2AE^2=2AD^2$; wherefore $AB^2+AC^2=2BD^2+2AD^2$. Therefore, &c. Q. E. D.

PROP. B. THEOR.

The sum of the squares of the diameters of any parallelogram is equal to the sum of the squares of the sides of the parallelogram.

Let ABCD be a parallelogram, of which the diameters are AC and BD; the sum of the squares of AC and BD is equal to the sum of the squares of AB, BC, CD, DA.

Let AC and BD intersect one another in E: and because the vertical angles AED, CEB are equal (15. 1.), and also the alternate angles EAD, ECB (29. 1.), the triangles ADE, CEB have two angles in the one equal to two angles in the other, each to each; but the sides AD and BC, which are opposite to equal angles in these triangles, are also equal (34. 1.); therefore the other sides which are opposite to the equal angles are also equal (26. 1.), viz. AE to EC, and ED to EB.

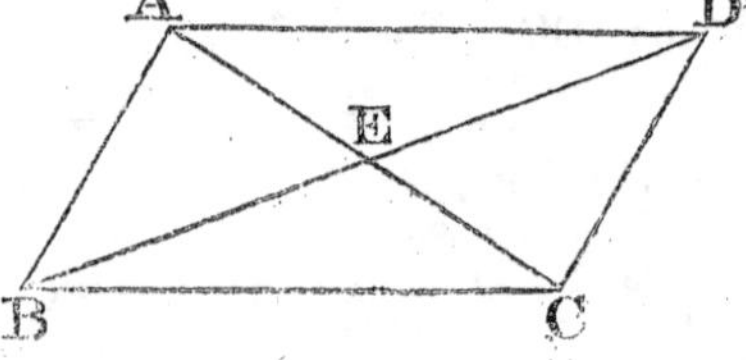

Since, therefore, BD is bisected in E, $AB^2+AD^2=$(A. 2.) $2BE^2+2AE^2$; and for the same reason, $CD^2+BC^2=2BE^2+2EC^2=2BE^2+2AE^2$, because $EC=AE$. Therefore $AB^2+AD^2+DC^2+BC^2=4BE^2+4AE^2$. But $4BE^2=BD^2$, and $4AE^2=AC^2$ (2. Cor. 8. 2.) because BD and AC are both bisected in E; therefore $AB^2+AC^2+CD^2+BC^2=BD^2+AC^2$. Therefore the sum of the squares, &c. Q. E. D.

Cor. From this demonstration, it is manifest that the diameters of every parallelogram bisect one another.

ELEMENTS

OF

GEOMETRY.

BOOK III.

DEFINITIONS.

A.

The radius of a circle is the straight line drawn from the centre to the circumference.

I.

A straight line is said to touch a circle, when it meets the circle, and being produced does not cut it.

II.

Circles are said to touch one another, which meet, but do not cut one another.

III.

Straight lines are said to be equally distant from the centre of a circle, when the perpendiculars drawn to them from the centre are equal.

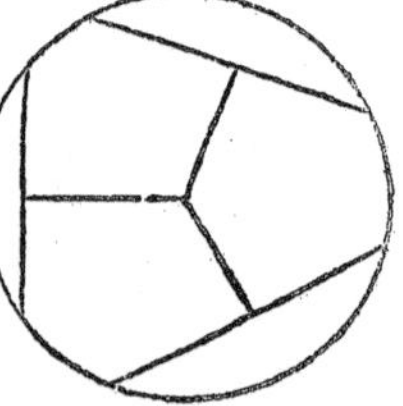

IV.

And the straight line on which the greater perpendicular falls, is said to be farther from the centre.

B.

An arch of a circle is any part of the circumference.

V.

A segment of a circle is the figure contained by a straight line, and the arch which it cuts off.

VI.

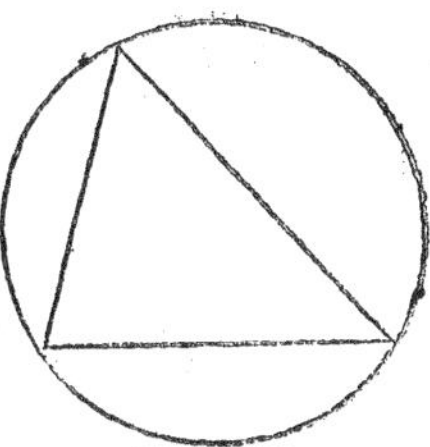

An angle in a segment is the angle contained by two straight lines drawn from any point in the circumference of the segment, to the extremities of the straight line which is the base of the segment.

VII.

And an angle is said to insist or stand upon the arch intercepted between the straight lines which contain the angle.

VIII.

The sector of a circle is the figure contained by two straight lines drawn from the centre, and the arch of the circumference between them.

IX.

Similar segments of a circle, are those in which the angles are equal, or which contain equal angles.

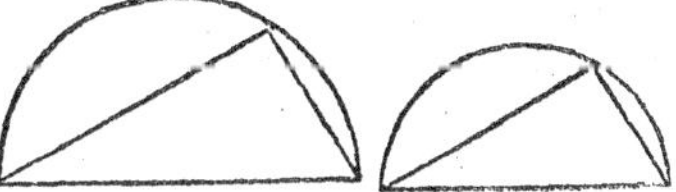

PROP. I. PROB.

To find the centre of a given circle.

Let ABC be the given circle; it is required to find its centre.

Draw within it any straight line AB, and bisect (10. 1.) it in D; from the point D draw (11. 1.) DC at right angles to AB, and produce it to E, and bisect CE in F: the point F is the centre of the circle ABC.

For, if it be not, let, if possible, G be the centre, and join GA, GD, GB: Then, because DA is equal to DB, and DG common to the two triangles ADG, BDG, the two sides AD, DG are equal to the two BD, DG, each to each; and the base GA is equal to the base GB, because they are radii of the same circle: therefore the angle ADG is equal (8. 1.) to the angle GDB: But when a straight line standing upon another straight line makes the adjacent angles equal to one another, each of the angles is a right angle (7. def. 1.) Therefore the angle GDB is a right angle: But FDB is likewise a right angle; wherefore the angle FDB is equal to the angle GDB, the greater

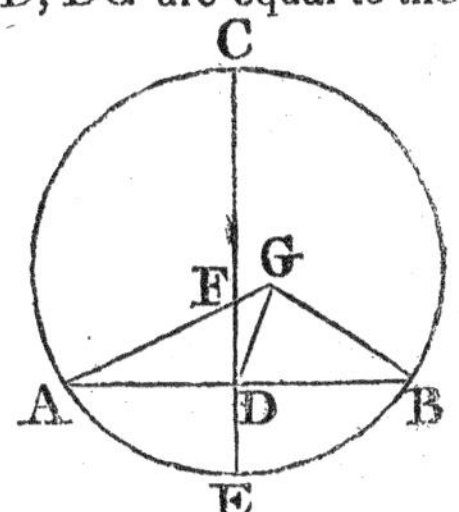

to the less, which is impossible: Therefore G is not the centre of the circle ABC: In the same manner, it can be shown, that no other point but F is the centre: that is, F is the centre of the circle ABC: Which was to be found.

COR. From this it is manifest that if in a circle a straight line bisect another at right angles, the centre of the circle is in the line which bisects the other.

PROP. II. THEOR.

If any two points be taken in the circumference of a circle, the straight line which joins them shall fall within the circle.

Let ABC be a circle, and A, B any two points in the circumference: the straight line drawn from A to B shall fall within the circle.

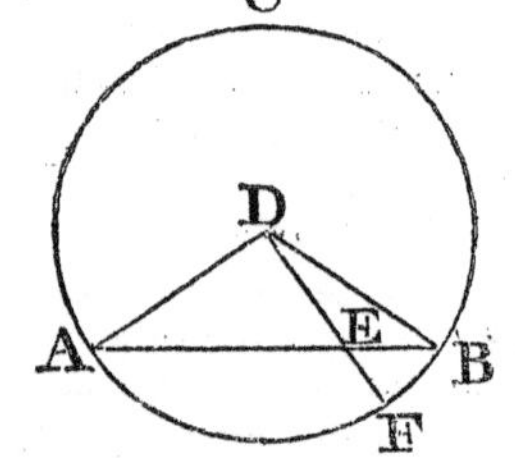

Take any point in AB as E; find D the centre of the circle ABC; join AD, DB and DE, and let DE meet the circumference in F. Then, because DA is equal to DB, the angle DAB is equal (5. 1.) to the angle DBA; and because AE, a side of the triangle DAE, is produced to B, the angle DEB is greater (16. 1.) than the angle DAE; but DAE is equal to the angle DBE; therefore the angle DEB is greater than the angle DBE: Now to the greater angle the greater side is opposite (19. 1.); DB is therefore greater than DE: but BD is equal to DF; wherefore DF is greater than DE, and the point E is therefore within the circle. The same may be demonstrated of any other point between A and B, therefore AB is within the circle. Wherefore, if any two points, &c. Q. E. D.

PROP. III. THEOR.

If a straight line drawn through the centre of a circle bisect a straight line in the circle, which does not pass through the centre, it will cut that line at right angles; and if it cut it at right angles, it will bisect it.

Let ABC be a circle, and let CD, a straight line drawn through the centre, bisect any straight line AB, which does not pass through the centre, in the point F: It cuts it also at right angles.

Take (1. 3.) E the centre of the circle, and join EA, EB. Then because AF is equal to FB, and FE common to the two triangles AFE, BFE, there are two sides in the one equal to two sides in the other:

but the base EA is equal to the base EB; therefore the angle AFE is equal (8. 1.) to the angle BFE. And when a straight line standing upon another makes the adjacent angles equal to one another, each of them is a right (7. Def. 1.) angle: Therefore each of the angles AFE, BFE is a right angle; wherefore the straight line CD, drawn through the centre bisecting AB, which does not pass through the centre, cuts AB at right angles.

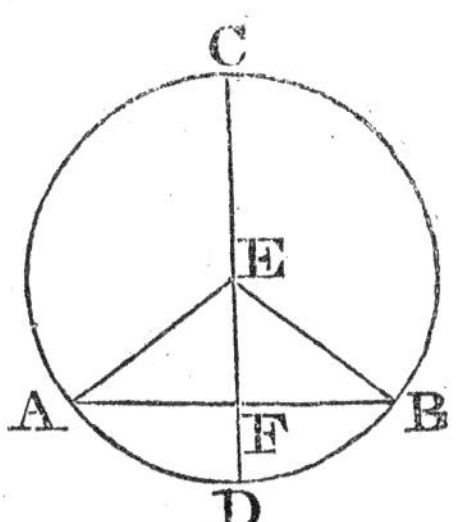

Again, let CD cut AB at right angles; CD also bisects AB, that is, AF is equal to FB.

The same construction being made, because the radii EA, EB are equal to one another, the angle EAF is equal (5. 1.) to the angle EBF; and the right angle AFE is equal to the right angle BFE: Therefore, in the two triangles EAF, EBF, there are two angles in one equal to two angles in the other; now the side EF, which is opposite to one of the equal angles in each, is common to both; therefore the other sides are equal (26. 1.); AF therefore is equal to FB. Wherefore, if a straight line, &c. Q. E. D.

PROP. IV. THEOR.

If in a circle two straight lines cut one another, which do not both pass through the centre, they do not bisect each other.

Let ABCD be a circle, and AC, BD two straight lines in it, which cut one another in the point E, and do not both pass through the centre: AC, BD do not bisect one another

For if it is possible, let AE be equal to EC, and BE to ED; If one of the lines pass through the centre, it is plain that it cannot be bisected by the other, which does not pass through the centre. But if neither of them pass through the centre, take (1. 3.) F the centre of the circle, and join EF: and because FE, a straight line through the centre, bisects another AC, which does not pass through the centre, it must cut it at right (3. 3.) angles; wherefore FEA is a right angle. Again, because the straight line FE bisects the straight line BD, which does not pass through the centre, it must cut it at right (3. 3) angles; wherefore FEB is a right angle: and FEA was shown to be a right angle: therefore FEA is equal to the angle FEB, the less to the greater, which is impossible: therefore AC, BD do not bisect one another. Wherefore, if in a circle, &c. Q. E. D.

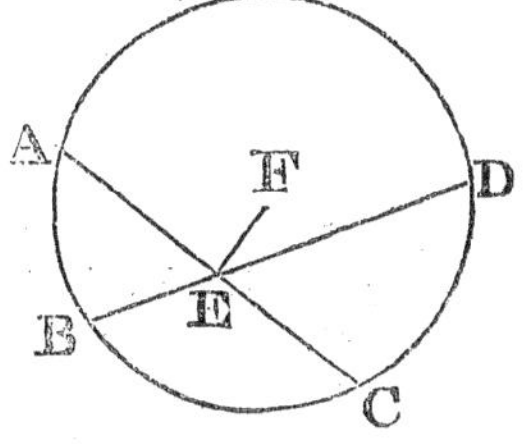

PROP. V. THEOR.

If two circles cut one another, they cannot have the same centre.

Let the two circles ABC, CDG cut one another in the points B, C; they have not the same centre.

For, if it be possible, let E be their centre: join EC, and draw any straight line EFG meeting the circles in F and G: and because E is the centre of the circle ABC, CE is equal to EF: Again, because E is the centre of the circle CDG, CE is equal to EG: but, CE was shown to be equal to EF, therefore EF is equal to EG, the less to the greater, which is impossible: therefore E is not the centre of the circles ABC, CDG. Wherefore, if two circles, &c. Q.E.D.

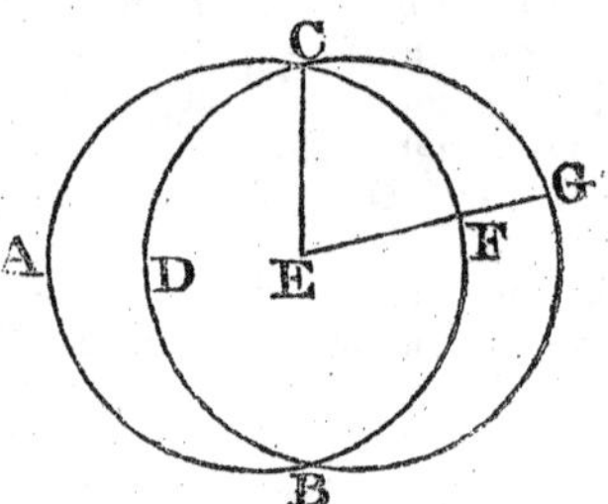

PROP. VI. THEOR.

If two circles touch one another internally, they cannot have the same centre.

Let the two circles ABC, CDE, touch one another internally in the point C: they have not the same centre.

For, if they have, let it be F; join FC, and draw any straight line FEB meeting the circles in E and B; and because F is the centre of the circle ABC, CF is equal to FB; also, because F is the centre of the circle CDE, CF is equal to FE: but CF was shown to be equal to FB; therefore FE is equal to FB, the less to the greater, which is impossible; wherefore F is not the centre of the circles ABC, CDE. Therefore, if two circles, &c. Q. E. D.

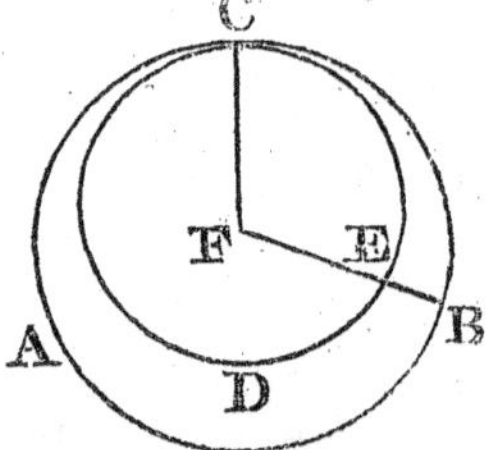

PROP. VII. THEOR.

If any point be taken in the diameter of a circle which is not the centre of all the straight lines which can be drawn from it to the circumference, the greatest is that in which the centre is, and the other part of that diameter is the least; and, of any others, that which is nearer to the line passing through the centre is always greater than one more remote from it; And from the same point there can be drawn only two straight lines that are equal to one another, one upon each side of the shortest line.

Let ABCD be a circle, and AD its diameter, in which let any point F be taken which is not the centre: let the centre be E; of all the straight lines FB, FC, FG, &c. that can be drawn from F to the circumference, FA is the greatest; and FD, the other part of the diameter AD, is the least; and of the others, FB is greater than FC, and FC than FG.

Join BE, CE, GE; and because two sides of a triangle are greater (20. 1.) than the third, BE, EF are greater than BF; but AE is equal to EB; therefore AE and EF, that is, AF, is greater than BF: again, because BE is equal to CE, and FE common to the triangles BEF, CEF, the two sides BE, EF are equal to the two CE, EF: but the angle BEF is greater than the angle CEF; therefore the base BF is greater (24. 1.) than the base FC; for the same reason, CF is greater than GF. Again, because GF, FE are greater (20. 1.) than EG, and EG is equal to ED; GF, FE are greater than ED; take away the common part FE, and the remainder GF is greater than the remainder FD: therefore FA is the greatest, and FD the least of all the straight lines from F to the circumference; and BF is greater than CF, and CF than GF.

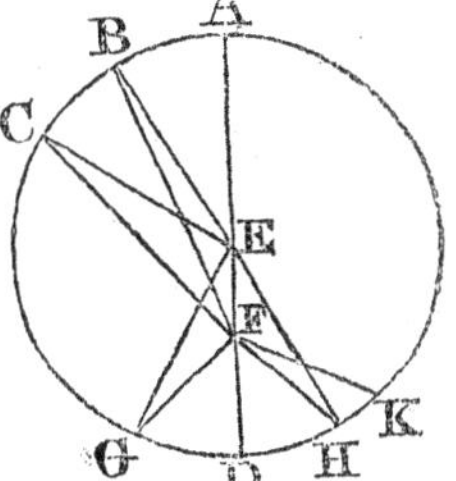

Also there can be drawn only two equal straight lines from the point F to the circumference, one upon each side of the shortest line FD: at the point E in the straight line EF, make (23. 1.) the angle FEH equal to the angle GEF, and join FH: Then, because GE is equal to EH, and EF common to the two triangles GEF, HEF; the two sides GE, EF are equal to the two HE, EF; and the angle GEF is equal to the angle HEF; therefore the base FG is equal (4. 1.) to the base FH: but besides FH, no straight line can be drawn from F to the circumference equal to FG: for, if there can, let it be FK; and because FK is equal to FG, and FG to FH, FK is equal to FH; that is, a line nearer to that which passes through the centre, is equal to one more remote, which is impossible. Therefore, if any point be taken, &c. Q. E. D.

PROP. VIII. THEOR.

If any point be taken without a circle, and straight lines be drawn from it to the circumference, whereof one passes through the centre; of those which fall upon the concave circumference, the greatest is that which passes through the centre; and of the rest that which is nearer to that through the centre is always greater than the more remote; But of those which fall upon the convex circumference, the least is that between the point without the circle, and the diameter; and of the rest, that which is nearer to the least is always less than the more remote: And only two equal straight lines can be drawn from the point unto the circumference, one upon each side of the least.

Let ABC be a circle, and D any point without it, from which let the straight lines DA, DE, DF, DC be drawn to the circumference, whereof DA passes through the centre. Of those which fall upon the concave part of the circumference AEFC, the greatest is AD, which passes through the centre; and the line nearer to AD is always greater than the more remote, viz. DE than DF, and DF than DC: but of those which fall upon the convex circumference HLKG, the least is DG, between the point D and the diameter AG; and the nearer to it is always less than the more remote, viz. DK than DL, and DL than DH.

Take (1. 3.) M the centre of the circle ABC, and join ME, MF, MC, MK, ML, MH: And because AM is equal to ME, if MD be added to each, AD is equal to EM and MD; but EM and MD are greater (20. 1.) than ED; therefore also AD is greater than ED. Again, because ME is equal to MF, and MD common to the triangles EMD, FMD; EM, MD are equal to FM, MD; but the angle EMD is greater than the angle FMD; therefore the base ED is greater (24. 1.) than the base FD. In like manner it may be shewn that FD is greater than CD. Therefore DA is the greatest; and DE greater than DF, and DF than DC.

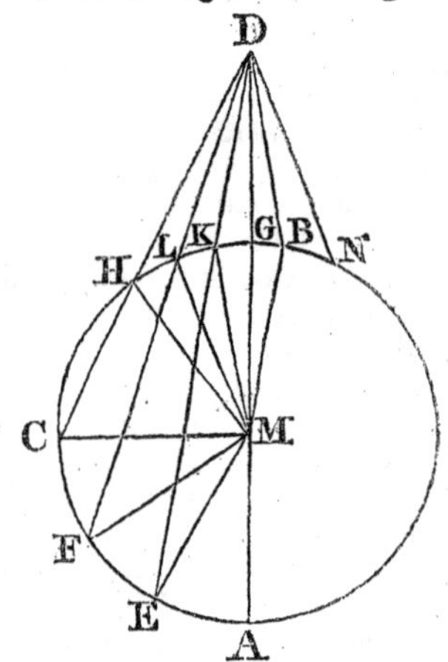

And because MK, KD are greater (20. 1.) than MD, and MK is equal to MG, the remainder KD is greater (5. Ax.) than the remainder GD, that is, GD is less than KD: And because MK, DK are drawn to the point K within the triangle MLD from M, D, the extremities of its side MD; MK, KD are less (21. 1.) than ML; LD, whereof MK is equal to ML; therefore the remainder DK is less than the remainder DL: In like manner, it may be shewn that DL is less than DH: Therefore DG is the least, and DK less than DL, and DL than DH.

Also there can be drawn only two equal straight lines from the point D to the circumference, one upon each side of the least: at the point M, in the straight line MD make the angle DMB equal to the angle DMK, and join DB; and because in the triangles KMD, BMD, the side KM is equal to the side BM, and MD common to both, and also the angle KMD equal to the angle BMD, the base DK is equal (4. 1.) to the base DB. But, besides DB, no straight line can be drawn from D to the circumference, equal to DK: for, if there can, let it be DN; then, because DN is equal to DK, and DK equal to DB, DB is equal to DN; that is, the line nearer to DG, the least, equal to the more remote, which has been shewn to be impossible. If, therefore, any point, &c. Q. E. D.

PROP. IX. THEOR.

If a point be taken within a circle, from which there fall more than two equal straight lines upon the circumference, that point is the centre of the circle.

Let the point D be taken within the circle ABC, from which there fall on the circumference more than two equal straight lines, viz. DA, DB, DC, the point D is the centre of the circle.

For, if not, let E be the centre, join DE, and produce it to the circumference in F, G; then FG is a diameter of the circle ABC: And because in FG, the diameter of the circle ABC, there is taken the point D which is not the centre, DG is the greatest line from it to the circumference, and DC greater (7. 3.) than DB, and DB than DA; but they are likewise equal, which is impossible: Therefore E is not the centre of the circle ABC: In like manner, it may be demonstrated, that no other point but D is the centre: D therefore is the centre. Wherefore, if a point be taken, &c. Q. E. D.

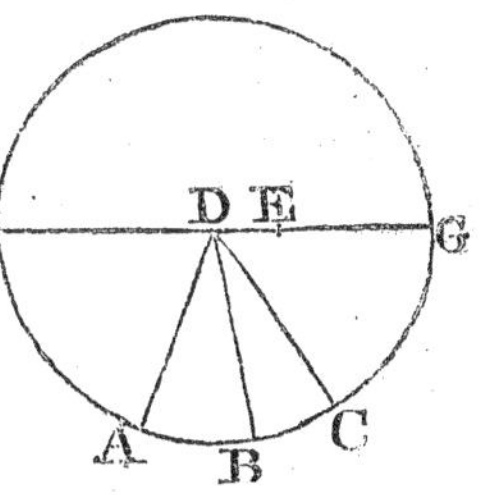

PROP. X. THEOR.

One circle cannot cut another in more than two points.

If it be possible, let the circumference FAB cut the circumference DEF in more than two points, viz, in B, G, F; take the centre K of the circle ABC, and join KB, KG, KF; and because within the circle DEF there is taken the point K, from which more than two equal straight lines, viz. KB, KG, KF, fall on the circumference DEF, the point K is (9. 3.) the centre of the circle DEF; but K is also the centre of the circle ABC; therefore the same point is the centre of two circles that cut one another, which is impossible (5. 3.). Therefore one circumference of a circle cannot cut another in more than two points. Q. E. D.

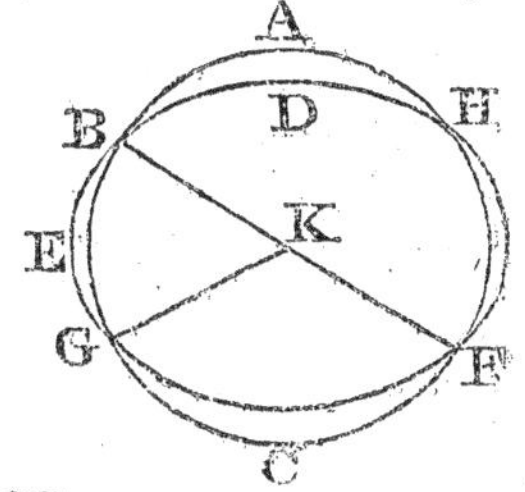

PROP. XI. THEOR.

If two circles touch each other internally, the straight line which joins their centres being produced, will pass through the point of contact.

Let the two circles ABC, ADE, touch each other internally in the point A, and let F be the centre of the circle ABC, and G the centre of the circle ADE; the straight line which joins the centres F, G, being produced, passes through the point A.

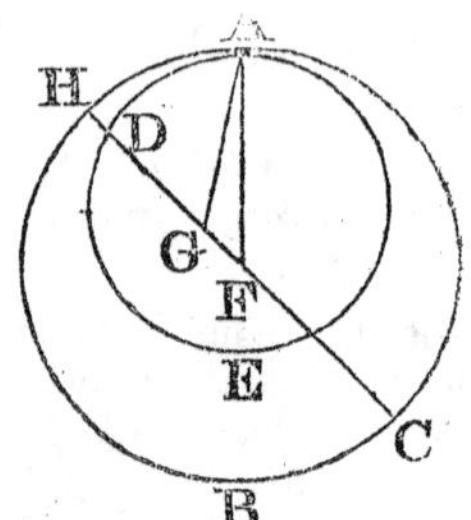

For, if not, let it fall otherwise, if possible, as FGDH, and join AF, AG: And because AG, GF are greater (20. 1.) than FA, that is, than FH, for FA is equal to FH, being radii of the same circle; take away the common part FG, and the remainder AG is greater than the remainder GH. But AG is equal to GD, therefore GD is greater than GH; and it is also less, which is impossible. Therefore the straight line which joins the points F and G cannot fall otherwise than on the point A; that is, it must pass through A. Therefore, if two circles, &c. Q. E. D.

PROP. XII. THEOR.

If two circles touch each other externally, the straight line which joins their centres will pass through the point of contact.

Let the two circles ABC, ADE, touch each other externally in the point A; and let F be the centre of the circle ABC, and G the centre of ADE; The straight line which joins the points F, G shall pass through the point of contact.

For, if not, let it pass otherwise, if possible, FCDG, and join FA, AG: and because F is the centre of the circle ABC, AF is equal to FC: Also because G is the centre of the circle, ADE, AG is equal to GD. Therefore FA, AG are equal to FC, DG; wherefore the whole FG is greater than FA, AG; but it is also less (20. 1.), which is impossible: Therefore the straight line which joins the points F, G cannot pass otherwise than through the point of contact A; that is, it passes through A. Therefore, if two circles, &c. Q. E. D.

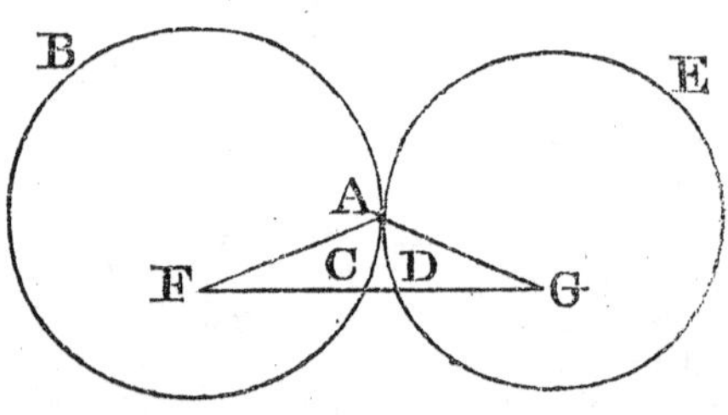

PROP. XIII. THEOR.

One circle cannot touch another in more points than one, whether it touches it on the inside or outside.

For, if it be possible, let the circle EBF touch the circle ABC in more points than one, and first on the inside, in the points B, D; join BD, and draw (10. 11. 1.) GH, bisecting BD at right angles: Therefore because the points B, D are in the circumference of each of the

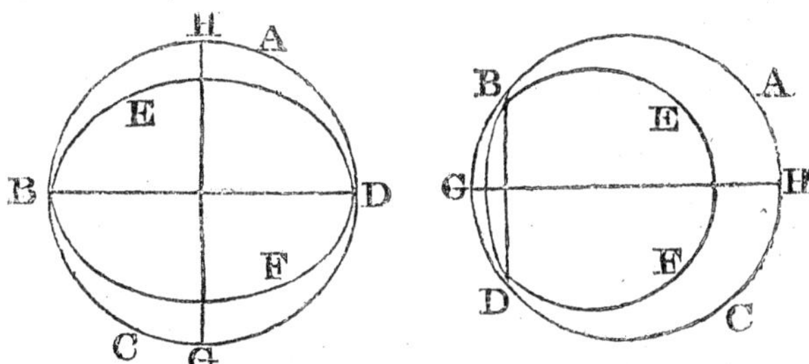

circles, the straight line BD falls within each (2. 3.) of them: and therefore their centres are (Cor. 1. 3.) in the straight line GH which bisects BD at right angles: Therefore GH passes through the point of contact (11. 3.); but it does not pass through it, because the points B, D are without the straight line GH, which is absurd: Therefore one circle cannot touch another in the inside in more points than one.

Nor can two circles touch one another on the outside in more than one point: For if it be possible, let the circle ACK touch the circle ABC in the points A, C, and join AC: Therefore, because the two points A, C are in the circumference of the circle ACK, the straight line AC which joins them shall fall within the circle ACK: And the circle ACK is without the circle ABC; and therefore the straight line AC is also without ABC; but, because the points A, C are in the circumference of the circle ABC, the straight line AC must be within (2. 3.) the same circle, which is absurd: Therefore a circle cannot touch another on the outside in more than one point: and it has been shewn, that a circle cannot touch another on the inside in more than one point. Therefore, one circle, &c. Q. E. D.

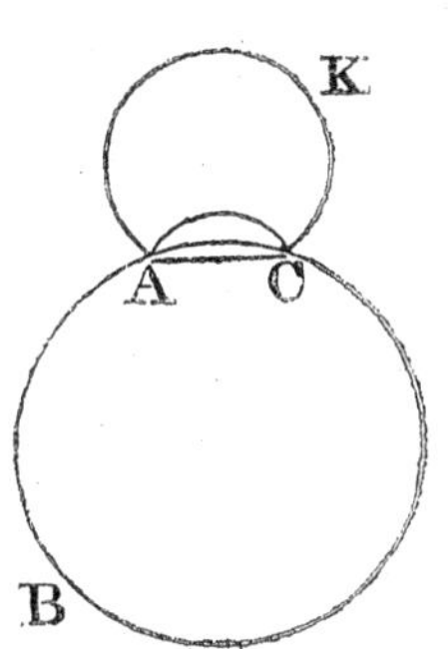

PROP. XIV. THEOR.

Equal straight lines in a circle are equally distant from the centre; and those which are equally distant from the centre, are equal to one another.

Let the straight lines AB, CD, in the circle ABDC, be equal to one another; they are equally distant from the centre.

Take E the centre of the circle ABDC, and from it draw EF, EG, perpendiculars to AB, CD; join AE and EC. Then, because the straight line EF passing through the centre, cuts the straight line AB, which does not pass through the centre at right angles, it also bisects (3. 3.) it: Wherefore AF is equal to FB, and AB double of AF. For the same reason, CD is double of CG: But AB is equal to CD; therefore AF is equal to CG: And because AE is equal to EC, the square of AE is equal to the square of EC: Now the squares of AF, FE are equal (47. 1.) to the square of AE, because the angle AFE is a right angle; and, for the like reason, the squares of EG, GC are equal to the square of EC: Therefore the squares of AF, FE are equal to the squares of CG, GE, of which the square of AF is equal to the square of CG, because AF is equal to CG; therefore the remaining square of FE is equal to the remaining square of EG, and the straight line EF is therefore equal to EG: But straight lines in a circle are said to be equally distant from the centre when the perpendiculars drawn to them from the centre are equal (3. Def. 3.): Therefore AB, CD are equally distant from the centre.

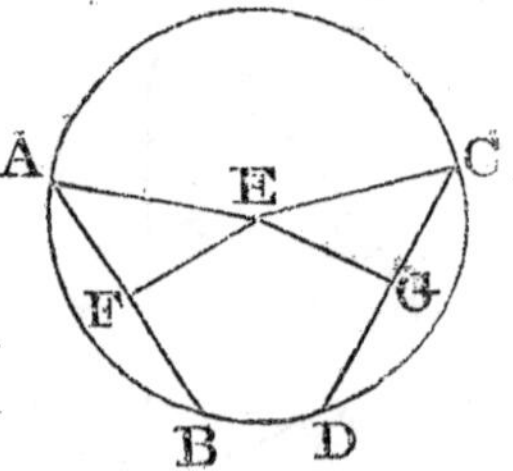

Next, if the straight lines AB, CD be equally distant from the centre, that is, if FE be equal to EG, AB is equal to CD. For, the same construction being made, it may, as before, be demonstrated, that AB is double of AF, and CD double of CG, and that the squares of EF, FA are equal to the squares of EG, GC; of which the square of FE is equal to the square of EG, because FE is equal to EG; therefore the remaining square of AF is equal to the remaining square of CG; and the straight line AF is therefore equal to CG: But AB is double of AF, and CD double of CG; wherefore AB is equal to CD. Therefore equal straight lines, &c. Q. E. D.

PROP XV. THEOR.

The diameter is the greatest straight line in a circle; and of all others, that which is nearer to the centre is always greater than one more remote; and the greater is nearer to the centre than the less.

Let ABCD be a circle, of which the diameter is AD, and the centre E; and let BC be nearer to the centre than FG; AD is greater than any straight line BC which is not a diameter, and BC greater than FG.

From the centre draw EH, EK perpendiculars to BC, FG, and join EB, EC, EF; and because AE is equal to EB, and ED to EC, AD is equal to EB, EC: But EB, EC are greater (20. 1.) than BC; wherefore, also AD is greater than BC.

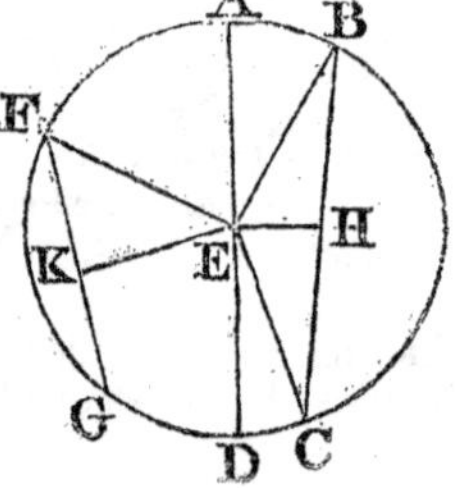

And, because BC is nearer to the centre than FG, EH is less (4. Def. 3.) than EK; But, as was demonstrated in the preceding, BC is double of BH, and FG double of FK, and the squares of EH, HB are equal to the squares of EK, KF, of which the square of EH, is less than the square of EK, because EH is less than EK; therefore the square of BH is greater than the square of FK, and the straight line BH greater than FK; and therefore BC is greater than FG.

Next, let BC be greater than FG; BC is nearer to the centre than FG: that is, the same construction being made, EH is less than EK; Because BC is greater than FG, BH likewise is greater than KF: but the squares of BH, HE are equal to the squares of FK, KE, of which the square of BH is greater than the square of FK, because BH is greater than FK; therefore the square of EH is less than the square of EK, and the straight line EH less than EK. Wherefore, the diameter, &c. Q. E. D.

PROP. XVI. THEOR.

The straight line drawn at right angles to the diameter of a circle, from the extremity of it, falls without the circle; and no straight line can be drawn between that straight line and the circumference, from the extremity of the diameter, so as not to cut the circle.

Let ABC be a circle, the centre of which is D, and the diameter AB: and let AE be drawn from A perpendicular to AB, AE shall fall without the circle.

In AE take any point F, join DF and let DF meet the circle in C. Because DAF is a right angle, it is greater than the angle AFD (32. 1.); but the greater angle of any triangle is subtended by the greater side (19. 1.), therefore DF is greater than DA; now DA is equal to DC, therefore DF is greater than DC, and the point F is therefore without the circle. And F is any point whatever in the line AE, therefore AE falls without the circle.

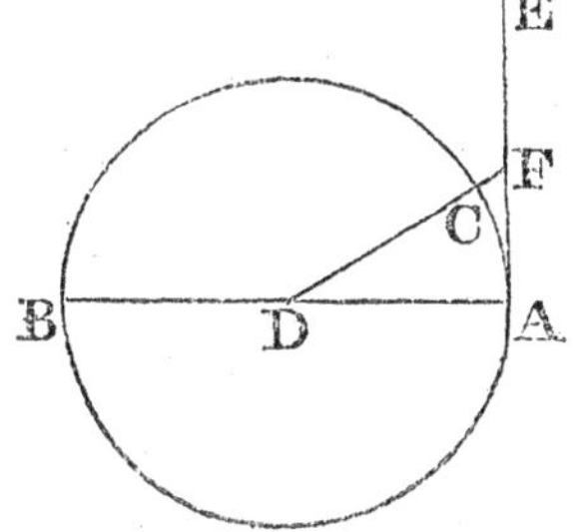

Again, between the straight line AE and the circumference, no straight line can be drawn from the point A, which does not cut the circle. Let AG be drawn in the angle DAE:

from D draw DH at right angles to AG; and because the angle DHA is a right angle, and the angle DAH less than a right angle, the side DH of the triangle DAH is less than the side DA (19. 1.). The point H, therefore, is within the circle, and therefore the straight line AG cuts the circle.

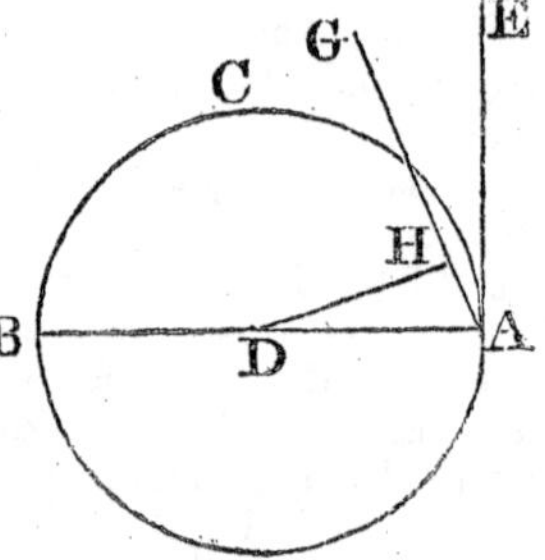

COR. From this it is manifest, that the straight line which is drawn at right angles to the diameter of a circle from the extremity of it, touches the circle; and that it touches it only in one point; because, if it did meet the circle in two, it would fall within it (2. 3.). Also it is evident that there can be but one straight line which touches the circle in the same point.

PROP. XVII. PROB.

To draw a straight line from a given point either without or in the circumference, which shall touch a given circle.

First, Let A be a given point without the given circle BCD; it is required to draw a straight line from A which shall touch the circle.

Find (1. 3.) the centre E of the circle, and join AE; and from the centre E, at the distance EA, describe the circle AFG; from the point D draw (11. 1.) DF at right angles to EA, join EBF, and draw AB. AB touches the circle BCD.

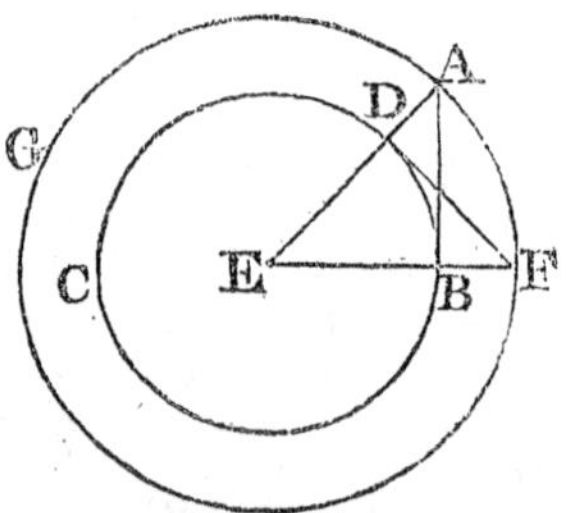

Because E is the centre of the circles BCD, AFG, EA is equal to EF, and ED to EB; therefore the two sides AE, EB are equal to the two FE, ED, and they contain the angle at E common to the two triangles AEB, FED; therefore the base DF is equal to the base AB, and the triangle FED to the triangle AEB, and the other angles to the other angles (4. 1.); Therefore the angle EBA is equal to the angle EDF; but EDF is a right angle, wherefore EBA is a right angle; and EB is a line drawn from the centre: but a straight line drawn from the extremity of a diameter, at right angles to it, touches the circle (Cor. 16. 3.): Therefore AB touches the circle; and is drawn from the given point A. Which was to be done.

But if the given point be in the circumference of the circle, as the point D, draw DE to the centre E, and DF at right angles to DE; DF touches the circle (Cor. 16. 3.)

PROP. XVIII. THEOR.

If a straight line touch a circle, the straight line drawn from the centre to the point of contact, is perpendicular to the line touching the circle.

Let the straight line DE touch the circle ABC in the point C; take the centre F, and draw the straight line FC: FC is perpendicular to DE.

For, if it be not, from the point F draw FBG perpendicular to DE; and because FGC is a right angle, GCF must be (17. 1.) an acute angle; and to the greater angle the greater (19. 1.) side is opposite: Therefore FC is greater than FG; but FC is equal to FB; therefore FB is greater than FG, the less than the greater, which is impossible; wherefore FG is not perpendicular to DE: In the same manner it may be shewn, that no other line but FC can be perpendicular to DE; FC is therefore perpendicular to DE. Therefore, if a straight line, &c. Q. E. D.

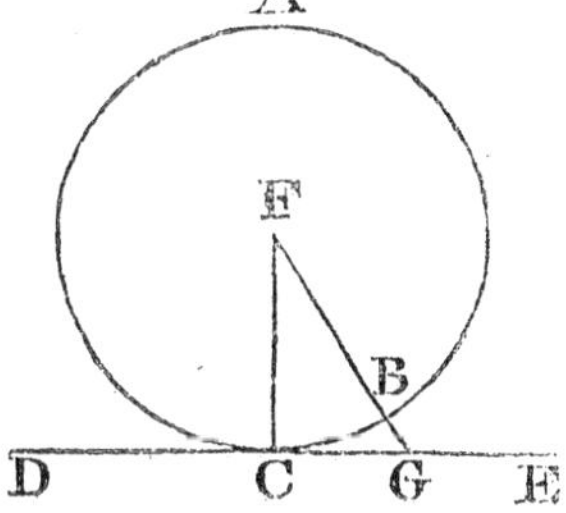

PROP. XIX. THEOR.

If a straight line touch a circle, and from the point of contact a straight line be drawn at right angles to the touching line, the centre of the circle is in that line.

Let the straight line DE touch the circle ABC, in C, and from C let CA be drawn at right angles to DE; the centre of the circle is in CA.

For, if not, let F be the centre, if possible, and join CF. Because DE touches the circle ABC, and FC is drawn from the centre to the point of contact, FC is perpendicular (13. 3.) to DE; therefore FCE is a right angle; But ACE is also a right angle; therefore the angle FCE is equal to the angle ACE, the less to the greater, which is impossible; Wherefore F is not the centre of the circle ABC: In the same manner it may be shewn, that no other point which is not in CA, is the centre; that is, the centre is in CA. Therefore, if a straight line, &c. Q. E. D.

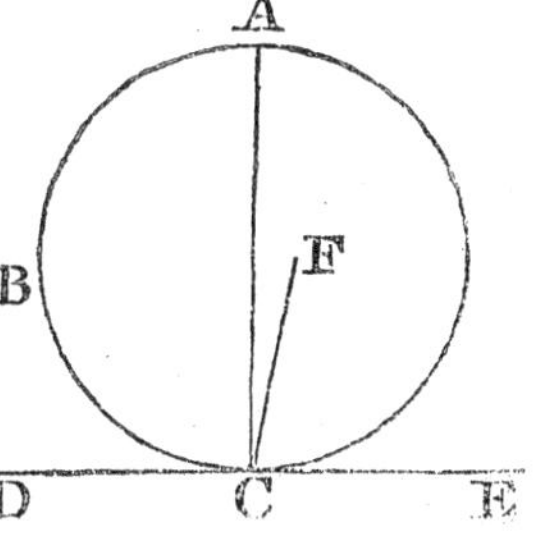

PROP. XX. THEOR.

The angle at the centre of a circle is double of the angle at the circumference, upon the same base, that is, upon the same part of the circumference.

Let ABC be a circle, and BDC an angle at the centre, and BAC an angle at the circumference which have the same circumference BC for the base; the angle BDC is double of the angle BAC.

First, let D, the centre of the circle, be within the angle BAC, and join AD, and produce it to E: Because DA is equal to DB, the angle DAB is equal (5. 1.) to the angle DBA: therefore the angles DAB, DBA together are double of the angle DAB; but the angle BDE is equal (32. 1.) to the angles DAB, DBA; therefore also the angle BDE is double of the angle DAB; For the same reason, the angle EDC is double of the angle DAC: Therefore the whole angle BDC is double of the whole angle BAC.

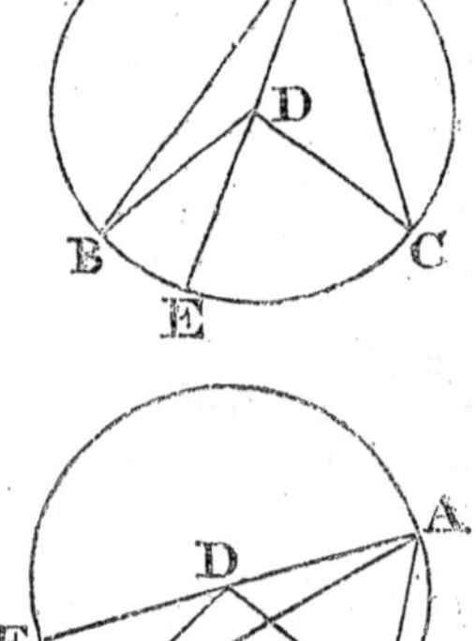

Again, let D, the centre of the circle, be without the angle BAC; and join AD and produce it to E. It may be demonstrated, as in the first case, that the angle EDC is double of the angle DAC, and that EDB a part of the first, is double of DAB, a part of the other; therefore the remaining angle BDC is double of the remaining angle BAC. Therefore the angle at the centre, &c. Q.E.D.

PROP. XXI. THEOR.

The angles in the same segment of a circle are equal to one another.

Let ABCD be a circle, and BAD, BED angles in the same segment BAED: The angles BAD, BED are equal to one another.

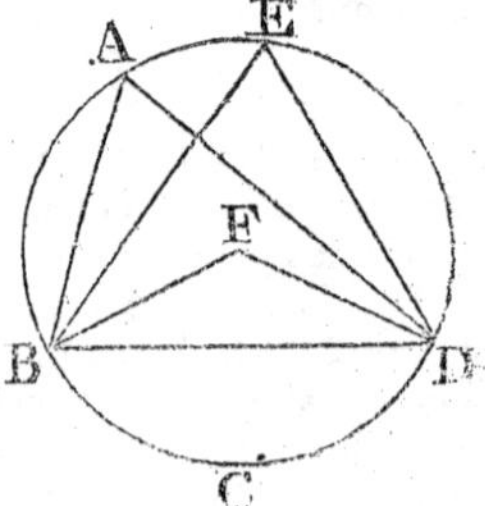

Take F the centre of the circle ABCD: And, first, let the segment BAED be greater than a semicircle, and join BF, FD: And because the angle BFD is at the centre, and the angle BAD at the circumference, both having the same part of the circumference, viz. BCD, for their base; therefore the angle BFD is double (20. 3.) of the angle BAD: for the same reason, the angle BFD is double of the angle BED: Therefore the angle BAD is equal to the angle BED.

But, if the segment BAED be not greater than a semicircle, let BAD, BED be angles in it; these also are equal to one another. Draw AF to the centre, and produce to C, and join CE: Therefore the segment BADC is greater than a semicircle; and the angles in it BAC, BEC are equal, by the first case: For the same reason, because CBED is greater than a semicircle, the angles CAD, CED are equal; Therefore the whole angle BAD is equal to the whole angle BED. Wherefore the angles in the same segment, &c. Q. E. D.

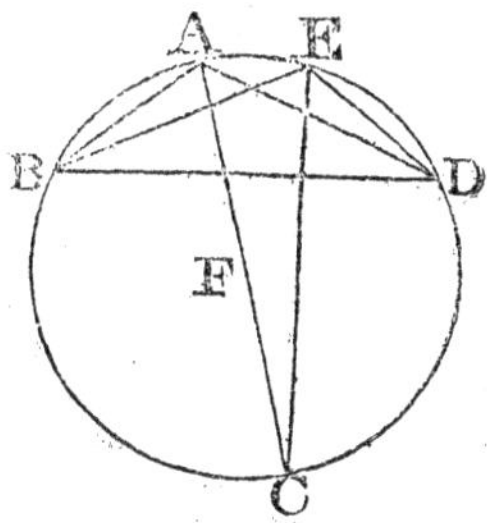

PROP. XXII. THEOR.

The opposite angles of any quadrilateral figure described in a circle, are together equal to two right angles.

Let ABCD be a quadrilateral figure in the circle ABCD; any two of its opposite angles are together equal to two right angles.

Join AC, BD. The angle CAB is equal (21. 3.) to the angle CDB, because they are in the same segment BADC, and the angle ACB is equal to the angle ADB, because they are in the same segment ADCB; therefore the whole angle ADC is equal to the angles CAB, ACB: To each of these equals add the angle ABC; and the angles ABC, ADC, are equal to the angles ABC, CAB, BCA. But ABC, CAB, BCA are equal to two right angles (32. 1.); therefore also the angles ABC, ADC are equal to two right angles: In the same manner, the angles BAD, DCB may be shewn to be equal to two right angles. Therefore the opposite angles, &c. Q. E. D.

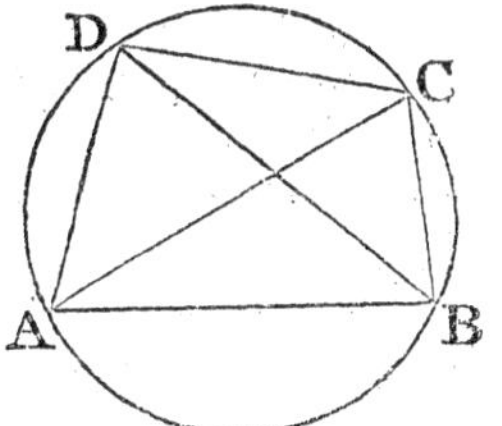

PROP. XXIII. THEOR.

Upon the same straight line, and upon the same side of it, there cannot be two similar segments of circles, not coinciding with one another.

If it be possible, let the two similar segments of circles, viz. ACB, ADB, be upon the same side of the same straight line AB, not coinciding with one another; then, because the circles ACB, ADB, cut one another in the two points A, B, they cannot cut one another in any other point (10. 3.): one of the segments must therefore fall within the other: let ACB fall within ADB, draw the straight line BCD, and join CA, DA: and because the segment ACB is similar to the segment ADB, and similar segments of circles contain (9. def. 3.) equal angles, the angle ACB is equal to the angle ADB, the exterior to the interior, which is impossible (16. 1.). Therefore, there cannot be two similar segments of circles upon the same side of the same line, which do not coincide. Q. E. D.

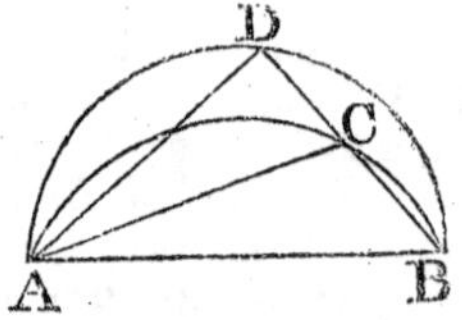

PROP. XXIV. THEOR.

Similar segments of circles upon equal straight lines are equal to one another.

Let AEB, CFD be similar segments of circles upon the equal straight lines AB, CD; the segment AEB is equal to the segment CFD.

For, if the segment AEB be applied to the segment CFD, so as the point A be on C, and the straight line AB upon CD, the point B shall coincide with the point D, because AB is equal to CD: Therefore the straight line AB coinciding with CD, the segment AEB must (23. 3.) coincide with the segment CFD, and therefore is equal to it. Wherefore, similar segments, &c. Q. E. D.

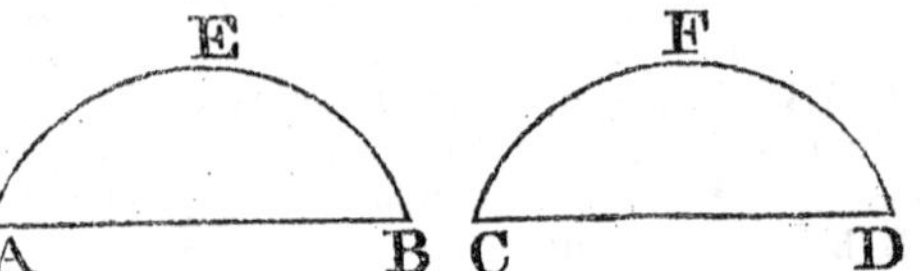

PROP. XXV. PROB.

A segment of a circle being given, to describe the circle of which it is the segment.

Let ABC be the given segment of a circle; it is required to describe the circle of which it is the segment.

Bisect (10. 1.) AC in D, and from the point D draw (11. 1.) DB at right angles to AC, and join AB: First, let the angles ABD, BAD be equal to one another; then the straight line BD is equal (6. 1.) to DA, and therefore to DC; and because the three straight lines DA, DB, DC, are all equal; D is the centre of the circle (9. 3.): from the centre D, at the distance of any of the three DA, DB, DC, describe a circle; this shall pass through the other points; and the circle of which ABC is a segment is described; and because the centre D is in AC, the segment ABC is a semicircle. Next let the angles ABD, BAD be unequal; at the point A, in the straight line AB make (23. 1.) the angle BAE equal to the angle ABD, and produce BD if neces-

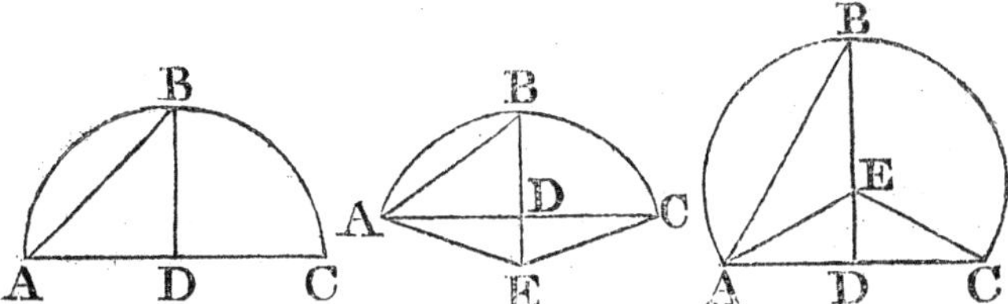

sary, to E, and join EC: and because the angle ABE is equal to the angle BAE, the straight line BE is equal (6. 1.) to EA: and because AD is equal to DC, and DE common to the triangles ADE, CDE, the two sides AD, DE are equal to the two CD, DE, each to each; and the angle ADE is equal to the angle CDE, for each of them is a right angle; therefore the base AE is equal (4. 1.) to the base EC: but AE was shewn to be equal to EB, wherefore also BE is equal to EC: and the three straight lines AE, EB, EC are therefore equal to one another; wherefore (9. 3.) E is the centre of the circle. From the centre E, at the distance of any of the three AE, EB, EC, describe a circle, this shall pass through the other points: and the circle of which ABC is a segment is described: also, it is evident, that if the angle ABD be greater than the angle BAD, the centre E falls without the segment ABC, which therefore is less than a semicircle; but if the angle ABD be less than BAD, the centre E falls within the segment ABC, which is therefore greater than a semicircle: Wherefore, a segment of a circle being given, the circle is described of which it is a segment. Which was to be done.

PROP. XXVI. THEOR.

In equal circles, equal angles stand upon equal arches, whether they be at the centres or circumferences.

Let ABC, DEF be equal circles, and the equal angles BGC, EHF, at their centres, and BAC, EDF at their circumferences: the arch BKC is equal to the arch ELF.

Join BC, EF; and because the circles ABC, DEF are equal, the straight lines drawn from their centres are equal: therefore the two

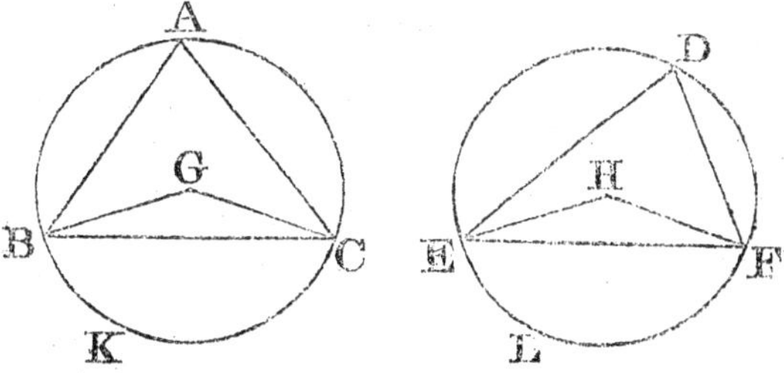

sides BG, GC, are equal to the two EH, HF; and the angle at G is equal to the angle at H; therefore the base BC is equal (4. 1.) to the base EF: and because the angle at A is equal to the angle at D, the segment BAC is similar (9. def. 3.) to the segment EDF; and they are upon equal straight lines BC, EF; but similar segments of circles upon equal straight lines are equal (24. 3.) to one another, therefore, the segment BAC is equal to the segment EDF: but the whole circle ABC is equal to the whole DEF; therefore the remaining segment BKC is equal to the remaining segment ELF, and the arch BKC to the arch ELF. Wherefore in equal circles, &c. Q. E. D.

PROP. XXVII. THEOR.

In equal circles, the angles which stand upon equal arches are equal to one another, whether they be at the centres or circumferences.

Let the angles BGC, EHF at the centres, and BAC, EDF at the circumferences of the equal circles ABC, DEF stand upon the equal arches BC, EF; the angle BGC is equal to the angle EHF, and the angle BAC to the angle EDF.

If the angle BGC be equal to the angle EHF, it is manifest(20. 3.) that the angle BAC is also equal to EDF. But, if not, one of them is the greater: let BGC be the greater, and at the point G, in the straight line BG, make the angle (23. 1.) BGK equal to the angle EHF. And because equal angles stand upon equal arches (26. 3.), when they are at the centre, the arch BK is equal to the arch EF: but EF is equal to BC; therefore also BK is equal to BC, the less to the greater, which is impossible. Therefore the angle BGC is not unequal to the angle EHF; that is, it is equal to it: and the angle at A is half the

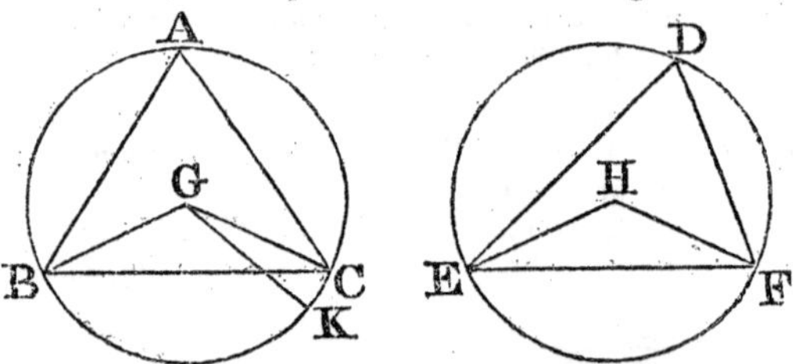

angle BGC, and the angle at D half of the angle EHF; therefore the angle at A is equal to the angle at D. Wherefore, in equal circles, &c. Q. E. D.

PROP. XXVIII. THEOR.

In equal circles, equal straight lines cut off equal arches, the greater equal to the greater, and the less to the less.

Let ABC, DEF be equal circles, and BC, EF equal straight lines in them, which cut off the two greater arches BAC, EDF, and the two less BGC, EHF: the greater BAC is equal to the greater EDF, and the less BGC to the less EHF.

Take (1. 3.) K, L, the centres of the circles, and join BK, KC, EL, LF; and because the circles are equal, the straight lines from their

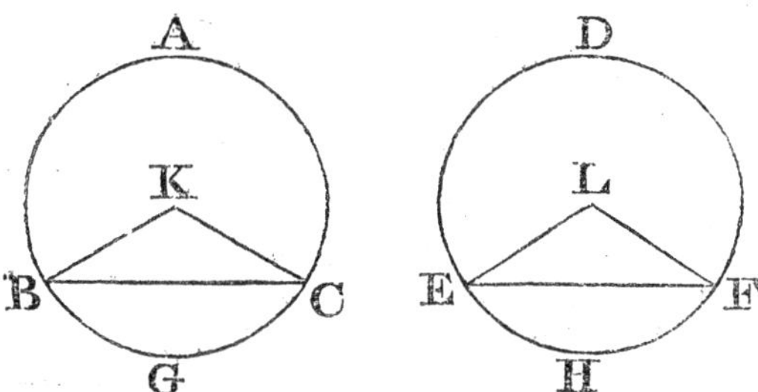

centres are equal; therefore BK, KC are equal to EL, LF; but the base BC is also equal to the base EF; therefore the angle BKC is equal (8. 1.) to the angle ELF: and equal angles stand upon equal (26. 3.) arches, when they are at the centres; therefore the arch BGC is equal to the arch EHF. But the whole circle ABC is equal to the whole EDF; the remaining part, therefore, of the circumference, viz. BAC, is equal to the remaining part EDF. Therefore, in equal circles, &c. Q. E. D.

PROP. XXIX. THEOR.

In equal circles equal arches are subtended by equal straight lines.

Let ABC, DEF be equal circles, and let the arches BGC, EHF also be equal; and join BC, EF: the straight line BC is equal to the straight line EF.

Take (1. 3.) K, L, the centres of the circles, and join BK, KC, EL, LF: and because the arch BGC is equal to the arch EHF, the angle BKC is equal (27. 3.) to the angle ELF: also because the circles ABC, DEF are equal, their radii are equal: therefore BK, KC are

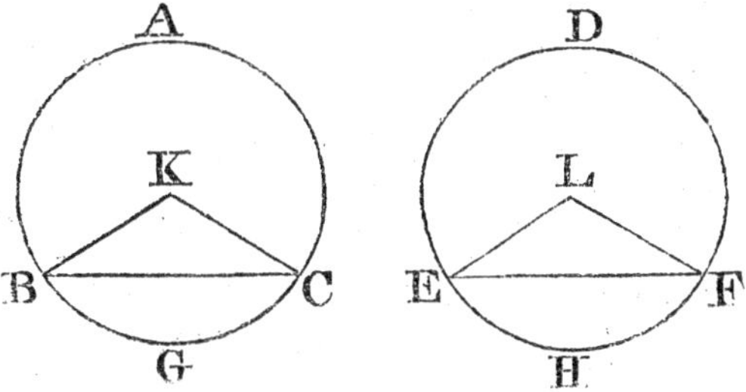

equal to EL, LF: and they contain equal angles; therefore the base BC is equal (4. 1.) to the base EF. Therefore, in equal circles, &c. Q. E. D.

PROP. XXX. PROB.

To bisect a given arch, that is, to divide it into two equal parts.

Let ADB be the given arch; it is required to bisect it.

Join AB, and bisect (10. 1.) it in C; from the point C draw CD at right angles to AB, and join AD, DB: the arch ADB is bisected in the point D.

Because AC is equal to CB, and CD common to the triangle ACD, BCD, the two sides AC, CD are equal to the two BC, CD; and the angle ACD is equal to the angle BCD, because each of them is a right angle; therefore the base AD is equal (4. 1.) to the base BD. But equal straight lines cut off equal (28. 3.) arches, the greater equal to the greater, and the less to the less; and AD, DB are each of them less than a semicircle, because DC passes through the centre (Cor. 1. 3.); wherefore the arch AD is equal to the arch DB; and therefore the given arch ADB is bisected in D. Which was to be done.

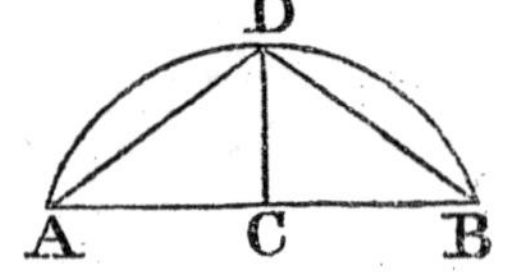

PROP. XXXI. THEOR.

In a circle, the angle in a semicircle is a right angle; but the angle in a segment greater than a semicircle is less than a right angle; and the angle in a segment less than a semicircle is greater than a right angle.

Let ABCD be a circle, of which the diameter is BC, and centre E; draw CA dividing the circle into segments ABC, ADC, and join BA, AD, DC; the angle in the semicircle BAC is a right angle; and the angle in the segment ABC, which is greater than a semicircle, is less than a right angle; and the angle in the segment ADC, which is less than a semicircle, is greater than a right angle.

Join AE, and produce BA to F; and because BE is equal to EA, the angle EAB is equal (5. 1.) to EBA: also, because AE is equal to EC, the angle EAC is equal to ECA; wherefore the whole angle BAC is equal to the two angles ABC, ACB. But FAC, the exterior angle of the triangle ABC, is also equal (32. 1.) to the two angles ABC, ACB; therefore the angle BAC is equal to the angle FAC, and each of them is therefore a right (7. def. 1.) angle; wherefore the angle BAC in a semicircle is a right angle.

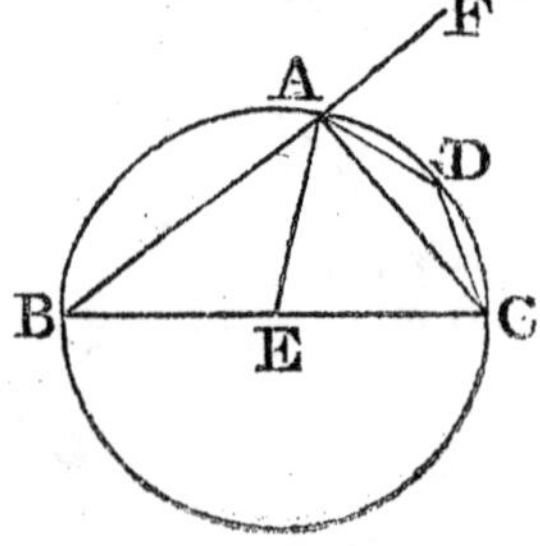

And because the two angles ABC, BAC of the triangle ABC are together less (17. 1.) than two right angles, and BAC is a right angle, ABC must be less than a right angle; and therefore the angle in a segment ABC, greater than a semicircle, is less than a right angle.

Also because ABCD is a quadrilateral figure in a circle, any two of its opposite angles are equal (22. 3.) to two right angles; therefore the angles ABC, ADC are equal to two right angles; and ABC is less than a right angle: wherefore the other ADC is greater than a right angle. Therefore, in a circle, &c. Q. E. D.

Cor. From this it is manifest, that if one angle of a triangle be equal to the other two, it is a right angle, because the angle adjacent to it is equal to the same two; and when the adjacent angles are equal, they are right angles.

PROP. XXXII. THEOR.

If a straight line touch a circle, and from the point of contact a straight line be drawn cutting the circle, the angles made by this line with the line which touches the circle, shall be equal to the angles in the alternate segments of the circle.

Let the straight line EF touch the circle ABCD in B, and from the point B let the straight line BD be drawn cutting the circle: The angles which BD makes with the touching line EF shall be equal to the angles in the alternate segments of the circle; that is, the angle FBD is equal to the angle which is in the segment DAB, and the angle DBE to the angle in the segment BCD.

From the point B draw (11. 1.) BA at right angles to EF, and take any point C in the arch BD, and join AD, DC, CB; and because the straight line EF touches the circle ABCD in the point B, and BA is drawn at right angles to the touching line, from the point of contact B, the centre of the circle is (19. 3.) in BA; therefore the angle ADB, in a semicircle, is a right (31. 3.) angle, and consequently the other two angles, BAD, ABD are equal (32. 1.) to a right angle: but ABF is likewise a right angle; therefore the angle ABF is equal to the angles BAD, ABD: take from these equals the common angle ABD; and there will remain the angle DBF equal to the angle BAD, which is in the alternate segment of the circle.

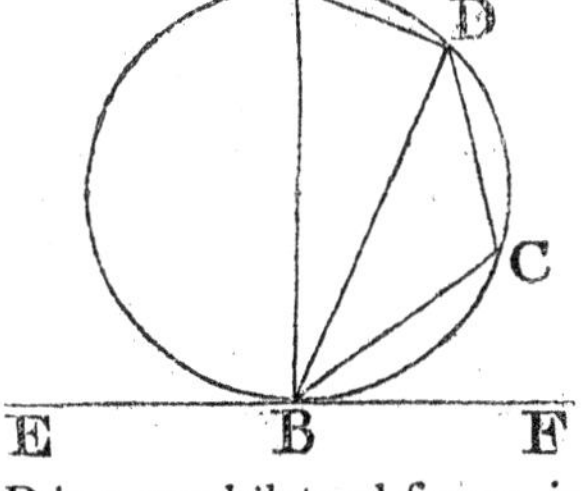

And because ABCD is a quadrilateral figure in a circle, the opposite angles BAD, BCD are equal (22. 3.) to two right angles; therefore the angles DBF, DBE, being likewise equal (13. 1.) to two right angles, are equal to the angles BAD, BCD; and DBF has been proved equal to BAD: therefore the remaining angle DBE is equal to the angle BCD in the alternate segment of the circle.

Wherefore, if a straight line, &c. Q. E. D.

PROP. XXXIII. PROB.

Upon a given straight line to describe a segment of a circle, containing an angle equal to a given rectilineal angle.

Let AB be the given straight line, and the angle at C the given rectilineal angle; it is required to describe upon the given straight line AB a segment of a circle, containing an angle equal to the angle C.

First, let the angle at C be a right angle; bisect (10. 1.) AB in F, and from the centre F, at the distance FB, describe the semicircle AHB; the angle AHB being in a semicircle is (31. 3.) equal to the right angle at C.

But if the angle C be not a right angle at the point A, in the straight line AB, make (23. 1.) the angle BAD equal to the angle C, and from the point A draw (11. 1.) AE at right angles to AD; bisect (10. 1.) AB in F, and from F draw (11. 1.) FG at right angles to AB, and join GB: Then because AF is equal to FB, and FG common to the triangles AFG, BFG, the two sides AF, FG are equal to the two BF, FG; but the angle AFG is also equal to the angle BFG; therefore the base AG is equal (4. 1.) to the base GB; and the circle described from the centre G, at the distance GA, shall pass through the point B; let this be the circle AHB: And because from the point A the extremity of the diameter AE, AD is drawn at right angles to AE, therefore AD (Cor. 16. 3.) touches the circle; and because AB, drawn from the point of contact A, cuts the circle, the angle DAB is equal to the angle in the alternate segment AHB (32. 3.); but the angle DAB is equal to the angle C, therefore also the angle C is equal to the angle in the segment AHB;

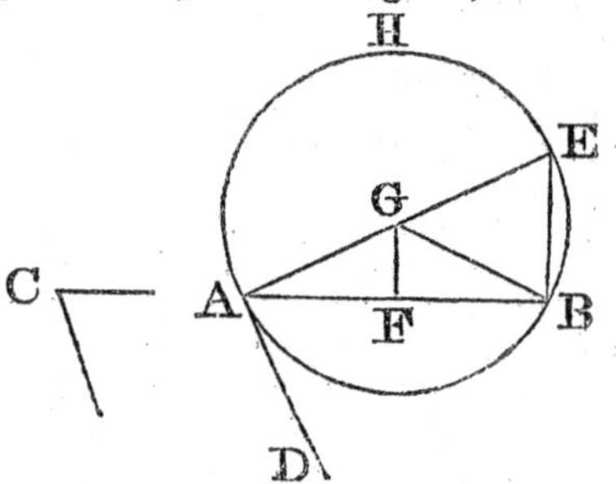

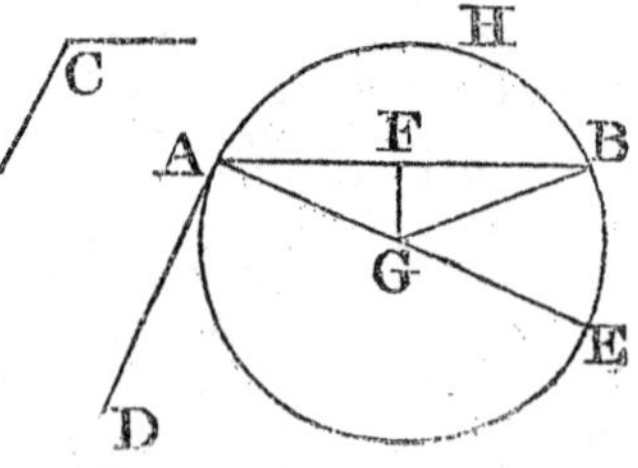

Wherefore, upon the given straight line AB the segment AHB of a circle is described which contains an angle equal to the given angle at C. Which was to be done.

PROP. XXXIV. PROB.

To cut off a segment from a given circle which shall contain an angle equal to a given rectilineal angle.

Let ABC be the given circle, and D the given rectilineal angle; it is required to cut off a segment from the circle ABC that shall contain an angle equal to the angle D.

Draw (17. 3.) the straight line EF touching the circle ABC in the point B, and at the point B, in the straight line BF, make (23. 1.) the angle FBC equal to the angle D; therefore, because the straight line EF touches the circle ABC, and BC is drawn from the point of contact B, the angle FBC is equal (32. 3.) to the angle in the alternate segment BAC; but the angle FBC is equal to the angle D: therefore the angle in the segment BAC is equal to the angle D: wherefore the segment BAC is cut off from the given circle ABC containing an angle equal to the given angle D. Which was to be done.

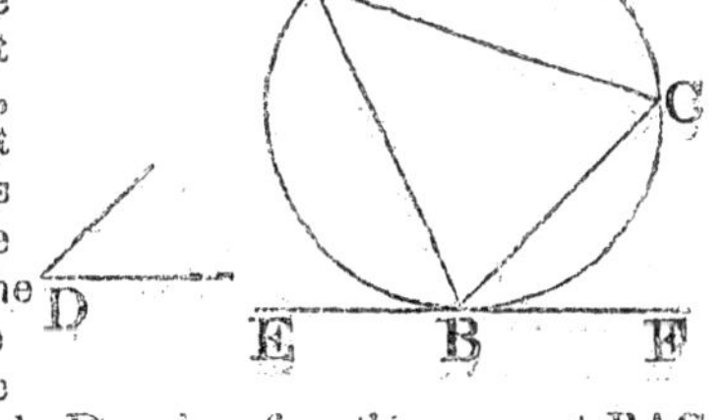

PROP. XXXV. THEOR.

If two straight lines within a circle cut one another, the rectangle contained by the segments of one of them is equal to the rectangle contained by the segments of the other.

Let the two straight lines AC, BD, within the circle ABCD, cut one another in the point E: the rectangle contained by AE, EC is equal to the rectangle contained by BE, ED.

If AC, BD pass each of them through the centre, so that E is the centre, it is evident that AE, EC, BE, ED, being all equal, the rectangle AE.EC is likewise equal to the rectangle BE.ED.

But let one of them, BD, pass through the centre, and cut the other AC, which does not pass through the centre, at right angles in the point E; then, if BD be bisected in F, F is the centre of the circle

ABCD; join AF: and because BD, which passes through the centre, cuts the straight line AC, which does not pass through the centre at right angles, in E, AE, EC are equal (3. 3.) to one another; and because the straight line BD is cut into two equal parts in the point F, and into two unequal, in the point E, BE.ED (5. 2.) + $EF^2 = FB^2 = AF^2$. But $AF^2 = AE^2 +$ (47. 1.) EF^2, therefore BE.ED + $EF^2 = AE^2 + EF^2$, and taking EF^2 from each, BE.ED $= AE^2 =$ AE.EC.

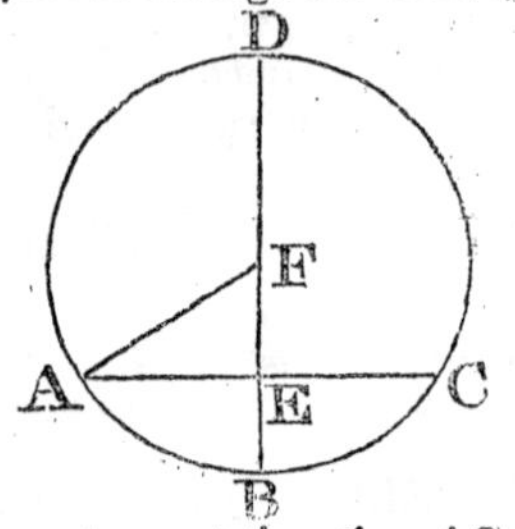

Next, Let BD, which passes through the centre, cut the other AC, which does not pass through the centre, in E, but not at right angles: then, as before, if BD be bisected in F, F is the centre of the circle. Join AF, and from F draw (12. 1.) FG perpendicular to AC; therefore AG is equal (3. 3.) to GC; wherefore AE.EC + (5. 2.) $EG^2 = AG^2$, and adding GF^2 to both, AE.EC + $EG^2 + GF^2 = AG^2 + GF^2$. Now $EG^2 + GF^2 = EF^2$, and $AG^2 + GF^2 = AF^2$; therefore, AE.EC + $EF^2 = AF^2 = FB^2$. But FB^2 = BE.ED + (5. 2.) EF^2, therefore AE.EC + EF^2 = BE.ED + EF^2, and taking EF^2 from both, AE.EC = BE.ED.

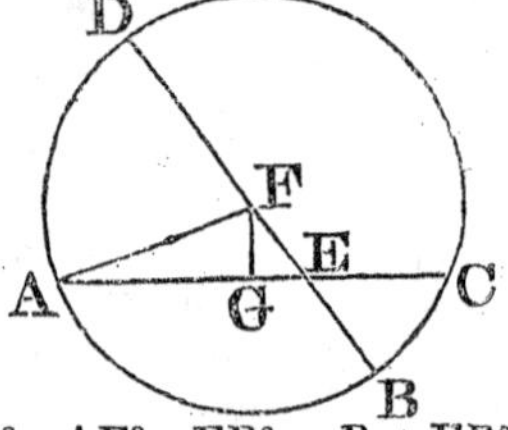

Lastly, Let neither of the straight lines AC, BD pass through the centre: take the centre F, and through E, the intersection of the straight lines AC, DB, draw the diameter GEFH: and because, as has been shown, AE.EC = GE.EH, and BE.ED = GE.EH; therefore AE.EC = BE.ED. Wherefore, if two straight lines, &c.

Q. E. D.

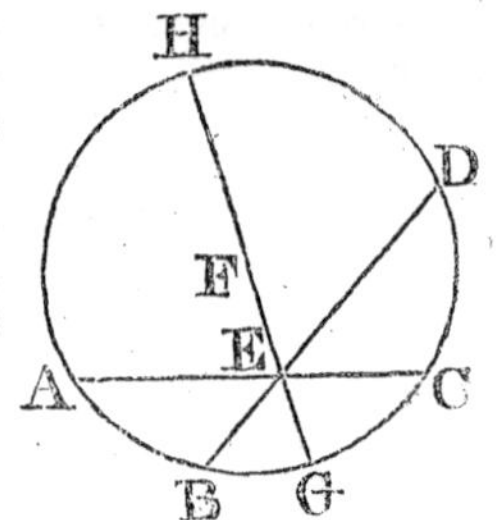

PROP. XXXVI. THEOR.

If from any point without a circle two straight lines be drawn, one of which cuts the circle, and the other touches it; the rectangle contained by the whole line which cuts the circle, and the part of it without the circle, is equal to the square of the line which touches it.

Let D be any point without the circle ABC, and DCA, DB two straight lines drawn from it, of which DCA cuts the circle, and DB touches it: the rectangle AD.DC is equal to the square of DB.

Either DCA passes through the centre, or it does not; first, Let it pass through the centre E, and join EB; therefore the angle EBD is a right (18. 3.) angle: and because the straight line AC is bisected in E, and produced to the point D, $AD.DC + EC^2 = ED^2$ (6. 2.). But EC=EB, therefore $AD.DC + EB^2 = ED^2$. Now $ED^2 =$ (47. 1.) $EB^2 + BD^2$, because EBD is a right angle; therefore $AD.DC + EB^2 = EB^2 + BD^2$, and taking EB^2 from each, $AD.DC = BD^2$.

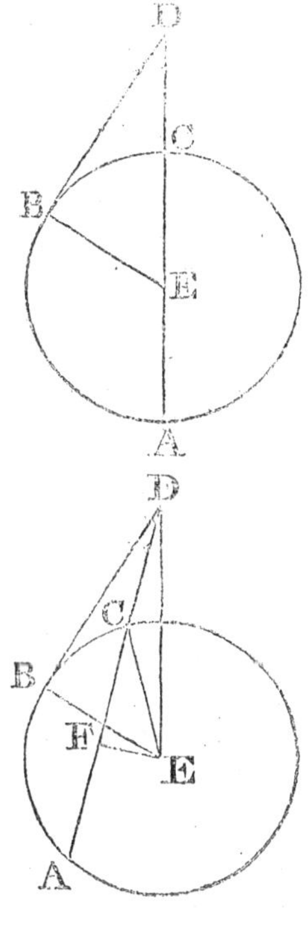

But, if DCA does not pass through the centre of the circle ABC, take (1. 3.) the centre E, and draw EF perpendicular (12. 1.) to AC, and join EB, EC, ED: and because the straight line EF, which passes through the centre, cuts the straight line AC, which does not pass through the centre, at right angles, it likewise bisects (3. 3.) it; therefore AF is equal to FC; and because the straight line AC is bisected in F, and produced to D (6. 2.), $AD.DC + FC^2 = FD^2$; add FE^2 to both, then $AD.DC + FC^2 + FE^2 = FD^2 + FE^2$. But (47. 1.) $EC^2 = FC^2 + FE^2$, and $ED^2 = FD^2 + FE^2$, because DFE is a right angle; therefore $AD.DC + EC^2 = ED^2$. Now, because EBD is a right angle, $ED^2 = EB^2 + BD^2 = EC^2 + BD^2$, and therefore, $AD.DC + EC^2 = EC^2 + BD^2$, and $AD.DC = BD^2$.

Wherefore, if from any point, &c. Q. E. D.

Cor. If from any point without a circle, there be drawn two straight lines cutting it, as AB, AC, the rectangles contained by the whole lines and the parts of them without the circle, are equal to one another, viz. BA.AE=CA.AF; for each of these rectangles is equal to the square of the straight line AD, which touches the circle.

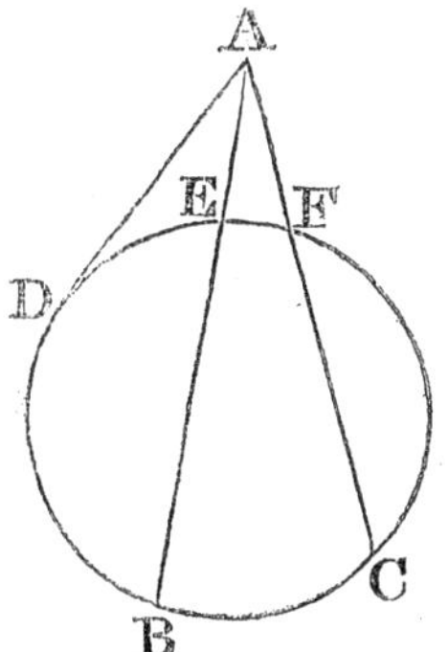

PROP. XXXVII. THEOR.

If from a point without a circle there be drawn two straight lines, one of which cuts the circle, and the other meets it; if the rectangle contained by the whole line, which cuts the circle, and the part of it without the circle, be equal to the square of the line which meets it, the line which meets shall touch the circle.

Let any point D be taken without the circle ABC, and from it let two straight lines DCA and DB be drawn, of which DCA cuts the circle, and DB meets it; if the rectangle AD.DC be equal to the square of DB, DB touches the circle.

Draw (17. 3.) the straight line DE touching the circle ABC; find the centre F, and join FE, FB, FD; then FED is a right (18. 3.) angle: and because DE touches the circle ABC, and DCA cuts it, the rectangle AD.DC is equal (36. 3.) to the square of DE; but the rectangle AD.DC is, by hypothesis, equal to the square of DB: therefore the square of DE is equal to the square of DB; and the straight line DE equal to the straight line DB: but FE is equal to FB, wherefore DE, EF are equal to DB, BF; and the base FD is common to the two triangles DEF, DBF; therefore the angle DEF is equal (8. 1.) to the angle DBF; and DEF is a right angle, therefore also DBF is a right angle: but FB, if produced, is a diameter, and the straight line which is drawn at right angles to a diameter, from the extremity of it, touches (16. 3.) the circle: therefore DB touches the circle ABC. Wherefore, if from a point, &c. Q. E. D.

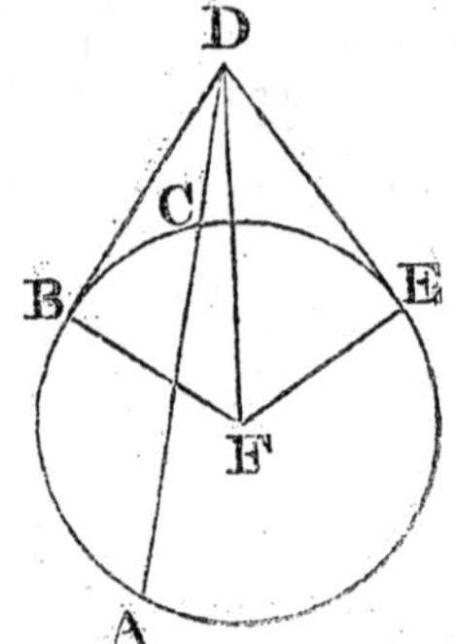

ELEMENTS

OF

GEOMETRY.

BOOK IV.

DEFINITIONS.

I.

A RECTILINEAL figure is said to be inscribed in another rectilineal figure, when all the angles of the inscribed figure are upon the sides of the figure in which it is inscribed, each upon each.

II.

In like manner, a figure is said to be described about another figure, when all the sides of the circumscribed figure pass through the angular points of the figure about which it is described, each through each.

III.

A rectilineal figure is said to be inscribed in a circle, when all the angles of the inscribed figure are upon the circumference of the circle.

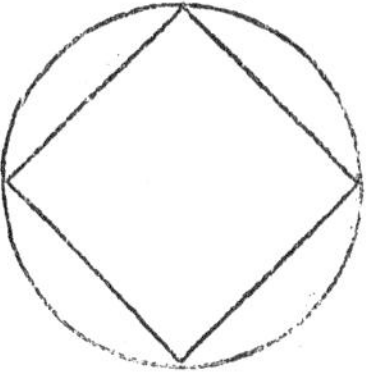

IV.

A rectilineal figure is said to be described about a circle, when each side of the circumscribed figure touches the circumference of the circle.

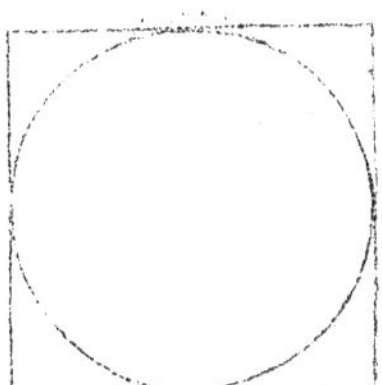

V.

In like manner, a circle is said to be inscribed in a rectilineal figure, when the circumference of the circle touches each side of the figure.

VI.

A circle is said to be described about a rectilineal figure, when the circumference of the circle passes through all the angular points of the figure about which it is described.

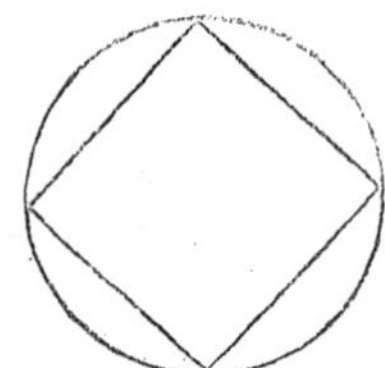

VII.

A straight line is said to be placed in a circle, when the extremities of it are in the circumference of the circle.

PROP. I. PROB.

In a given circle to place a straight line equal to a given straight line, not greater than the diameter of the circle.

Let ABC be the given circle, and D the given straight line, not greater than the diameter of the circle.

Draw BC the diameter of the circle ABC; then, if BC is equal to D, the thing required is done; for in the circle ABC a straight line BC is placed equal to D: But, if it is not, BC is greater that D; make CE equal (3. 1.) to D, and from the centre C, at the distance CE, describe the circle AEF, and join CA: Therefore, because C is the centre of the circle AEF, CA is equal to CF; but D is equal to CE; therefore D is equal to CA: Wherefore, in the circle ABC, a straight line is placed, equal to the given straight line D, which is not greater than the diameter of the circle. Which wasto be done.

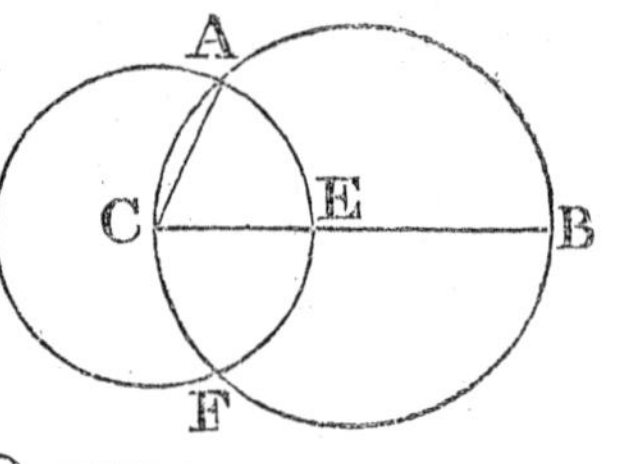

PROP. II. PROB.

In a given circle to inscribe a triangle equiangular to a given triangle.

Let ABC be the given circle, and DEF the given triangle; it is required to inscribe in the circle ABC a triangle equiangular to the triangle DEF.

Draw (17. 3.) the straight line GAH touching the circle in the point A, and at the point A, in the straight line AH, make (23. 1.) the angle HAC equal to the angle DEF: and at the point A, in the straight line

AG, make the angles GAB equal to the angle DFE, and join BC. Therefore, because HAG touches the circle ABC and AC is drawn from the point of contact, the angle HAC is equal (32. 3.) to the angle ABC in the alternate segment of the circle: But HAC is equal to the angle DEF; therefore also the angle ABC is equal to DEF; for the same reason, the angle ACB is equal to the angle DFE; therefore the remaining angle BAC is equal (32. 1.) to the remaining angle EDF: Wherefore the triangle ABC is equiangular to the triangle DEF, and it is inscribed in the circle ABC. Which was to be done.

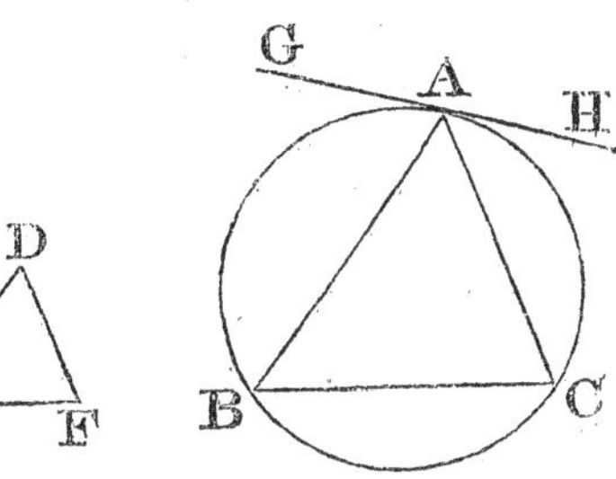

PROP. III. PROB.

About a given circle to describe a triangle equiangular to a given triangle.

Let ABC be the given circle, and DEF the given triangle; it is required to describe a triangle about the circle ABC equiangular to the triangle DEF.

Produce EF both ways to the points G, H, and find the centre K of the circle ABC, and from it draw any straight line KB; at the point K in the straight line KB, make (23. 1.) the angle BKA equal to the angle DEG, and the angle BKC equal to the angle DFH; and through the points A, B, C, draw the straight lines LAM, MBN, NCL, touching (17. 3.) the circle ABC: Therefore, because LM, MN, NL touch the circle ABC in the points A, B, C, to which from the centre are drawn KA, KB, KC, the angles at the points A, B, C, are right (18. 3.) angles. And because the four angles of the quadrilateral figure AMBK are equal to four right angles, for it can be divided into two triangles; and because two of them, KAM, KBM are right angles

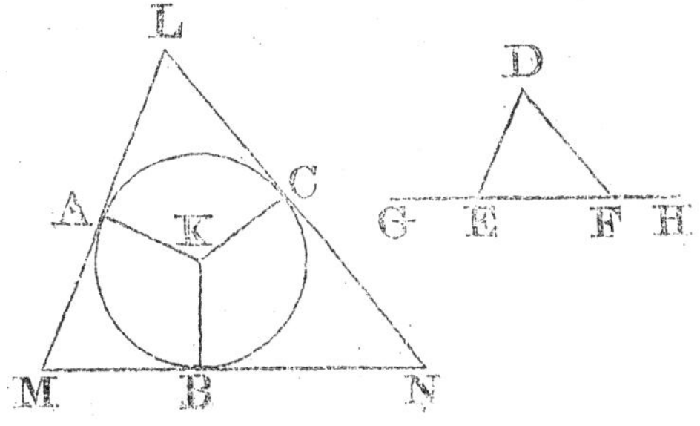

the other two AKB, AMB are equal to two right angles: But the angles DEG, DEF are likewise equal (13. 1.) to two right angles; therefore the angles AKB, AMB are equal to the angles DEG, DEF, of which AKB is equal to DEG; wherefore the remaining angle AMB is equal to the remaining angle DEF. In like manner, the angle LMN may be demonstrated to be equal to DFE; and therefore the remaining angle MLN is equal (32. 1.) to the remaining angle EDF: Wherefore the triangle LMN is equiangular to the triangle DEF: And it is described about the circle ABC. Which was to be done.

PROP. IV. PROB.

To inscribe a circle in a given triangle.

Let the given triangle be ABC; it is required to inscribe a circle in ABC.

Bisect (9. 1.) the angles ABC, BCA by the straight lines BD, CD meeting one another in the point D, from which draw (12. 1.) DE, DF, DG perpendiculars to AB, BC, CA.

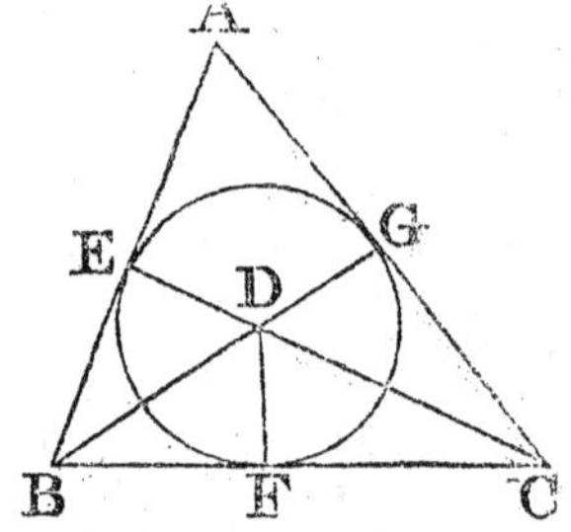

Then because the angle EBD is equal to the angle FBD, the angle ABC being bisected by BD; and because the right angle BED, is equal to the right angle BFD, the two triangles EBD, FBD have two angles of the one equal to two angles of the other; and the side BD, which is opposite to one of the equal angles in each, is common to both; therefore their other sides are equal (26. 1.); wherefore DE is equal to DF. For the same reason, DG is equal DF; therefore the three straight lines DE, DF, DG, are equal to one another, and the circle described from the centre D, at the distance of any of them, will pass through the extremities of the other two, and will touch the straight lines AB, BC, CA, because the angles at the points E, F, G, are right angles, and the straight line which is drawn from the extremity of a diameter at right angles to it, touches (Cor. 16. 3) the circle. Therefore the straight lines AB, BC, CA, do each of them touch the circle, and the circle EFG is inscribed in the triangle ABC. Which was to be done.

PROP. V. PROB.

To describe a circle about a given triangle.

Let the given triangle be ABC; it is required to describe a circle about ABC.

Bisect (10. 1.) AB, AC in the points D, E, and from these points draw DF, EF at right angles (11. 1.) to AB, AC; DF, EF produced

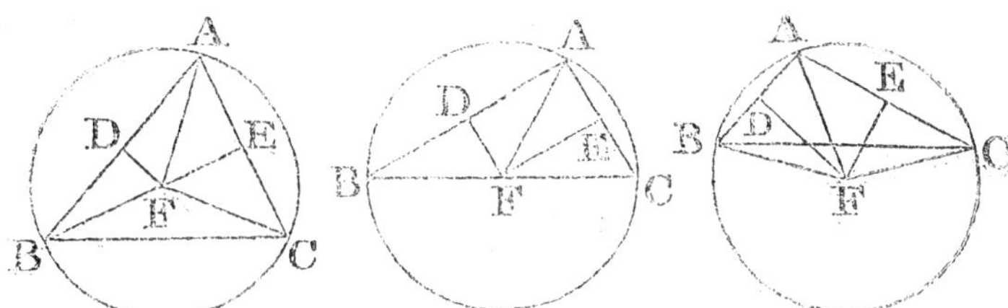

will meet one another; for, if they do not meet, they are parallel, wherefore, AB, AC, which are at right angles to them, are parallel, which is absurd; Let them meet in F, and join FA; also, if the point F be not in BC, join BF, CF: then, because AD is equal to BD, and DF common, and at right angles to AB, the base AF is equal (4. 1.) to the base FB. In like manner, it may be shewn that CF is equal to FA; and therefore BF is equal to FC; and FA, FB, FC are equal to one another; wherefore the circle described from the centre F, at the distance of one of them, will pass through the extremities of the other two, and be described about the triangle ABC, which was to be done.

COR. When the centre of the circle falls within the triangle, each of its angles is less than a right angle, each of them being in a segment greater than a semicircle; but when the centre is in one of the sides of the triangle, the angle opposite to this side, being in a semicircle, is a right angle: and if the centre falls without the triangle, the angle opposite to the side beyond which it is, being in a segment less than a semicircle, is greater than a right angle. Wherefore, if the given triangle be acute angled, the centre of the circle falls within it: if it be a right angled triangle, the centre is in the side opposite to the right angle; and if it be an obtuse angled triangle, the centre falls without the triangle, beyond the side opposite to the obtuse angle.

PROP. VI. PROB.

To inscribe a square in a given circle.

Let ABCD be the given circle; it is required to inscribe a square in ABCD.

Draw the diameters AC, BD at right angles to one another, and join AB, BC, CD, DA; because BE is equal to ED, E being the centre, and because EA is at right angles to BD, and common to the triangles ABE, ADE; the base BA is equal (4. 1.) to the base AD; and, for the same reason, BC, CD are each of them equal to BA or AD; therefore the quadrilateral figure ABCD is equilateral. It is also rectangular; for the straight line BD being a diameter of the circle ABCD, BAD is a semicircle; wherefore the angle BAD is a right (31. 3.) angle; for the same reason each

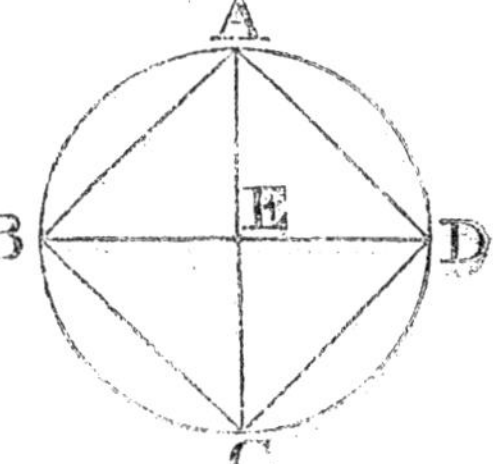

of the angles ABC, BCD, CDA is a right angle; therefore the quadrilateral figure ABCD is rectangular, and it has been shewn to be equilateral; therefore it is a square; and it is inscribed in the circle ABCD. Which was to be done.

PROP. VII. PROB.

To describe a square about a given circle.

Let ABCD be the given circle; it is required to describe a square about it.

Draw two diameters AC, BD of the circle ABCD, at right angles to one another, and through the points A, B, C, D draw (17. 3.) FG, GH, HK, KF touching the circle; and because FG touches the circle ABCD, and EA is drawn from the centre E to the point of contact A, the angles at A are right (18. 3.) angles; for the same reason, the angles at the points B, C, D are right angles; and because the angle AEB is a right angle, as likewise is EBG, GH is parallel (28. 1.) to AC; for the same reason, AC is parallel to FK, and in like manner, GF, HK may each of them be demonstrated to be parallel to BED; therefore the figures GK, GC, AK, FB, BK are parallelograms; and GF is therefore equal (34. 1.) to HK, and GH to FK; and because AC is equal to BD, and also to each of the two GH, FK; and BD to each of the two GF, HK: GH, FK are each of them equal to GF or HK; therefore the quadrilateral figure FGHK is equilateral. It is also rectangular; for, GBEA being a parallelogram, and AEB a right angle, AGB (34. 1.) is likewise a right angle: In the same manner, it may be shewn that the angles at H, K, F are right angles; therefore the quadrilateral figure FGHK is rectangular; and it was demonstrated to be equilateral; therefore it is a square; and it is described about the circle ABCD. Which was to be done.

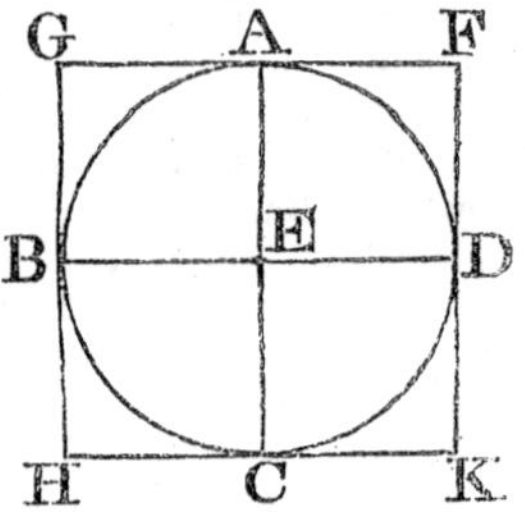

PROP. VIII. PROB.

To inscribe a circle in a given square.

Let ABCD be the given square; it is required to inscribe a circle in ABCD.

Bisect (10. 1.) each of the sides AB, AD, in the points F, E, and through E draw (31. 1.) EH parallel to AB or DC, and through F draw FK parallel to AD or BC: therefore each of the figures, AK, KB, AH, HD, AG, GC, BG, GD is a parallelogram, and their opposite sides are equal (34. 1.); and because that AD is equal to AB, and that AE is the half of AD, and AF the half of AB, AE is equal to AF; wherefore the sides opposite to these are equal, viz. FG to GE; in the same manner, it may be demonstrated, that GH, GK, are each of them equal to FG or GE; therefore the four straight lines, GE, GF, GH, GK, are equal to one another; and the circle described from the centre G, at the distance of one of them, will pass through the extremities of the other three; and will also touch the straight lines AB, BC, CD, DA, because the angles at the points E, F, H, K, are right (29. 1.) angles, and because the straight line which is drawn from the extremity of a diameter at right angles to it, touches the circle (16. 3.); therefore each of the straight lines AB, BC, CD, DA touches the circle, which is therefore inscribed in the square ABCD. Which was to be done.

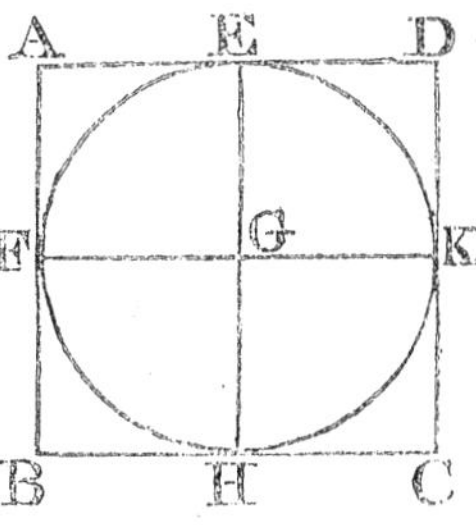

PROP. IX. PROB.

To describe a circle about a given square.

Let ABCD be the given square; it is required to describe a circle about it.

Join AC, BD, cutting one another in E; and because DA is equal to AB, and AC common to the triangles DAC, BAC, the two sides DA, AC are equal to the two BA, AC, and the base DC is equal to the base BC; wherefore the angle DAC is equal (8. 1.) to the angle BAC, and the angle DAB is bisected by the straight line AC. In the same manner, it may be demonstrated, that the angles ABC, BCD, CDA are severally bisected by the straight lines BD, AC; therefore, because the angle DAB is equal to the angle ABC, and the angle EAB is the half of DAB, and EBA the half of ABC; the angle EAB is equal to the angle EBA: and the side EA (6. 1.) to the side EB. In the same manner, it may be demonstrated, that the straight lines EC, ED are each of them equal to EA, or EB; therefore the four straight lines EA, EB, EC, ED are equal to one another; and the circle described from the centre E, at the distance of one of them must pass through the extremities of the other three, and be described about the square ABCD. Which was to be done.

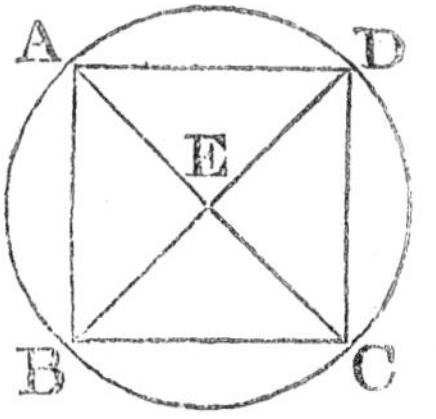

PROP. X. PROB.

To describe an isosceles triangle, having each of the angles at the base double of the third angle.

Take any straight line AB, and divide (11. 2.) it in the point C, so that the rectangle AB, BC may be equal to the square of AC; and from the centre A, at the distance AB, describe the circle BDE, in which place (1. 4.) the straight line BD equal to AC, which is not greater than the diameter of the circle BDE; join DA, DC, and about the triangle ADC describe (5. 4.) the circle ACD; the triangle ABD is such as is required, that is, each of the angles ABD, ADB is double of the angle BAD.

Because the rectangle AB.BC is equal to the square of AC, and AC equal to BD, the rectangle AB.BC is equal to the square of BD; and because from the point B without the circle ACD two straight lines BCA, BD are drawn to the circumference, one of which cuts, and the other meets the circle, and the rectangle AB.BC contained by the whole of the cutting line, and the part of it without the circle, is equal to the square of BD, which meets it; the straight line BD touches (37. 3.) the circle ACD. And because BD touches the circle, and DC is drawn from the point of contact D, the angle BDC is equal (32. 3.) to the angle DAC in the alternate segment of the circle; to each of these add the angle CDA: therefore the whole angle BDA is equal to the two angles CDA, DAC; but the exterior angle BCD is equal (32. 1.) to the angles CDA, DAC; therefore also BDA is equal to BCD; but BDA is equal (5. 1.) to CDB, because the side AD is equal to the side AB; therefore CBD, or DBA is equal to BCD; and consequently the three angles BDA, DBA, BCD, are equal to one another. And because the angle DBC is equal to the angle BCD, the side BD is equal (6. 1.) to the side DC; but BD was made equal to CA; therefore also CA is equal to CD, and the angle CDA equal (5. 1.) to the angle DAC; therefore the angles CDA, DAC together, are double of the angle DAC; but BCD is equal to the angles CDA, DAC (32. 1.); therefore also BCD is double of DAC. But BCD is equal to each of the angles BDA, DBA, and therefore each of the angles BDA, DBA, is double of the angle DAB; wherefore an isosceles triangle ABD is described, having each of the angles at the base double of the third angle. Which was to be done.

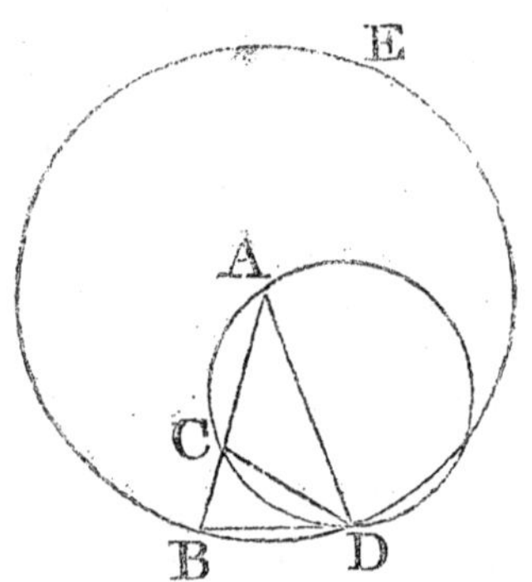

"Cor. 1. The angle BAD is the fifth part of two right angles. "For since each of the angles ABD and ADB is equal to twice the "angle BAD, they are together equal to four times BAD, and there- "fore all the three angles ABD, ADB, BAD, taken together, are "equal to five times the angle BAD. But the three angles ABD, "ADB, BAD are equal to two right angles, therefore five times the "angle BAD is equal to two right angles; or BAD is the fifth part of "two right angles."

"Cor. 2. Because BAD is the fifth part of two, or the tenth part "of four right angles, all the angles about the centre A are together "equal to ten times the angle BAD, and may therefore be divided "into ten parts each equal to BAD. And as these ten equal angles at "the centre, must stand on ten equal arches, therefore the arch BD "is one-tenth of the circumference; and the straight line BD, that is, "AC, is therefore equal to the side of an equilateral decagon inscrib- "ed in the circle BDE."

PROP. XI. PROB.

To inscribe an equilateral and equiangular pentagon in a given circle.

Let ABCDE be the given circle, it is required to inscribe an equilateral and equiangular pentagon in the circle ABCDE.

Describe (10. 4.) an isosceles triangle FGH, having each of the angles at G, H, double of the angle at F; and in the circle ABCDE inscribe (2. 4.) the triangle ACD equiangular to the triangle FGH, so that the angle CAD be equal to the angle at F, and each of the angles ACD, CDA equal to the angle at G or H; wherefore each of the angles ACD, CDA is double of the angle CAD. Bisect (9. 1.) the angle ACD, CDA by the straight lines CE, DB; and join AB, BC, DE, EA. ABCDE is the pentagon required.

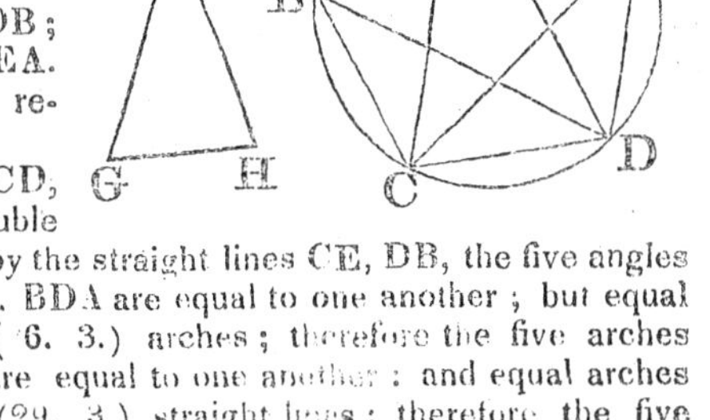

Because the angles ACD, CDA are each of them double of CAD, and are bisected by the straight lines CE, DB, the five angles DAC, ACE, ECD, CDB, BDA are equal to one another; but equal angles stand upon equal (26. 3.) arches; therefore the five arches AB, BC, CD, DE, EA are equal to one another: and equal arches are subtended by equal (29. 3.) straight lines; therefore the five straight lines AB, BC, CD, DE, EA are equal to one another. Wherefore the pentagon ABCDE is equilateral. It is also equiangular; because the arch AB is equal to the arch DE; if to each be added BCD, the whole ABCD is equal to the whole EDCB: and the angle AED stands on the arch ABCD, and the angle BAE on the arch

EDCB; therefore the angle BAE is equal (27. 3.) to the angle AED: for the same reason, each of the angles ABC, BCD, CDE is equal to the angle BAE or AED: therefore the pentagon ABCDE is equiangular: and it has been shown that it is equilateral. Wherefore, in the given circle, an equilateral and equiangular pentagon has been inscribed. Which was to be done.

Otherwise.

" Divide the radius of the given circle, so that the rectangle con-
" tained by the whole and one of the parts may be equal to the square
" of the other (11. 2.) Apply in the circle, on each side of a given
" point, a line equal to the greater of these parts; then (2. Cor. 10. 4.),
" each of the arches cut off will be one-tenth of the circumference, and
" therefore the arch made up of both will be one-fifth of the circum-
" ference; and if the straight line subtending this arch be drawn, it
" will be the side of an equilateral pentagon inscribed in the circle."

PROP. XII. PROB.

To describe an equilateral and equiangular pentagon about a given circle.

Let ABCDE be the given circle, it is required to describe an equilateral and equiangular pentagon about the circle ABCDE.

Let the angles of a pentagon, inscribed in the circle, by the last proposition, be in the points A, B, C, D, E, so that the arches AB, BC, CD, DE, EA are equal (11. 4.) and through the points A, B, C, D, E draw GH, HK, KL, LM, MG, touching (17. 3.) the circle; take the centre F, and join FB, FK, FC, FL, FD. And because the straight line KL touches the circle ABCDE in the point C, to which FC is drawn from the centre F, FC is perpendicular (18. 3.) to KL; therefore each of the angles at C is a right angle; for the same reason, the angles at the points B, D are right angles; and because FCK is a right angle, the square of FK is equal (47. 1.) to the squares of FC, CK. For the same reason, the square of FK is equal to the squares of FB, BK: therefore the squares of FC, CK are equal to the squares of FB, BK, of which the square of FC is equal to the square of FB; the remaining square of CK is therefore equal to the remaining square of BK, and the straight line CK equal to BK: and because FB is equal to FC, and FK common to the triangles BFK, CFK, the two BF, FK are equal to the two CF, FK; and the base BK is equal to the base KC: therefore the angle BFK is equal (8. 1.) to the angle KFC, and the angle BKF to FKC; wherefore the angle BFC is double of the angle KFC, and BKC double of FKC: for the same reason, the angle CFD is double of the angle CFL, and CLD double of CLF: and because the arch BC is equal to the arch CD, the angle BFC is equal (27. 3.) to the angle CFD; and BFC is double of the angle KFC, and CFD double of CFL; therefore the angle KFC is equal to the angle CFL; now the right angle FCK is equal to the right angle FCL; and therefore in the two triangles FKC, FLC, there are two angles of one equal to two angles of the other, each to each, and the side FC, which is adjacent to the equal angles in each, is common to both; therefore the other sides are equal (26. 1.) to the other sides, and the third angle to the third angle; therefore the straight line KC is equal to CL, and the angle FKC to the angle FLC: and because KC is equal to CL, KL is double of KC; in the same manner, it may be shown that HK is double of BK; and because BK is equal to KC, as was demonstrated, and KL is double of KC, and HK double of BK, HK is equal to KL; in like manner, it may be shown that GH, GM, ML, are each of them equal to HK or KL: therefore the pentagon GHKLM is equilateral. It is also equiangular; for, since the angle FKC is equal to the angle FLC, and the angle HKL double of the angle FKC and KLM double of FLC, as was before demonstrated, the angle HKL is equal to KLM; and in like manner it may be shown, that each of the angles KHG, HGM, GML is equal to the angle HKL or KLM; therefore the five angles GHK, HKL, KLM, LMG, MGH being equal to one another, the pentagon GHKLM is equiangular: and it is equilateral as was demonstrated; and it is described about the circle ABCDE. Which was to be done.

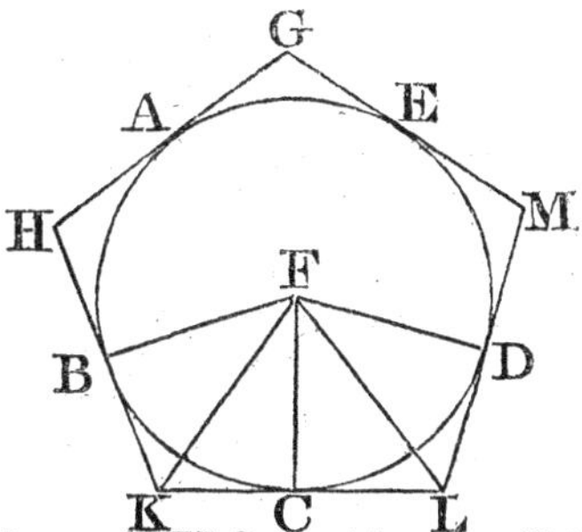

PROP. XIII. PROB.

To inscribe a circle in a given equilateral and equiangular pentagon.

Let ABCDE be the given equilateral and equiangular pentagon; it is required to inscribe a circle in the pentagon ABCDE.

Bisect (9. 1.) the angles BCD, CDE by the straight lines CF, DF, and from the point F, in which they meet, draw the straight lines FB, FA, FE; therefore, since BC is equal to CD, and CF common to the triangles BCF, DCF, the two sides BC, CF are equal to the two DC, CF; and the angle BCF is equal to the angle DCF: therefore the base BF is equal (4. 1.) to the base FD, and the other angles to the other angles, to which the equal sides are opposite: therefore the angle CBF is equal to the angle CDF; and because the angle CDE is double of CDF, and CDE equal to CBA, and CDF to CBF; CBA is also double of the angle CBF; therefore the angle ABF is equal to the angle CBF; wherefore the angle ABC is bisected by the straight line BF: In the same manner, it may be demonstrated that the angles BAE, AED, are bisected by the straight lines AF, EF: from the point F draw (12. 1.) FG, FH, FK, FL, FM perpendiculars to the straight lines AB, BC, CD, DE, EA: and because the angle HCF is equal to KCF, and the right angle FHC, equal to the right angle FKC; in the triangles FHC, FKC there are two angles of one equal to two angles of the other, and the side FC, which is opposite to one of the equal angles in each, is common to both; therefore, the other sides shall be equal (26. 1.), each to each; wherefore the perpendicular FH is equal to the perpendicular FK; in the same manner it may be demonstrated, that FL, FM, FG are each of them equal to FH or FK; therefore the five straight lines FG, FH, FK, FL, FM are equal to one another; wherefore the circle described from the centre F, at the distance of one of these five, will pass through the extremities of the other four, and touch the straight lines AB, BC, CD, DE, EA, because that the angles at the points G, H, K, L, M are right angles, and that a straight line drawn from the extremity of the diameter of a circle at right angles to it, touches (Cor. 16. 3.) the circle: therefore each of the straight lines AB, BC, CD, DE, EA touches the circle; wherefore the circle is inscribed in the pentagon ABCDE. Which was to be done.

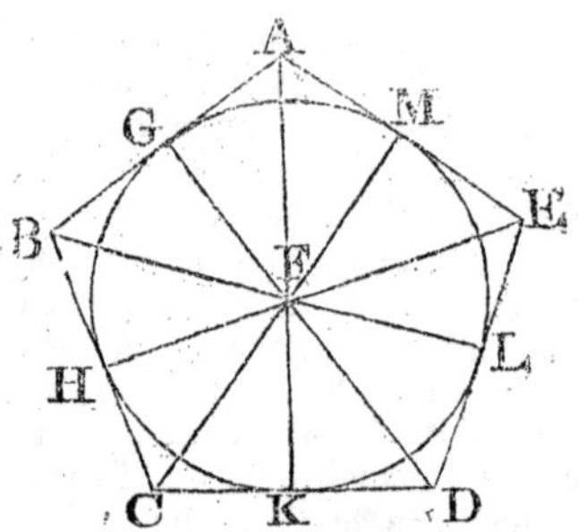

PROP. XIV. PROB.

To describe a circle about a given equilateral and equiangular pentagon.

Let ABCDE be the given equilateral and equiangular pentagon; it is required to describe a circle about it.

Bisect (9. 1. the angles BCD, CDE by the straight lines CF, FD, and from the point F, in which they meet, draw the straight lines FB, FA, FE to the points B, A, E. It may be demonstrated, in the same manner as in the preceding proposition, that the angles CBA, BAE, AED are bisected by the straight lines FB, FA, FE: and because that the angle BCD is equal to the angle CDE, and that FCD is the half of the angle BCD, and CDF the half of CDE; the angle FCD is equal

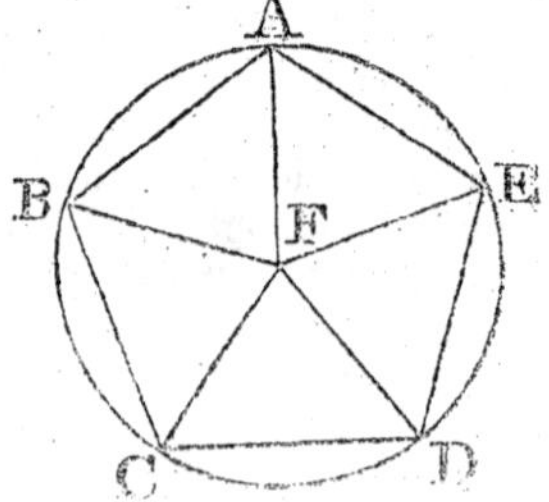

to FDC ; wherefore the side CF is equal (6. 1.) to the side FD : In like manner it may be demonstrated, that FB, FA, FE are each of them equal to FC, or FD : therefore the five straight lines FA, FB, FC, FD, FE are equal to one another ; and the circle described from the centre F, at the distance of one of them, will pass through the extremities of the other four, and be described about the equilateral and equiangular pentagon ABCDE. Which was to be done.

PROP. XV. PROB.

To inscribe an equilateral and equiangular hexagon in a given circle.

Let ABCDEF be the given circle ; it is required to inscribe an equilateral and equiangular hexagon in it.

Find the centre G of the circle ABCDEF, and draw the diameter AGD : and from D, as a centre, at the distance DG, describe the circle EGCH, join EG, CG, and produce them to the points B, F ; and join AB, BC, CD, DE, EF, FA : the hexagon ABCDEF is equilateral and equiangular.

Because G is the centre of the circle ABCDEF, GE is equal to GD : and because D is the centre of the circle EGCH, DE is equal to DG ; wherefore GE is equal to ED, and the triangle EGD is equilateral ; and therefore its three angles EGD, GDE, DEG are equal to one another (Cor. 5. 1.) ; and the three angles of a triangle are equal (32. 1.) to two right angles ; therefore the angle EGD is the third part of two right angles : In the same manner it may be demonstrated that the angle DGC is also the third part of two right angles : and because the straight line GC makes with EB the adjacent angles EGC, CGB equal (13. 1.) to two right angles : the remaining angle CGB is the third part of two right angles : therefore the angles EGD, DGC, CGB, are equal to one another : and also the angles vertical to them, BGA, AGF, FGE (15. 1.); therefore the six angles EGD, DGC, CGB, BGA, AGF, FGE are equal to one another. But equal angles at the centre stand upon equal (26. 3.) arches ; therefore the six arches AB, BC, CD, DE, EF, FA are equal to one another : and equal arches are subtended by equal (29. 3.) straight lines ; therefore the six straight lines are equal to one another, and the hexagon ABCDEF is equilateral. It is also equiangular ; for, since the arch AF is equal to ED, to each of these add the arch ABCD ; therefore the whole arch FABCD shall be equal to the whole EDCBA : and the angle FED stands upon the arch FABCD, and the angle AFE upon EDCBA ; therefore the angle

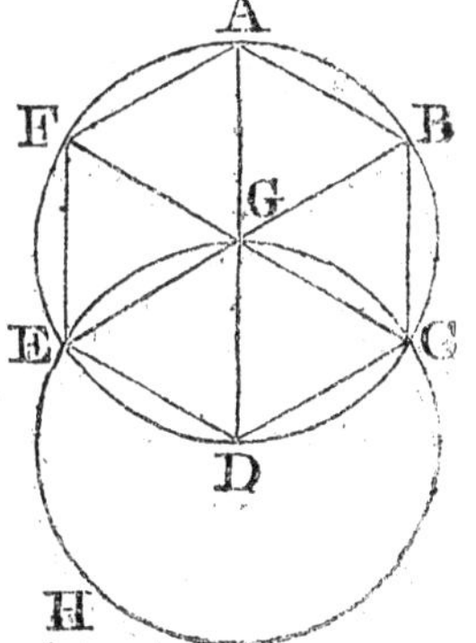

AFE is equal to FED: in the same manner it may be demonstrated, that the other angles of the hexagon ABCDEF are each of them equal to the angle AFE or FED; therefore the hexagon is equiangular; it is also equilateral, as was shown; and it is inscribed in the given circle ABCDEF. Which was to be done.

Cor. From this it is manifest, that the side of the hexagon is equal to the straight line from the centre, that is, to the radius of the circle.

And if through the points A, B, C, D, E, F, there be drawn straight lines touching the circle, an equilateral and equiangular hexagon shall be described about it, which may be demonstrated from what has been said of the pentagon; and likewise a circle may be inscribed in a given equilateral and equiangular hexagon, and circumscribed about it, by a method like to that used for the pentagon.

PROP. XVI. PROB.

To inscribe an equilateral and equiangular quindecagon in a given circle.

Let ABCD be the given circle; it is required to inscribe an equilateral and equiangular quindecagon in the circle ABCD.

Let AC be the side of an equilateral triangle inscribed (2. 4.) in the circle, and AB the side of an equilateral and equiangular pentagon inscribed (11. 4.) in the same; therefore, of such equal parts as the whole circumference ABCDF contains fifteen, the arch ABC, being the third part of the whole, contains five; and the arch AB, which is the fifth part of the whole, contains three; therefore BC their difference contains two of the same parts: bisect (30. 3.) BC in E; therefore BE, EC are, each of them, the fifteenth part of the whole circumference ABCD: therefore if the straight lines BE, EC be drawn, and straight lines equal to them be placed (1. 4.) around in the whole circle, an equilateral and equiangular quindecagon will be inscribed in it. Which was to be done.

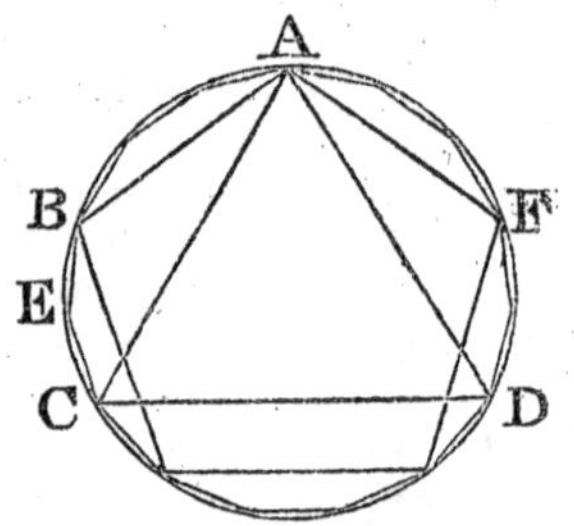

And in the same manner as was done in the pentagon, if through the points of division made by inscribing the quindecagon, straight lines be drawn touching the circle, an equilateral and equiangular quindecagon may be described about it: And likewise, as in the pentagon, a circle may be inscribed in a given equilateral and equiangular quindecagon, and circumscribed about it.

ELEMENTS

OF

GEOMETRY.

BOOK V.

In the demonstrations of this book there are certain "*signs* or *characters* which it has been found convenient to employ.

" 1. The letters A, B, C, &c. are used to denote magnitudes of any " kind. The letters m, n, p, q, are used to denote numbers only.

" 2. The sign + (plus), written between two letters, that denote " magnitudes or numbers, signifies the sum of those magnitudes or " numbers. Thus, A + B is the sum of the two magnitudes denoted " by the letters A and B; $m+n$ is the sum of the numbers denoted " by m and n.

" 3. The sign – (minus), written between two letters, signifies the " excess of the magnitude denoted by the first of these letters, which " is supposed the greatest, above that which is denoted by the other. " Thus, A – B signifies the excess of the magnitude A above the " magnitude B.

" 4. When a number, or a letter denoting a number, is written " close to another letter denoting a magnitude of any kind, it sig- " nifies that the magnitude is multiplied by the number. Thus, " 3A signifies three times A; mB, m times B, or a multiple of B by " m. When the number is intended to multiply two or more mag- " nitudes that follow, it is written thus, m(A+B), which signifies " the sum of A and B taken m times; m(A – B) is m times the ex- " cess of A above B.

" Also, when two letters that denote numbers are written close to " one another, they denote the product of those numbers, when mul- " tiplied into one another. Thus, mn is the product of m into n; and " mnA is A multiplied by the product of m into n.

" 5. The sign = signifies the equality of the magnitudes denoted "by the letters that stand on the opposite sides of it; A=B signifies "that A is equal to B; A+B=C−D signifies that the sum of A and "B is equal to the excess of C above D.

"6. The sign 7 is used to signify that the magnitudes between "which it is placed are unequal, and that the magnitude to which the "opening of the lines is turned is greater than the other. Thus A "7 B signifies that A is greater than B; and A ∠ B signifies that A "is less than B."

DEFINITIONS.

I.

A less magnitude is said to be a part of a greater magnitude, when the less measures the greater, that is, when the less is contained a certain number of times, exactly, in the greater.

II.

A greater magnitude is said to be a multiple of a less, when the greater is measured by the less, that is, when the greater contains the less a certain number of times exactly.

III.

Ratio is a mutual relation of two magnitudes, of the same kind, to one another, in respect of quantity.

IV.

Magnitudes are said to be of the same kind, when the less can be multiplied so as to exceed the greater; and it is only such magnitudes that are said to have a ratio to one another.

V.

If there be four magnitudes, and if any equimultiples whatsoever be taken of the first and third, and any equimultiples whatsoever of the second and fourth, and if, according as the multiple of the first is greater than the multiple of the second, equal to it, or less, the multiple of the third is also greater than the multiple of the fourth, equal to it, or less; then the first of the magnitudes is said to have to the second the same ratio that the third has to the fourth.

VI.

Magnitudes are said to be proportionals, when the first has the same ratio to the second that the third has to the fourth; and the third to the fourth the same ratio which the fifth has to the sixth, and so on whatever be their number.

"When four magnitudes, A, B, C, D are proportionals, it is usual to "say that A is to B as C to D, and to write them thus, A : B :: "C : D, or thus, A : B = C : D."

VII.

When of the equimultiples of four magnitudes, taken as in the fifth definition, the multiple of the first is greater than that of the second, but the multiple of the third is not greater than the multiple of the fourth: then the first is said to have to the second a greater ratio than the third magnitude has to the fourth: and, on the contrary, the third is said to have to the fourth a less ratio than the first has to the second.

VIII.

When there is any number of magnitudes greater than two, of which the first has to the second the same ratio that the second has to the third, and the second to the third the same ratio which the third has to the fourth, and so on, the magnitudes are said to be continual proportionals.

IX.

When three magnitudes are continual proportionals, the second is said to be a mean proportional between the other two.

X.

When there is any number of magnitudes of the same kind, the first is said to have to the last the ratio compounded of the ratio which the first has to the second, and of the ratio which the second has to the third, and of the ratio which the third has to the fourth, and so on unto the last magnitude.

For example, if A, B, C, D, be four magnitudes of the same kind, the first A is said to have to the last D, the ratio compounded of the ratio of A to B, and of the ratio of B to C, and of the ratio of C to D; or, the ratio of A to D is said to be compounded of the ratios of A to B, B to C, and C to D.

And if A : B :: E : F; and B : C :: G : H, and C : D :: K : L, then, since by this definition A has to D the ratio compounded of the ratios of A to B, B to C, C to D; A may also be said to have to D the ratio compounded of the ratios which are the same with the ratios of E to F, G to H, and K to L.

In like manner, the same things being supposed, if M has to N the same ratio which A has to D, then, for shortness' sake, M is said to have to N a ratio compounded of the same ratios, which compound the ratio of A to D; that is, a ratio compounded of the ratios of E to F, G to H, and K to L.

XI.

If three magnitudes are continual proportionals, the ratio of the first to the third is said to be duplicate of the ratio of the first to the second.

"Thus, if A be to B as B to C, the ratio of A to C is said to be duplicate of the ratio of A to B. Hence, since by the last definition, "the ratio of A to C is compounded of the ratios of A to B, and B "to C, a ratio, which is compounded of two equal ratios, is duplicate of either of these ratios."

XII.

If four magnitudes are continual proportionals, the ratio of the first to the fourth is said to be triplicate of the ratio of the first to the second, or of the ratio of the second to the third, &c.

"So also, if there are five continual proportionals; the ratio of the first to the fifth is called quadruplicate of the ratio of the first to the second; and so on, according to the number of ratios. Hence, a ratio compounded of three equal ratios, is triplicate of any one of those ratios; a ratio compounded of four equal ratios quadruplicate," &c.

XIII.

In proportionals, the antecedent terms are called homologous to one another, as also the consequents to one another.

Geometers make use of the following technical words to signify certain ways of changing either the order or magnitude of proportionals, so as that they continue still to be proportionals.

XIV.

Permutando, or alternando, by permutation, or alternately; this word is used when there are four proportionals, and it is inferred, that the first has the same ratio to the third which the second has to the fourth; or that the first is to the third as the second to the fourth: See Prop. 16. of this Book.

XV.

Invertendo, by inversion: When there are four proportionals, and it is inferred, that the second is to the first, as the fourth to the third. Prop. A. Book 5.

XVI.

Componendo, by composition: When there are four proportionals, and it is inferred, that the first, together with the second, is to the second as the third, together with the fourth, is to the fourth. 18th Prop. Book 5.

XVII.

Dividendo, by division: when there are four proportionals, and it is inferred, that the excess of the first above the second, is to the second, as the excess of the third above the fourth, is to the fourth. 17th Prop. Book 5.

XVIII.

Convertendo, by conversion; when there are four proportionals, and it is inferred, that the first is to its excess above the second, as the third to its excess above the fourth. Prop. D. Book 5.

XIX.

Ex æquali (sc. distantia), or ex æquo, from equality of distance; when there is any number of magnitudes more than two, and as many others, so that they are proportionals when taken two and two of each rank, and it is inferred, that the first is to the last of the first rank of magnitudes, as the first is to the last of the others; Of this there are the two following kinds which arise from the different order in which the magnitudes are taken two and two.

XX.

Ex æquali, from equality; this term is used simply by itself, when the first magnitude is to the second of the first rank, as the first to the second of the other rank; and as the second is to the third of the first rank, so is the second to the third of the other; and so on in order, and the inference is as mentioned in the preceding definition; whence this is called ordinate proportion.

It is demonstrated in the 22d Prop. Book 5.

XXI.

Ex æquali, in proportione perturbata, seu inordinata: from equality, in perturbate, or disorderly proportion; this term is used when the first magnitude is to the second of the first rank, as the last but one is to the last of the second rank; and as the second is to the third of the first rank, so is the last but two to the last but one of the second rank; and as the third is to the fourth of the first rank, so is the third from the last, to the last but two, of the second rank; and so on in a cross, or *inverse*, order; and the inference is as in the 19th definition. It is demonstrated in the 23d Prop. of Book 5.

AXIOMS.

I.

Equimultiples of the same, or of equal magnitudes, are equal to one another.

II.

Those magnitudes of which the same, or equal magnitudes, are equimultiples, are equal to one another.

III.

A multiple of a greater magnitude is greater than the same multiple of a less.

IV.

That magnitude of which a multiple is greater than the same multiple of another, is greater than that other magnitude.

PROP. I. THEOR.

If any number of magnitudes be equimultiples of as many others, each of each, what multiple soever any one of the first is of its part, the same multiple is the sum of all the first of the sum of all the rest.

Let any number of magnitudes A, B, and C be equimultiples of as many others, D, E, and F, each of each, $A+B+C$ is the same multiple of $D+E+F$, that A is of D.

Let A contain D, B contain E, and C contain F, each the same number of times, as, for instance, three times,

Then, because A contains D three times, $A=D+D+D$.
For the rame reason, $B=E+E+E$;
And also, $C=F+F+F$.

Therefore, adding equals to equals (Ax. 2. 1.), $A+B+C$ is equal to $D+E+F$, taken three times. In the same manner, if A, B, and C were each any other equimultiple of D, E, and F, it would be shown that $A+B+C$ was the same multiple of $D+E+F$. Therefore, &c. Q. E. D.

Cor. Hence, if m be any number, $mD+mE+mF=m(D+E+F)$. For mD, mE, and mF are multiples of D, E, and F by m, therefore their sum is also a multiple of $D+E+F$ by m.

PROP. II. THEOR.

If to a multiple of a magnitude by any number, a multiple of the same magnitude by any number be added, the sum will be the same multiple of that magnitude that the sum of the two numbers is of unity.

Let $A=mC$, and $B=nC$; $A+B=(m+n)C$.

For, since $A=mC$, $A=C+C+C+$ &c. C being repeated m times.
For the same reason, $B=C+C+$ &c. C being repeated n times.
Therefore, adding equals to equals, $A+B$ is equal to C taken $m+n$ times; that is, $A+B=(m+n)C$. Therefore $A+B$ contains C as oft as there are units in $m+n$. Q. E. D.

Cor. 1. In the same way, if there be any number of multiples whatsoever, as $A=mE$, $B=nE$, $C=pE$, it is shown, that $A+B+C=(m+n+p)E$.

Cor. 2. Hence also, since $A+B+C=(m+n+p)E$, and since $A=mE$, $B=nE$, and $C=pE$, $mE+nE+pE=(m+n+p)E$.

PROP. III. THEOR.

If the first of three magnitudes contain the second as oft as there are units in a certain number, and if the second contain the third also, as often as there are units in a certain number, the first will contain the third as oft as there are units in the product of these two numbers.

Let $A=mB$, and $B=nC$; then $A=mnC$.

Since $B=nC$, $mB=nC+nC+$&c. repeated m times. But $nC+nC$ &c. repeated m times is equal to C (2. Cor. 2. 5.), multiplied by $n+n+$&c. n being added to itself m times; but n added to itself m times, is n multiplied by m, or mn. Therefore $nC+nC+$&c. repeated m times $=mnC$; whence also $mB=mnC$, and by hypothesis $A=mB$, therefore $A=mnC$. Therefore, &c. Q. E. D.

PROP. IV. THEOR.

If the first of four magnitudes has the same ratio to the second which the third has to the fourth, and if any equimultiples whatever be taken of the first and third, and any whatever of the second and fourth; the multiple of the first shall have the same ratio to the multiple of the second, that the multiple of the third has to the multiple of the fourth.

Let $A : B :: C : D$, and let m and n be any two numbers; $mA : nB :: mC : nD$.

Take of mA and mC equimultiples by any number p, and of nB and nD equimultiples by any number q. Then the equimultiples of mA, and mC by p, are equimultiples also of A and C, for they contain A and C as oft as there are units in pm (3. 5.), and are equal to pmA and pmC. For the same reason the multiples of nB and nD by q, are qnB, qnD. Since, therefore, $A : B :: C : D$, and of A and C there are taken any equimultiples, viz. pmA and pmC, and of B and D, any equimultiples qnB, qnD, if pmA be greater than qnB, pmC must be greater than qnD (def. 5. 5.); if equal, equal; and if less, less. But pmA, pmC are also equimultiples of mA and mC, and qnB, qnD are equimultiples of nB and nD, therefore (def. 5. 5.), $mA : nB :: mC : nD$. Therefore, &c. Q. E. D.

Cor. In the same manner it may be demonstrated, that if $A : B :: C : D$, and of A and C equimultiples be taken by any number m, viz. mA and mC, $mA : B :: mC : D$. This may also be considered as included in the proposition, and as being the case when $n=1$.

PROP. V. THEOR.

If one magnitude be the same multiple of another, which a magnitude taken from the first is of a magnitude taken from the other; the remainder is the same multiple of the remainder, that the whole is of the whole.

Let mA and mB be any equimultiples of the two magnitudes A and B, of which A is greater than B; $mA-mB$ is the same multiple of $A-B$ that mA is of A, that is, $mA-mB=m(A-B)$.

Let D be the excess of A above B, then $A-B=D$, and adding B to both, $A=D+B$. Therefore (1. 5.) $mA=mD+mB$; take mB from both, and $mA-mB=mD$; but $D=A-B$, therefore $mA-mB=m(A-B)$. Therefore, &c. Q. E. D.

PROP. VI. THEOR.

If from a multiple of a magnitude by any number a multiple of the same magnitude by a less number be taken away, the remainder will be the same multiple of that magnitude that the difference of the numbers is of unity.

Let mA and nA be multiples of the magnitude A, by the numbers m and n, and let m be greater than n; $m\text{A}-n\text{A}$ contains A as oft as $m-n$ contains unity, or $m\text{A}-n\text{A}=(m-n)\text{A}$.

Let $m-n=q$; then $m=n+q$. Therefore (2. 5.) $m\text{A}=n\text{A}+q\text{A}$; take nA from both, and $m\text{A}-n\text{A}=q\text{A}$. Therefore $m\text{A}-n\text{A}$ contains A as oft as there are units in q, that is, in $m-n$, or $m\text{A}-n\text{A}=(m-n)$ A. Therefore, &c. Q. E. D.

Cor. When the difference of the two numbers is equal to unity, or $m-n=1$, then $m\text{A}-n\text{A}=\text{A}$.

PROP. A. THEOR.

If four magnitudes be proportionals, they are proportionals also when taken inversely.

If A : B :: C : D, then also B : A :: D : C.

Let mA and mC be any equimultiples of A and C; nB and nD any equimultiples of B and D. Then, because A : B :: C : D, if mA be less than nB, mC will be less than nD (def. 5. 5.), that is, if nB be greater than mA, nD will be greater than mC. For the same reason, if $n\text{B}=m\text{A}$, $n\text{D}=m\text{C}$, and if $n\text{B}\angle m\text{A}$, $n\text{D}\angle m\text{C}$. But nB, nD are any equimultiples of B and D, and mA, mC any equimultiples of A and C, therefore (def. 5. 5.), B : A :: D : C. Therefore, &c. Q. E. D.

PROP. B. THEOR.

If the first be the same multiple of the second, or the same part of it, that the third is of the fourth; the first is to the second as the third to the fourth.

First, if mA, mB be equimultiples of the magnitudes A and B, mA : A :: mB : B.

Take of mA and mB equimultiples by any number n; and of A and B equimultiples by any number p; these will be nmA (3. 5.), pA, nmB (3. 5.), pB. Now, if nmA be greater than pA, nm is also greater than p; and if nm is greater than p, nmB is greater than pB, therefore, when nmA is greater than pA, nmB is greater than pB. In the same manner, if $nm\text{A}=p\text{A}$, $nm\text{B}=p\text{B}$, and if $nm\text{A}\angle p\text{A}$, $nm\text{B}\angle p\text{B}$. Now, nmA, nmB are any equimultiples of mA and mB; and

pA, pB are any equimultiples of A and B, therefore $mA : A :: mB : B$ (def. 5. 5.).

Next, Let C be the same part of A that D is of B; then A is the same multiple of C that B is of D, and therefore, as has been demonstrated, $A : C :: B : D$, and inversely (A. 5.) $C : A :: D : B$. Therefore, &c. Q. E. D.

PROP. C. THEOR.

If the first be to the second as the third to the fourth; and if the first be a multiple or a part of the second, the third is the same multiple or the same part of the fourth.

Let $A : B :: C : D$, and first, let A be a multiple of B, C is the same multiple of D, that is, if $A=mB$, $C=mD$.

Take of A and C equimultiples by any number as 2, viz. 2A and 2C; and of B and D, take equimultiples by the number $2m$, viz. $2mB$, $2mD$ (3. 5.); then, because $A=mB$, $2A=2mB$; and since $A : B :: C : D$, and since $2A=2mB$, therefore $2C=2mD$ (def. 5. 5.), and $C=mD$, that is, C contains D m times, or as often as A contains B.

Next, Let A be a part of B, C is the same part of D. For, since $A : B :: C : D$, inversely (A. 5.), $B : A :: D : C$. But A being a part of B, B is a multiple of A; and therefore, as is shewn above, D is the same multiple of C, and therefore C is the same part of D that A is of B. Therefore, &c. Q. E. D.

PROP. VII. THEOR.

Equal magnitudes have the same ratio to the same magnitude; and the same has the same ratio to equal magnitudes.

Let A and B be equal magnitudes, and C any other; $A : C :: B : C$.

Let mA, mB, be any equimultiples of A and B; and nC any multiple of C.

Because $A=B$, $mA=mB$ (Ax. 1. 5.); wherefore, if mA be greater than nC, mB is greater than nC; and if $mA=nC$, $mB=nC$; or, if $mA \angle nC$, $mB \angle nC$. But mA and mB are any equimultiples of A and B, and nC is any multiple of C, therefore (def. 5. 5.) $A : C :: B : C$.

Again, if $A = B$, $C : A :: C : B$; for, as has been proved, $A : C :: B : C$, and inversly (A. 5.), $C : A :: C : B$. Therefore, &c. Q. E. D.

PROP. VIII. THEOR.

Of unequal magnitudes, the greater has a greater ratio to the same than the less has; and the same magnitude has a greater ratio to the less than it has to the greater.

Let $A+B$ be a magnitude greater than A, and C a third magnitude,

A+B has to C a greater ratio than A has to C; and C has a greater ratio to A than it has to A+B.

Let m be such a number that mA and mB are each of them greater than C; and let nC be the least multiple of C that exceeds mA+mB; then nC − C, that is, (n − 1)C (1. 5.) will be less than mA+mB, or mA+mB, that is, m(A+B) is greater than (n − 1.)C. But because nC is greater than mA+mB, and C less than mB, nC − C is greater than mA, or mA is less than nC − C, that is, than (n—1.)C. Therefore the multiple of A+B by m exceeds the multiple of C by n − 1, but the multiple of A by m does not exceed the multiple of C by n—1; therefore A+B has a greater ratio to C that A has to C (def. 7. 5.).

Again, because the multiple of C by n—1, exceeds the multiple of A by m, but does not exceed the multiple of A+B by m, C has a greater ratio to A than it has to A+B (def. 7. 5.). Therefore, &c. Q. E. D.

PROP. IX. THEOR.

Magnitudes which have the same ratio to the same magnitude are equal to one another; and those to which the same magnitude has the same ratio are equal to one another.

If A : C : : B : C, A=B.

For, if not, let A be greater than B; then, because A is greater than B, two numbers, m and n, may be found, as in the last proposition, such that mA shall exceed nC, while mB does not exceed nC. But because A : C : : B : C; and if mA exceed nC, mB must also exceed nC (def. 5. 5.); and it is also shewn that mB does not exceed nC, which is impossible. Therefore A is not greater than B; and in the same way it is demonstrated that B is not greater than A; therefore A is equal to B.

Next, let C : A : : C : B, A=B. For by inversion (A. 5.) A : C : : B : C; and therefore by the first case, A=B.

PROP. X. THEOR.

That magnitude, which has a greater ratio than another has to the same magnitude, is the greatest of the two: And that magnitude, to which the same has a greater ratio than it has to another magnitude, is the least of the two.

If the ratio of A to C be greater than that of B to C, A is greater than B.

Because A : C 7 B : C, two numbers m and n may be found, such that mA 7 nC, and mB ∠ nC (def. 7. 5.). Therefore also mA 7 mB, and A 7 B (Ax. 4. 5.).

Again, let C : B 7 C : A; B ∠ A. For two numbers, m and n may be found, such that mC 7 nB, and mC ∠ nA (def. 7. 5.) Therefore, since nB is less, and nA greater than the same magnitude mC, nB ∠ nA, and therefore B ∠ A. Therefore, &c. Q. E. D.

PROP. XI. THEOR.

Ratios that are equal to the same ratio are equal to one another.

If $A : B :: C : D$; and also $C : D :: E : F$; then $A : B :: E : F$.

Take mA, mC, mE, any equimultiples of A, C, and E; and nB, nD, nF any equimultiples of B, D, and F. Because $A : B :: C : D$, if $mA > nB$, $mC > nD$ (def. 5. 5); but if $mC > nD$, $mE > nF$ (def. 5. 5.), because $C : D :: E : F$; therefore if $mA > nB$, $mE > nF$. In the same manner, if $mA = nB$, $mE = nF$; and if $mA < nB$, $mE < nF$. Now, mA, mE are any equimultiples whatever of A and E; and nB, nF any whatever of B and F; therefore $A : B :: E : F$ (def. 5. 5.). Therefore, &c. Q. E. D.

PROP. XII. THEOR.

If any number of magnitudes be proportionals, as one of the antecedents is to its consequent, so are all the antecedents, taken together, to all the consequents.

If $A : B :: C : D$, and $C : D :: E : F$; then also, $A : B :: A+C+E : B+D+F$.

Take mA, mC, mE any equimultiples of A, C, and E; and nB, nD, nF, any equimultiples of B, D, and F. Then, because $A : B :: C : D$, if $mA > nB$, $nC > nD$ (def. 5. 5.); and when $mC > nD$, $mE > nF$, because $C : D :: E : F$. Therefore, if $mA > nB$, $mA+mC+mE > nB+nD+nF$: In the same manner, if $mA = nB$, $mA+mC+mE = nB+nD+nF$; and if $mA < nB$, $mA+mC+mE < nB+nD+nF$. Now, $mA+mC+mE = m(A+C+E)$ (Cor. 1. 5.), so that mA and $mA+mC+mE$ are any equimultiples of A, and of $A+C+E$. And for the same reason nB, and $nB+nD+nF$ are any equimultiples of B, and of $B+D+F$; therefore (def. 5. 5.) $A : B :: A+C+E : B+D+F$. Therefore, &c. Q. E. D.

PROP. XIII. THEOR.

If the first have to the second the same ratio which the third has to the fourth, but the third to the fourth a greater ratio than the fifth has to the sixth; the first has also to the second a greater ratio than the fifth has to the sixth.

If $A : B :: C : D$; but $C : D > E : F$; then also, $A : B > E : F$.

Because $C : D > E : F$, there are two numbers m and n, such that $mC > nD$, but $mE < nF$ (def. 7. 5.). Now, if $mC > nD$, $mA > nB$, because $A : B :: C : D$. Therefore $mA > nB$, and $mE < nF$, wherefore, $A : B > E : F$ (def. 7. 5.). Therefore, &c. Q. E. D.

PROP. XIV. THEOR.

If the first have to the second the same ratio which the third has to the fourth, and if the first be greater than the third, the second shall be greater than the fourth; if equal, equal; and if less, less.

If $A : B :: C : D$; then if $A > C$, $B > D$; if $A = C$, $B = D$; and if $A < C$, $B < D$.

First, let $A > C$; then $A : B > C : B$ (8. 5.), but $A : B :: C : D$, therefore $C : D > C : B$ (13. 5.), and therefore $B > D$ (10. 5.)

In the same manner, it is proved, that if $A = C$, $B = D$; and if $A < C$, $B < D$. Therefore, &c. Q. E. D.

PROP. XV. THEOR.

Magnitudes have the same ratio to ome another which their equimultiples have.

If A and B be two magnitudes, and m any number, $A : B :: mA : mB$.

Because $A : B :: A : B$ (7. 5.); $A : B :: A+A : B+B$ (12. 5.), or $A : B :: 2A : 2B$. And in the same manner since $A : B :: 2A : 2B$, $A : B :: A+2A : B+2B$ (12. 5.), or $A : B :: 3A : 3B$; and so on, for all the equimultiples of A and B. Therefore, &c. Q. E. D.

PROP. XVI. THEOR.

If four magnitudes of the same kind be proportionals, they will also be proportionals when taken alternately.

If $A : B :: C : D$, then alternately, $A : C :: B : D$.

Take mA, mB any equimultiples of A and B, and nC, nD any equimultiples of C and D. Then (15. 5.) $A : B :: mA : mB$; now $A : B :: C : D$, therefore (11. 5.) $C : D :: mA : mB$. But $C : D :: nC : nD$ (15. 5.); therefore $mA : mB :: nC : nD$ (11. 5): wherefore if $mA > nC$, $mB > nD$ (14. 5.); if $mA = nC$, $mB = nD$, or if $mA < nC$, $mB < nD$; therefore (def. 5. 5.) $A : C :: B : D$. Therefore, &c. Q. E. D.

PROP. XVII. THEOR.

If magnitudes, taken jointly, be proportionals, they will also be proportionals when taken separately; that is, if the first, together with the second, have to the second the same ratio which the third, together with the fourth, has to the fourth, the first will have to the second the same ratio which the third has to the fourth.

If $A+B : B :: C+D : D$, then by division $A : B :: C : D$.

Take mA and nB any multiples of A and B, by the numbers m and n; and first let $mA > nB$: to each of them add mB, then $mA+mB > mB+nB$. But $mA+mB=m(A+B)$ (Cor. 1. 5.), and $mB+nB=(m+n)B$ (2. Cor. 2. 5.), therefore $m(A+B) > (m+n)B$.

And because $A+B : B :: C+D : D$, if $m(A+B) > (m+n)B$, $m(C+D) > (m+n)D$, or $mC+mD > mD+nD$, that is, taking mD from both, $mC > nD$. Therefore, when mA is greater than nB, mC is greater than nD. In like manner, it is demonstrated, that if $mA=nB$, $mC=nD$, and if $mA < nB$, that $mD < nD$; therefore $A : B :: C : D$ (def. 5. 5.). Therefore, &c. Q. E. D.

PROP. XVIII. THEOR.

If magnitudes, taken separately, be proportionals, they will also be proportionals when taken jointly, that is, if the first be to the second as the third to the fourth, the first and second together will be to the second as the third and fourth together to the fourth.

If $A : B :: C : D$, then, by composition, $A+B : B :: C+D : D$.

Take $m(A+B)$, and nB any multiples whatever of $A+B$ and B : and first, let m be greater than n. Then, because $A+B$ is also greater than B, $m(A+B) > nB$. For the same reason, $m(C+D) > nD$. In this case, therefore, that is, when $m > n$, $m(A+B)$ is greater than nB, and $m(C+D)$ is greater than nD. And in the same manner it may be proved, that when $m=n$, $m(A+B)$ is greater than nB, and $m(C+D)$ greater than nD.

Next, let $m < n$, or $n > m$, then $m(A+B)$ may be greater than nB, or may be equal to it, or may be less; first, let $m(A+B)$ be greater than nB; then also, $mA+mB > nB$; take mB, which is less than nB, from both, and $mA > nB-mB$, or $mA > (n-m)B$ (6. 5.). But if $mA > (n-m)B$, $mC > (n-m)D$, because $A : B :: C : D$. Now, $(n-m)D=nD-mD$ (6. 5.), therefore, $mC > nD-mD$, and adding mD to both, $mC+mD > nD$, that is (1. 5.), $m(C+D) > nD$. If, therefore, $m(A+B) > nB$, $m(C+D) > nD$.

In the same manner it will be proved, that if $m(A+B)=nB$, $m(C+D)=nD$; and if $m(A+B) < nB$, $m(C+D) < nD$; therefore (def. 5. 5.), $A+B : B :: C+D : D$. Therefore, &c. Q. E. D.

PROP. XIX. THEOR.

If a whole magnitude be to a whole, as a magnitude taken from the first is to a magnitude taken from the other; the remainder will be to the remainder as the whole to the whole.

If $A : B :: C : D$, and if C be less than A, $A-C : B-D :: A : B$.

Because $A : B :: C : D$, alternately (16. 5.), $A : C :: B : D$; and therefore by division (17. 5.) $A-C : C :: B-D : D$. Wherefore, again alternately, $A-C : B-D :: C : D$; but $A : B :: C : D$, therefore (11. 5.) A—C : B—D : : A : B. Therefore, &c. Q. E. D.

Cor. A—C : B—D : : C : D.

PROP. D. THEOR.

If four magnitudes be proportionals, they are also proportionals by conversion, that is, the first is to its excess above the second, as the third to its excess above the fourth.

If A : B : : C : D, by conversion, A : A—B : : C : C—D.

For, since A : B : : C : D, by division (17. 5.), A—B : B : : C—D : D, and inversely (A. 5.), B : A—B : : D : C—D ; therefore, by composition (18. 5.), A : A—B : : C : C—D. Therefore, &c. Q. E. D.

Cor. In the same way, it may be proved that A : A+B : : C : C+D.

PROP. XX. THEOR.

If there be three magnitudes, and other three, which taken two and two, have the same ratio ; if the first be greater than the third, the fourth is greater than the sixth ; if equal, equal ; and if less, less.

If there be three magnitudes, A, B, and C, and other three D, E, and F ; and if A : B : : D : E ; and also B : C : : E : F, then if A7C, D7F ; if A=C, D=F : and if A∠C, D∠F.

A,	B,	C,
D,	E,	F.

First, let A7C ; then A : B7C : B (8. 5.). But A : B : : D : E, therefore also D : E7C : B (13. 5.). Now B : C : : E : F, and inversely (A. 5.), C : B : : F : E ; and it has been shewn that D : E7C : B, therefore D : E7F : E (13. 5.), and consequently D7F (10. 5.)

Next, let A=C ; then A : B : : C : B (7. 5.), but A : B : : D : E ; therefore, C : B : : D : E, but C : B : : F : E, therefore, D : E : : F : E (11. 5.), and D=F (9. 5.). Lastly, let A∠C. Then C7A, and because, as was already shewn, C : B : : F : E, and B : A : : E : D ; therefore, by the first case, if C7A, F7D, that is, if A∠C, D∠F. Therefore, &c. Q. E. D.

PROP. XXI. THEOR.

If there be three magnitudes, and other three, which have the same ratio taken two and two, but in a cross order ; if the first magnitude be greater than the third, the fourth is greater than the sixth ; if equal, equal ; and if less, less.

If there be three magnitudes, A, B, C, and other three, D, E, and F, such that A : B : : E : F, and B : C : : D : E ; if A7C, D7F if A=C, D=F, and if A∠C, D∠F.

First, let A 7 C. Then A : B 7 C : B (8. 5.), but A : B :: E : F, therefore F : F 7 C : B (13. 5). Now, B : C :: D : E, and inversely, C : B :: E : D ; therefore, E : F 7 E : D (13. 5.), wherefore, D 7 F (10. 5.).

A,	B,	C,
D,	E,	F.

Next, let A=C. Then (7. 5.) A : B :: C : B ; but A : B :: E : F, therefore, C : B :: E : F (11. 5.) ; but B : C :: D : E, and inversely, C : B :: E : D, therefore (11. 5.), E : F :: E : D, and, consequently, D=F (9. 5.).

Lastly, let A ∠ C. Then C 7 A, and, as was already proved, C : B :: E : D ; and B : A :: F : E, therefore, by the first case, since C 7 A, F 7 D, that is, D ∠ F. Therefore, &c. Q. E. D.

PROP. XXII. THEOR.

*If there be any number of magnitudes, and as many others, which, taken two and two in order, have the same ratio ; the first will have to the last of the first magnitudes, the same ratio which the first of the other has to the last.**

First, let there be three magnitudes, A, B, C, and other three, D, E, F, which, taken two and two, in order, have the same ratio, viz. A : B :: D : E, and B : C :: E : F ; then A : C :: D : F.

Take of A and D any equimultiples whatever, mA, mD ; and of B and D any whatever, nB, nE : and of C and F any whatever, qC, qF. Because A : B :: D : E, mA : nB :: mD : nE (4. 5.) ; and for the same reason, nB : qC :: nE : qF. Therefore (20. 5.), according as mA is greater than qC, equal to it, or less, mD is greater than qF, equal to it or less ; but mA, mD are any equimultiples of A and D ; and qC, qF are any equimultiples of C and F ; therefore (def. 5. 5.), A : C :: D : F.

A,	B,	C,
D,	E,	F,
mA,	nB,	qC,
mD,	nE,	qF.

Again, let there be four magnitudes, and other four which, taken two and two in order, have the same ratio, viz. A : B :: E : F ; B : C :: F : G ; C : D :: G : H, then A : D :: E : H.

For, since A, B, C are three magnitudes, and E, F, G other three, which, taken two and two, have the same ratio, by the foregoing case, A : C :: E : G. And because also C : D :: G : H, by that same case, A : D :: E : H. In the same manner is the demonstration extended to any number of magnitudes. Therefore, &c. Q. E. D.

A,	B,	C,	D,
E,	F,	G,	H,

* N. B. This proposition is usually cited by the words " ex æquali," or " ex æquo."

PROP. XXIII. THEOR.

*If there be any number of magnitudes, and as many others, which, taken two and two, in a cross order, have the same ratio; the first will have to the last of the first magnitudes the same ratio which the first of the others has to the last.**

First, let there be three magnitudes, A, B, C, and other three, D, E, and F, which, taken two and two in a cross order, have the same ratio, viz. A : B :: E : F, and B : C :: D : E, then A : C :: D : F. Take of A, B, and D, any equimultiples mA, mB, mD; and of C, E, F any equimultiples nC, nE nF.

Because A : B :: E : F, and because also A : B :: mA : mB (15. 5), and E : F :: nE : nF; therefore, mA : mB :: nE : nF (11. 5.). Again, because B : C :: D : E, mB : nC :: mD : nE (4. 5.); and it has been just shewn that mA : mB :: nE : nF; therefore, if $mA > nC$, $mD > nF$ (21. 5); if $mA = nC$, $mD = nF$; and if $mA < nC$, $mD < nF$. Now, mA and mD are any equimultiples of A and D, and nC, nF any equimultiples of C and F; therefore, A : C :: D : F (def. 5. 5.).

A,	B,	C,
D,	E,	F,
mA,	mB,	nC,
mD,	nB,	nF.

Next, Let there be four magnitudes, A, B, C, and D, and other four, E, F, G, and H, which taken two and two, in a cross order, have the same ratio, viz. A : B :: G : H; B : C :: F : G, and C : D :: E : F, then A : D :: E : H. For, since A, B, C, are three magnitudes, and F, G, H other three, which taken two and two, in a cross order, have the same ratio, by the first case, A : C :: F : H. But C : D :: E : F, therefore, again, by the first case, A : D :: E : H. In the same manner may the demonstration be extended to any number of magnitudes. Therefore, &c. Q. E. D.

A,	B,	C,	D,
E,	F,	G,	H.

PROP. XXIV. THEOR.

If the first has to the second the same ratio which the third has to the fourth; and the fifth to the second, the same ratio which the sixth has to the fourth; the first and fifth together, shall have to the second, the same ratio which the third and sixth together, have to the fourth.

Let A : B :: C : D, and also E : B :: F : D, then A+E : B :: C+F : D.

Because E : B :: F : D, by inversion, B : E :: D : F. But by hypothesis, A : B :: C : D, therefore, ex æquali (22. 5.), A : E :: C : F; and by composition (18. 5.), A+E : E :: C+F : F. And again by hypothesis, E : B :: F : D, therefore, ex æquali (22. 5.), A+E : B :: C+F : D. Therefore, &c. Q. E. D.

* N. B. This proposition is usually cited by the words "ex æquali in proportione perturbata; or, "ex æquo inversely."

PROP. E. THEOR.

If four magnitudes be proportionals, the sum of the first is to their difference as the sum of the other two to their difference.

Let $A : B :: C : D$; then if $A > B$,

$A+B : A-B :: C+D : C-D$; or if $A < B$

$A+B : B-A :: C+D : D-C$.

For, if $A > B$, then because $A : B :: C : D$, by division (17. 5.),

$A-B : B :: C-D : D$, and by inversion (A. 5.),

$B : A-B :: D : C-D$. But, by composition (18. 5.),

$A+B : B :: C+D : D$, therefore, ex æquali (22. 5.),

$A+B : A-B :: C+D : C-D$.

In the same manner, if $B > A$, it is proved, that

$A+B : B-A :: C+D : D-C$. Therefore, &c.

Q. E. D.

PROP. F. THEOR.

Ratios which are compounded of equal ratios, are equal to one another.

Let the ratios of A to B, and of B to C, which compound the ratio of A to C, be equal, each to each, to the ratios of D to E, and E to F, which compound the ratio of D to F, $A : C :: D : F$.

For, first, if the ratio of A to B be equal to that of D to E, and the ratio of B to C equal to that of E to F, ex æquali (22. 5.), $A : C :: D : F$.

A,	B,	C,
D,	E,	F.

And next, if the ratio of A to B be equal to that of E to F, and the ratio of B to C equal to that of D to E, ex æquali inversely (23. 5.), $A : C :: D : F$. In the same manner may the proposition be demonstrated, whatever be the number of ratios. Therefore, &c. Q. E. D.

ELEMENTS

OF

GEOMETRY.

BOOK VI.

DEFINITIONS.

I.

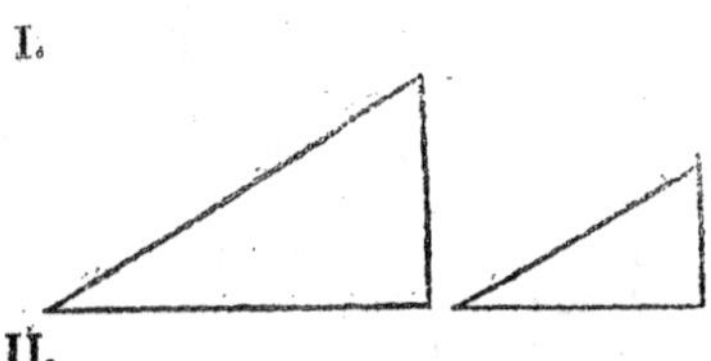

SIMILAR rectilineal figures are those which have their several angles equal, each to each, and the sides about the equal angles proportionals.

II.

Two sides of one figure are said to be reciprocally proportional to two sides of another, when one of the sides of the first is to one of the sides of the second, as the remaining side of the second is to the remaining side of the first.

III.

A straight line is said to be cut in extreme and mean ratio, when the whole is to the greater segment, as the greater segment is to the less.

IV.

The altitude of any figure is the straight line drawn from its vertex perpendicular to the base.

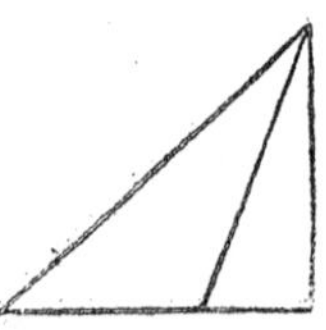

PROP. I. THEOR.

Triangles and parallelograms, of the same altitude, are one to another as their bases.

Let the triangles ABC, ACD, and the parallelograms EC, CF have the same altitude, viz. the perpendicular drawn from the point A to BD: Then, as the base BC, is to the base CD, so is the triangle

ABC to the triangle ACD, and the parallelogram EC to the parallelogram CF.

Produce BD both ways to the points H, L, and take any number of straight lines BG, GH, each equal to the base BC; and DK, KL, any number of them, each equal to the base CD; and join AG, AH, AK, AL. Then, because CB, BG, GH are all equal, the triangles AHG, AGB, ABC are all equal (38. 1.); Therefore, whatever multiple the base HC is of the base BC, the same multiple is the triangle AHC of the triangle ABC. For the same reason, whatever the base LC is of the base CD, the same multiple is the triangle ALC of the triangle ADC. But if the base HC be equal to the base CL, the triangle AHC is also equal to the triangle ALC (38. 1.): and if the base HC be greater than the base CL, likewise the triangle AHC is greater than the triangle ALC; and if less, less. Therefore, since there are four magnitudes, viz. the two bases BC, CD, and the two triangles ABC, ACD; and of the base BC and the triangle ABC, the first and third, any equimultiples whatever have been taken, viz. the base HC, and the triangle AHC; and of the base CD and triangle ACD, the second and fourth, have been taken any equimultiples whatever, viz. the base CL and triangle ALC; and since it has been shown, that if the base HC be greater than the base CL, the triangle AHC is greater than the triangle ALC; and if equal, equal; and if less, less; Therefore (def. 5. 5.), as the base BC is to the base CD, so is the triangle ABC to the triangle ACD.

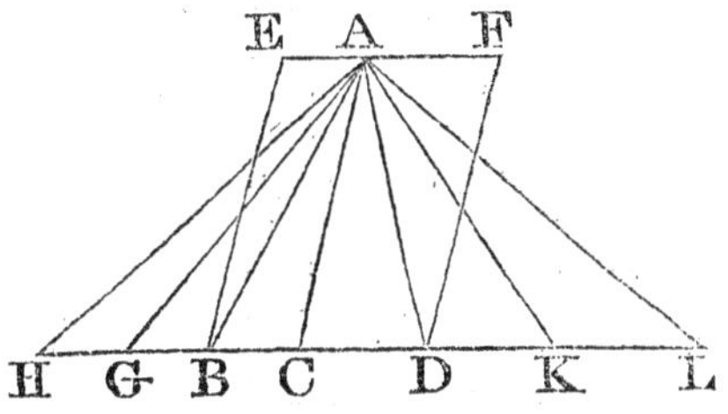

And because the parallelogram CE is double of the triangle ABC (41. 1.), and the parallelogram CF double of the triangle ACD, and because magnitudes have the same ratio which their equimultiples have (15. 5.); as the triangle ABC is to the triangle ACD, so is the parallelogram EC to the parallelogram CF. And because it has been shown, that, as the base BC is to the base CD, so is the triangle ABC to the triangle ACD; and as the triangle ABC to the triangle ACD, so is the parallelogram EC to the parallelogram CF; therefore, as the base BC is to the base CD, so is (11. 5.) the parallelogram EC to the parallelogram CF. Wherefore triangles, &c. Q. E. D,

Cor. From this it is plain, that triangles and parallelograms that have equal altitudes, are to one another as their bases.

Let the figures be placed so as to have their bases in the same straight line; and having drawn perpendiculars from the vertices of the triangles to the bases, the straight line which joins the vertices is parallel to that in which their bases are (33. 1.), because the perpendiculars are both equal and parallel to one another. Then, if the same construction be made as in the proposition, the demonstration will be the same.

PROP. II. THEOR.

If a straight line be drawn parallel to one of the sides of a triangle, it will cut the other sides, or the other sides produced, proportionally: And if the sides, or the sides produced, be cut proportionally, the straight line which joins the points of section will be parallel to the remaining side of the triangle.

Let DE be drawn parallel to BC, one of the sides of the triangle ABC: BD is to DA as CE to EA.

Join BE, CD; then the triangle BDE is equal to the triangle CDE (37. 1.), because they are on the same base DE and between the same parallels DE, BC: but ADE is another triangle, and equal magnitudes have, to the same, the same ratio (7. 5.); therefore, as the triangle BDE to the triangle ADE, so is the triangle CDE to the triangle ADE; but as the triangle BDE to the triangle ADE, so is (1. 6.) BD to DA, because having the same altitude, viz. the perpendicular drawn from the point E to AB, they are to one another as their bases; and for the same reason, as the triangle CDE to the triangle ADE, so is CE to EA. Therefore, as BD to DA, so is CE to EA (11. 5.)

Next, let the sides AB, AC of the triangle ABC, or these sides

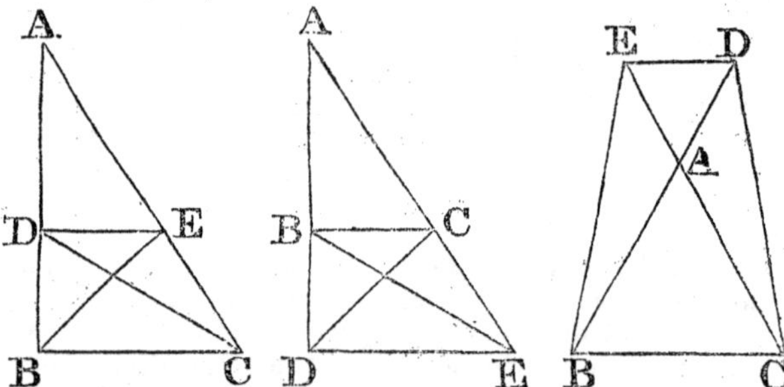

produced, be cut proportionally in the points D, E, that is, so that BD be to DA, as CE to EA, and join DE; DE is parallel to BC.

The same construction being made, because as BD to DA, so is CE to EA; and as BD to DA, so is the triangle BDE to the triangle ADE (1. 6.): and as CE to EA, so is the triangle CDE to the triangle ADE; therefore the triangle BDE, is to the triangle ADE, as the triangle CDE to the triangle ADE; that is, the triangles BDE, CDE have the same ratio to the triangle ADE; and therefore (9. 5.) the triangle BDE is equal to the triangle CDE: And they are on the same base DE; but equal triangles on the same base are between the same parallels (39. 1.) ; therefore DE is parallel to BC. Wherefore, if a straight line, &c. Q. E. D.

PROP. III. THEOR.

If the angle of a triangle be bisected by a straight line which also cuts the base ; the segments of the base shall have the same ratio which the other sides of the triangle have to one another ; And if the segments of the base have the same ratio which the other sides of the triangle have to one another, the straight line drawn from the vertex to the point of section, bisects the vertical angle.

Let the angle BAC, of any triangle ABC, be divided into two equal angles, by the straight line AD ; BD is to DC as BA to AC.

Through the point C draw CE parallel (31. 1.) to DA, and let BA produced meet CE in E. Because the straight line AC meets the parallels AD, EC, the angle ACE is equal to the alternate angle CAD (29. 1.) : But CAD, by the hypothesis, is equal to the angle BAD ; wherefore BAD is equal to the angle ACE. Again, because the straight line BAE meets the parallels AD, EC, the exterior angle BAD is equal to the interior and opposite angle AEC : But the angle ACE has been proved equal to the angle BAD ; therefore also ACE is equal to the angle AEC, and consequently the side AE is equal to the side (6. 1.) AC. And because AD is drawn parallel to one of the sides of the triangle BCE, viz. to EC, BD is to DC, as BA to AE (2. 6.) ;

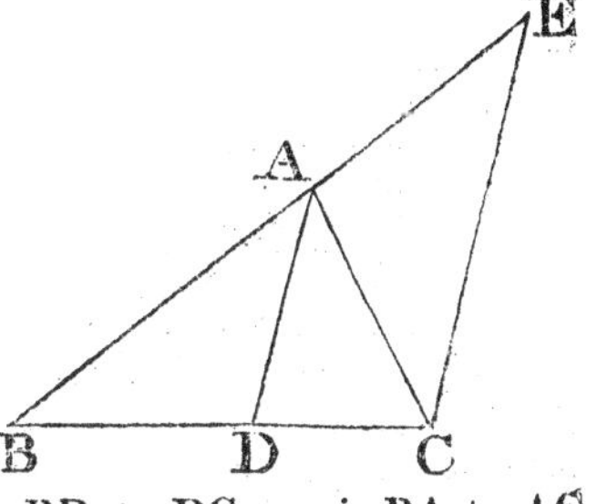

but AE is equal to AC ; therefore, as BD to DC, so is BA to AC (7. 5.).

Next, let BD be to DC, as BA to AC, and join AD ; the angle BAC is divided into two equal angles, by the straight line AD.

The same construction being made ; because, as BD to DC, so is BA to AC ; and as BD to DC, so is BA to AE (2. 6.), because AD is parallel to EC ; therefore AB is to AC, as AB to AE (11. 5.) : Consequently AC is equal to AE (9. 5.), and the angle AEC is therefore equal to the angle ACE (5. 1.) But the angle AEC is equal to the exterior and opposite angle BAD ; and the angle ACE is equal to the alternate an-

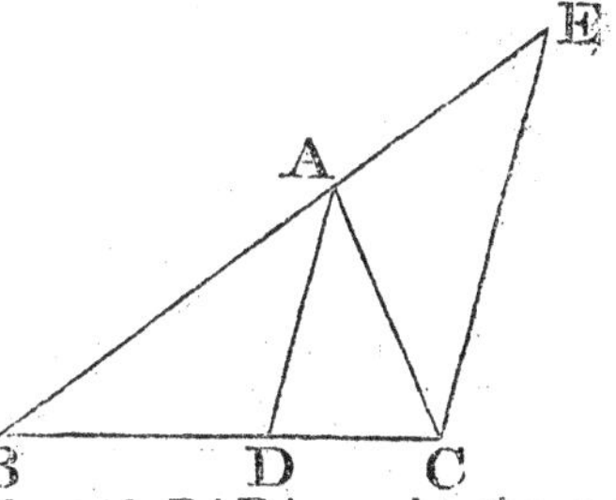

gle CAD (29. 1.) : Wherefore also the angle BAD is equal to the angle CAD : Therefore the angle BAC is cut into two equal angles by the straight line AD. Therefore, if the angle, &c. Q. E. D.

PROP. A. THEOR.

If the exterior angle of a triangle be bisected by a straight line which also cuts the base produced; the segments between the bisecting line and the extremities of the base have the same ratio which the other sides of the triangles have to one another: And if the segments of the base produced have the same ratio which the other sides of the triangle have, the straight line, drawn from the vertex to the point of section, bisect the exterior angle of the triangle.

Let the exterior angle CAE, of any triangle ABC, be bisected by the straight line AD which meets the base produced in D; BD is to DC, as BA to AC.

Through C draw CF parallel to AD (31. 1.): and because the straight line AC meets the parallels AD, FC, the angle ACF is equal to the alternate angle CAD (29. 1.): But CAD is equal to the angle DAE (Hyp.): therefore also DAE is equal to the angle ACF. Again, because the straight line FAE meets the parallels AD, FC the exterior angle DAE is equal to the interior and opposite angle CFA: But the angle ACF has been proved to be equal to the angle DAE; therefore also the angle ACF is equal to the angle CFA, and consequently the side AF is equal to the side AC (6. 1.); and, because AD is parallel to FC, a side of the triangle BCF, BD is to DC, as BA to AF (2. 6.); but AF is equal to AC; therefore as BD is to DC, so is BA to AC.

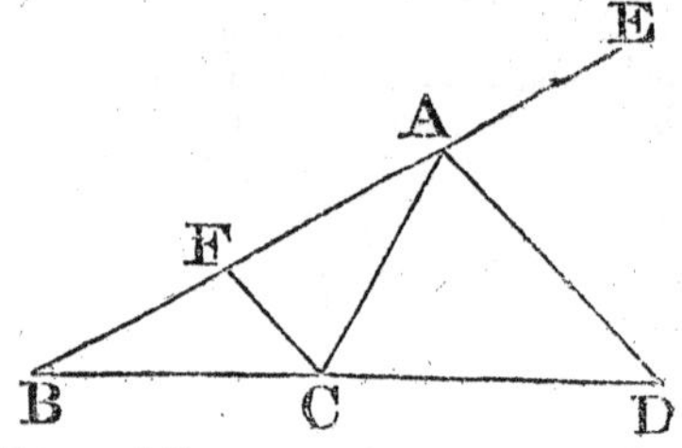

Now let BD be to DC, as BA to AC, and join AD; the angle CAD is equal to the angle DAE.

The same construction being made, because BD is to DC, as BA to AC; and also BD to DC, BA to AF (2. 6.); therefore BA is to AC, as BA to AF (11. 5.); wherefore AC is equal to AF (9. 5.), and the angle AFC equal (5. 1.) to the angle ACF: but the angle AFC is equal to the exterior angle EAD, and the angle ACF to the alternate angle CAD; therefore also EAD is equal to the angle CAD. Wherefore, if the exterior, &c. Q. E. D.

PROP. IV. THEOR.

The sides about the equal angles of equiangular triangles are proportionals; and those which are opposite to the equal angles are homologous sides, that is, are the antecedents or consequents of the ratios.

Let ABC, DCE, be equiangular triangles, having the angle ABC equal to the angle DCE, and the angle ACB to the angle DEC, and consequently (32. 1.) the angle BAC equal to the angle CDE. The sides about the equal angles of the triangles ABC, DCE are proportionals; and those are the homologous sides which are opposite to the equal angles.

Let the triangle DCE be placed, so that its side CE may be contiguous to BC, and in the same straight line with it: And because the angles ABC, ACB are together less than two right angles (17. 1.), ABC and DEC, which is equal to ACB, are also less than two right angles: wherefore BA, ED produced shall meet (Cor. 29. 1.); let them be produced and meet in the point F; and because the angle ABC is equal to the angle DCE, BF is parallel (28. 1.) to CD. Again, because the angle ACB is equal to the angle DEC, AC is parallel to FE (28. 1.); Therefore FACD is a parallelogram; and consequently AF is equal to CD, and AC to FD (34. 1.):

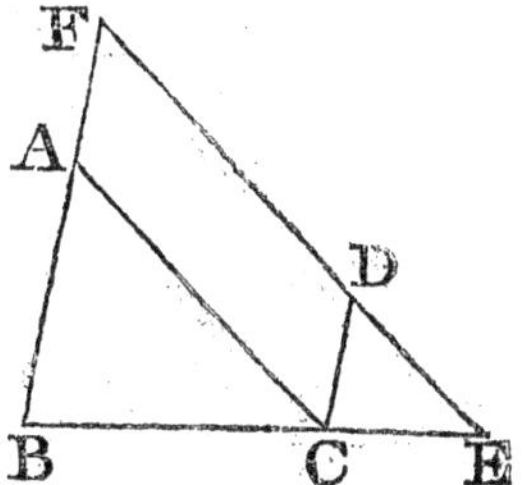

And because AC is parallel to FE, one of the sides of the triangle FBE, BA : AF :: BC : CE (2. 6.): but AF is equal to CD; therefore (7. 5.), BA : CD :: BC : CE; and alternately, BA : BC :: DC : CE (16. 5.): Again, because CD is parallel to BF, BC : CE :: FD : DE (2. 6.); but FD is equal to AC; therefore BC : CE :: AC : DE; and alternately, BC : CA :: CE : ED. Therefore, because it has been proved that AB : BC :: DC : CE; and BC : CA :: CE : ED, ex æquali, BA : AC :: CD : DE. Therefore the sides, &c. Q. E. D.

PROP. V. THEOR.

If the sides of two triangles, about each of their angles, be proportionals, the triangles shall be equiangular, and have their equal angles opposite to the homologous sides.

Let the triangles ABC, DEF have their sides proportionals, so that AB is to BC, as DE to EF; and BC to CA, as EF to FD; and consequently, ex æquali, BA to AC, as ED to DF; the triangle ABC is equiangular to the triangle DEF, and their equal angles are opposite to the homologous sides, viz. the angle ABC being equal to the angle DEF, and BCA to EFD, and also BAC to EDF.

At the points E, F, in the straight line EF, make (23. 1.) the angle FEG equal to the angle ABC, and the angle EFG equal to BCA, wherefore the remaining angle BAC is equal to the remaining angle EGF (32. 1.), and the triangle ABC is therefore equiangular to the triangle GEF; and consequently they have their sides opposite to the equal angles proportionals (4. 6.). Wherefore,

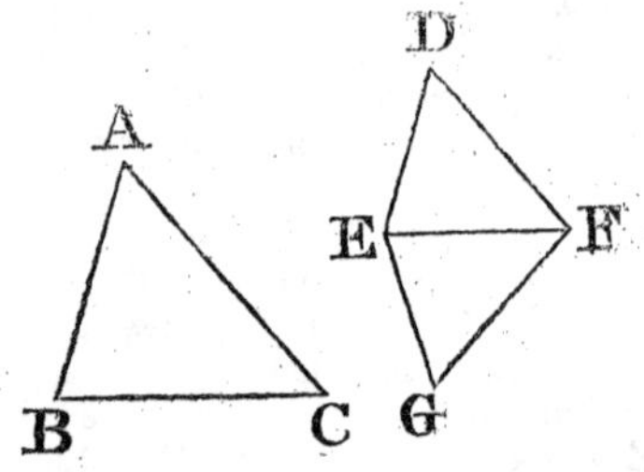

AB : BC :: GE : EF; but by supposition,
AB : BC :: DE : EF, therefore,
DE : EF :: GE : EF. Therefore (11. 5.) DE and GE have the same ratio to EF, and consequently are equal (9. 5.). For the same reason, DF is equal to FG: And because, in the triangles DEF, GEF, DE is equal to EG, and EF common, and also the base DF equal to the base GF; therefore the angle DEF is equal (8. 1.) to the angle GEF, and the other angles to the other angles, which are subtended by the equal sides (4. 1.) Wherefore the angle DFE is equal to the angle GFE, and EDF to EGF: and because the angle DEF is equal to the angle GEF, and GEF to the angle ABC; therefore the angle ABC is equal to the angle DEF: For the same reason, the angle ACB is equal to the angle DFE, and the angle at A to the angle at D. Therefore the triangle ABC is equiangular to the triangle DEF. Wherefore, if the sides, &c. Q. E. D.

PROP. VI. THEOR.

If two triangles have one angle of the one equal to one angle of the other, and the sides about the equal angles proportionals, the triangles shall be equiangular, and shall have those angles equal which are opposite to the homologous sides.

Let the triangles ABC, DEF have the angle BAC in the one equal to the angle EDF in the other, and the sides about those angles proportionals; that is, BA to AC, as ED to DF; the triangles ABC, DEF are equiangular, and have the angle ABC equal to the angle DEF, and ACB to DFE.

At the points D, F, is the straight line DF, make (23. 1.) the angle FDG equal to either of the angles BAC, EDF; and the angle DFG equal to the angle ACB; wherefore the remaining angle at B is equal to the remaining one at G (32. 1.), and consequently

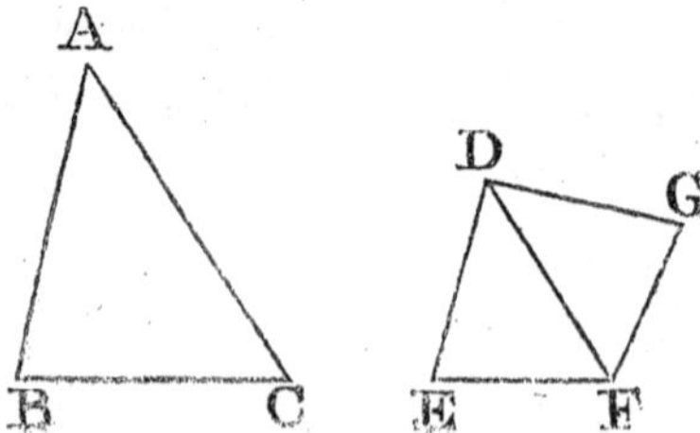

the triangle ABC is equiangular to the triangle DGF; and therefore

BA : AC :: GD (4. 6.): DF. But by hypothesis,

BA : AC :: ED : DF; and therefore

ED : DF :: GD : (11. 5.) DF; wherefore ED is equal (9. 5.) to DG: and DF is common to the two triangles EDF, GDF; therefore the two sides ED, DF are equal to the two sides GD, DF; but the angle EDF is also equal to the angle GDF; wherefore the base EF is equal to the base FG (4. 1.), and the triangle EDF to the triangle GDF, and the remaining angles to the remaining angles, each to each, which are subtended by the equal sides: Therefore the angle DFG is equal to the angle DFE, and the angle at G to the angle at E: But the angle DFG is equal to the angle ACB; therefore the angle ACB is equal the angle DFE, and the angle BAC is equal to the angle EDF (Hyp.); wherefore also the remaining angle at B is equal to the remaining angle at E. Therefore the triangle ABC is equiangular to the triangle DEF. Wherefore, if two triangles, &c. Q. E. D.

PROP. VII. THEOR.

If two triangles have one angle of the one equal to one angle of the other, and the sides about two other angles proportionals, then, if each of the remaining angles be either less, or not less, than a right angle, the triangles shall be equiangular, and have those angles equal about which the sides are proportionals.

Let the two triangles ABC, DEF have one angle in the one equal to one angle in the other, viz. the angle BAC to the angle EDF, and the sides about two other angles ABC, DEF proportionals, so that AB is to BC, as DE to EF; and, in the first case, let each of the remaining angles at C, F be less than a right angle. The triangle ABC is equiangular to the triangle DEF, that is, the angle ABC is equal to the angle DEF, and the remaining angle at C to the remaining angle at F.

For, if the angles ABC, DEF be not equal, one of them is greater than the other: Let ABC be the greater, and at the point B, in the straight line AB, make the angle ABG equal to the angle (23. 1.) DEF: and because the angle at A is equal to the angle at D, and the angle ABG to the angle DEF; the remaining angle AGB is equal (32. 1.) to the remaining angle DFE: Therefore the triangle ABG is equiangular to the triangle DEF;

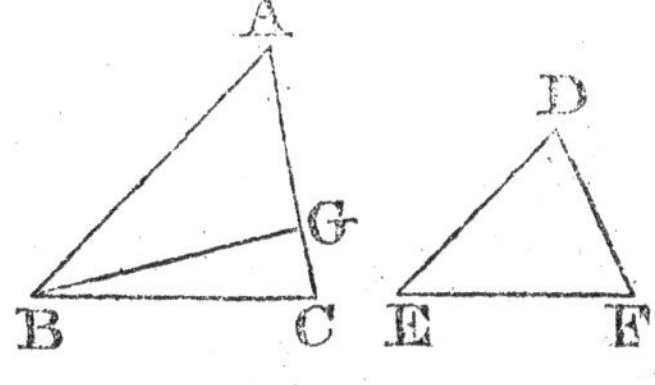

wherefore (4. 6.), AB : BG :: DE : EF; but,

by hypothesis, DE : EF :: AB : BC,

therefore AB : BC :: AB : BG (11. 5.)

and because AB has the same ratio to each of the lines BC, BG; BC is equal (9. 5.) to BG, and therefore the angle BGC is equal to the angle BCG (6. 1.): But the angle BCG is, by hypothesis, less than a right angle; therefore also the angle BGC is less than a right angle, and the adjacent angle AGB must be greater than a right angle (13. 1.). But it was proved that the angle AGB is equal to the angle at F; therefore the angle at F is greater than a right angle: But by the hypothesis, it is less than a right angle; which is absurd. Therefore the angles ABC, DEF are not unequal, that is, they are equal: And the angle at A is equal to the angle at D; wherefore the remaining angle at C is equal to the remaining angle at F: Therefore the triangle ABC is equiangular to the triangle DEF.

Next, let each of the angles at C, F be not less than a right angle; the triangle ABC is also, in this case, equiangular to the triangle DEF.

The same construction being made, it may be proved, in like manner, that BC is equal to BG, and the angle at C equal to the angle BGC: But the angle at C is not less than a right angle; therefore the angle BGC is not less than a right angle: Wherefore, two angles of the triangle BGC are together not less than two right angles, which is impossible (17. 1.); and therefore the triangle ABC may be proved to be equiangular to the triangle DEF, as in the first case.

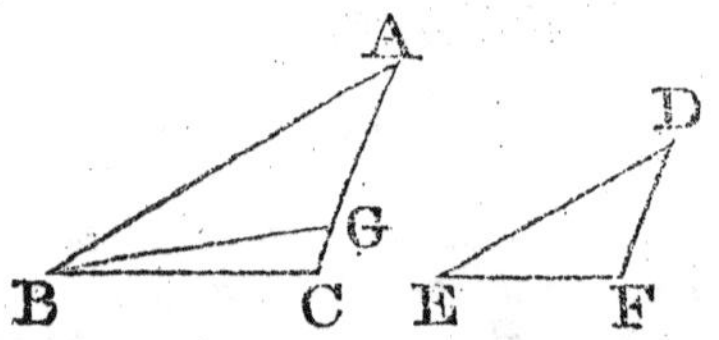

PROP. VIII. THEOR.

In a right angled triangle if a perpendicular be drawn from the right angle to the base; the triangles on each side of it are similar to the whole triangle, and to one another.

Let ABC be a right angled triangle, having the right angle BAC; and from the point A let AD be drawn perpendicular to the base BC: the triangles ABD, ADC are similar to the whole triangle ABC, and to one another.

Because the angle BAC is equal to the angle ADB, each of them being a right angle, and the angle at B common to the two triangles ABC, ABD; the remaining angle ACB is equal to the remaining angle BAD (32. 1.): therefore the triangle ABC is equiangular to the triangle ABD, and the sides about their equal angles are proportionals (4. 6.); wherefore the triangles are similar (def. 1. 6.). In the like manner, it may be demonstrated, that the triangle ADC is equian-

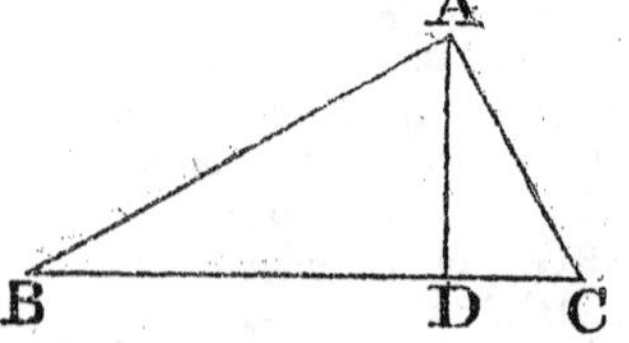

gular and similar to the triangle ABC: and the triangles ABD, ADC, being each equiangular and similar to ABC, and equiangular and similar to one another. Therefore, in a right angled, &c. Q. E. D.

Cor. From this it is manifest, that the perpendicular, drawn from the right angle of a right angled triangle, to the base, is a mean proportional between the segments of the base; and also that each of the sides is a mean proportional between the base, and its segment adjacent to that side. For in the triangles BDA, ADC, BD : DA :: DA : DC (4. 6.); and in the triangles ABC, BDA, BC : BA :: BA : BD (4. 6.); and in the triangles ABC, ACD, BC : CA :: CA : CD (4. 6.).

PROP. IX. PROB.

From a given straight line cut off any part required, that is, a part which shall be contained in it a given number of times.

Let AB be the given straight line; it is required to cut off from AB, a part which shall be contained in it a given number of times.

From the point A draw a straight line AC making any angle with AB; and in AC take any point D, and take AC such that it shall contain AD, as oft as AB is to contain the part, which is to be cut off from it; join BC, and draw DE parallel to it: then AE is the part required to be cut off.

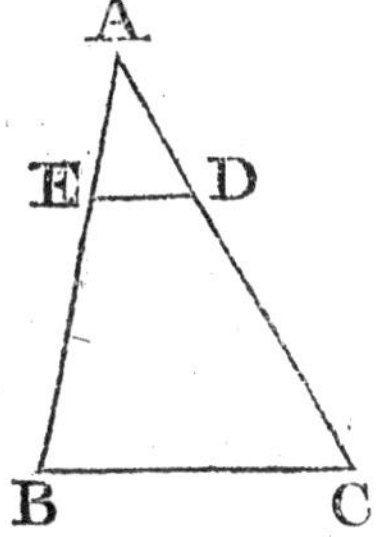

Because ED is parallel to one of the sides of the triangle ABC, viz. to BC, CD : DA :: DE : EA (2. 6.); and by composition (18. 5.), CA : AD :: BA : AE; But CA is a multiple of AD; therefore (C. 5.) BA is the same multiple of AE, or contains AE the same number of times that AC contains AD; and therefore, whatever part AD is of AC, AE is the same of AB; wherefore, from the straight line AB the part required is cut off. Which was to be done.

PROP. X. PROB.

To divide a given straight line similarly to a given divided straight line, that is, into parts that shall have the same ratios to one another which the parts of the divided given straight line have.

Let AB be the straight line given to be divided, and AC the divided line: it is required to divide AB similarly to AC.

Let AC be divided in the points D, E: and let AB, AC be placed so as to contain any angle, and join BC, and through the points D, E, draw (31. 1.) DF, EG, parallel to BC; and through D draw DHK, parallel to AB; therefore each of the figures FH, HB, is a parallelogram: wherefore DH is equal (34. 1.) to FG, and HK to GD; and because HE is parallel to KC, one of the sides of the triangle DKC, CE : ED :: (2. 6.) KH : HD; But KH=BG, and HD=GF; therefore CE : ED :: BG : GF: Again, because FD is parallel to EG, one of the sides of the triangle AGE, ED : DA :: GF : FA: But it has been proved that CE : ED :: BG : GF; therefore the given straight line AB is divided similarly to AC. Which was to be done.

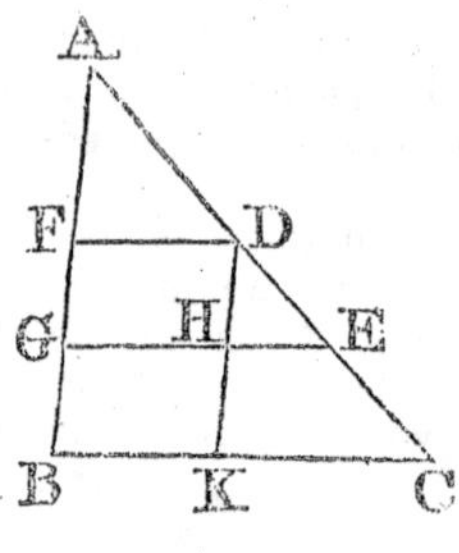

PROP. XI. PROB.

To find a third proportional to two given straight lines.

Let AB, AC be the two given straight lines, and let them be placed so as to contain any angle; it is required to find a third proportional to AB, AC.

Produce AB, AC to the points D, E; and make BD equal to AC; and having joined BC, through D draw DE parallel to it (31. 1.).

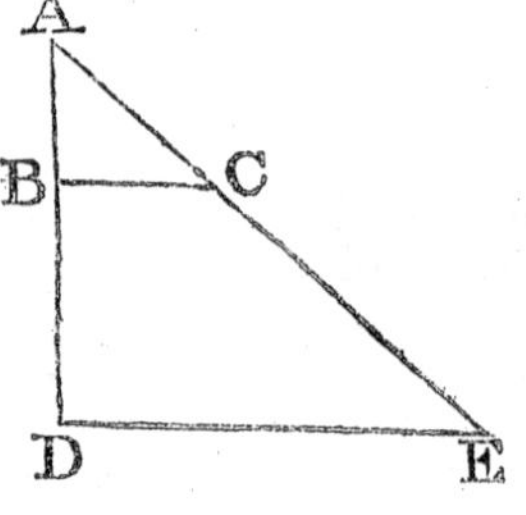

Because BC is parallel to DE, a side of the triangle ADE, AB: (2. 6.) BD :: AC : CE; but BE=AC: therefore AB : AC :: AC : CE. Wherefore to the two given straight lines AB, AC a third proportional, CE is found. Which was to be done.

PROP. XII. PROB.

To find a fourth proportional to three given straight lines.

Let A, B, C be the three given straight lines; it is required to find a fourth proportional to A, B, C.

Take two straight lines DE, DF, containing any angle EDF; and upon these make DG equal to A, GE equal to B, and DH equal to C; and having joined GH, draw EF parallel (31. 1.) to it through the

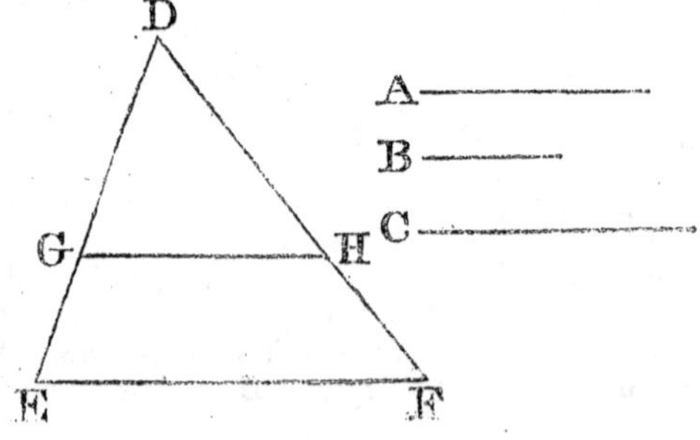

point E. And because GH is parallel to EF, one of the sides of the triangle DEF, DG : GE :: DH : HF (2. 6.); but DG=A, GE=B, and DH=C; and therefore A : B :: C : HF. Wherefore to the three given straight lines, A, B, C, a fourth proportional HF is found. Which was to be done.

PROP. XIII. PROB.

To find a mean proportional between two given straight lines.

Let AB, BC be the two given straight lines; it is required to find a mean proportional between them.

Place AB, BC in a straight line, and upon AC describe the semicircle ADC, and from the point B (11. 1.) draw BD at right angles to AC, and join AD, DC.

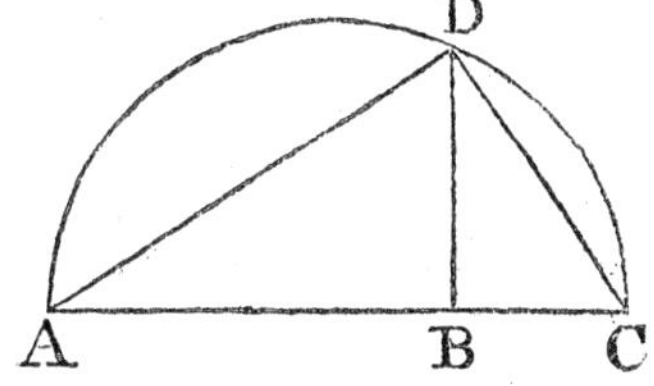

Because the angle ADC in a semicircle is a right angle (31. 3.) and because in the right angled triangle ADC, DB is drawn from the right angle, perpendicular to the base, DB is a mean proportional between AB, BC, the segments of the base (Cor. 8. 6.); therefore between the two given straight lines AB, BC, a mean proportional DB is found. Which was to be done.

PROP. III. PROB.

Equal parallelograms which have one angle of the one equal to one angle of the other, have their sides about the equal angles reciprocally proportional: And parallelograms which have one angle of the one equal to one angle of the other, and their sides about the equal angles reciprocally proportional, are equal to one another.

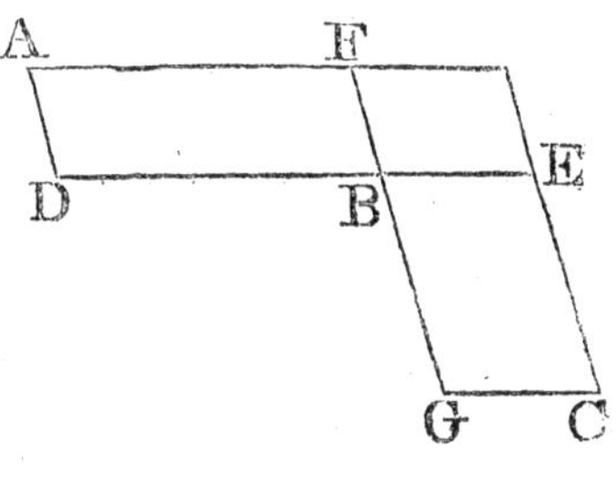

Let AB, BC be equal parallelograms, which have the angles at B equal, and let the sides DB, BE be placed in the same straight line; wherefore also FB, BG are in one straight line (14. 1.): the sides of the parallelograms AB, BC, about the equal angles, are reciprocally proportional; that is, DB is to BE, as GB to BF.

Complete the parallelogram FE; and because the parallelograms AB, BC are equal, and FE is another parallelogram,

AB : FE :: BC : FE (7. 5.);

but because the parallelograms AB, FE have the same altitude,

AB : FE :: DB : BE (1. 6.), also,
BC : FE :: GB : BF (1. 6.); therefore
DB : BE :: GB : BF (11. 5.). Wherefore, the sides of the parallelograms AB, BC about their equal angles are reciprocally proportional.

But, let the sides about the equal angles be reciprocally proportional, viz. as DB to BE, so GB to BF; the parallelogram AB is equal to the parallelogram BC.

Because, DB : BE :: GB : BF, and DB : BE :: AB : FE, and GB : BF :: BC : EF, therefore, AB : FE :: BC : FE (11. 5.); Wherefore the parallelogram AB is equal (9. 5.) to the parallelogram BC. Therefore equal parallelograms, &c. Q. E. D.

PROP. XV. THEOR.

Equal triangles which have one angle of the one equal to one angle of the other have their sides about the equal angles reciprocally proportional: And triangles which have one angle in the one equal to one angle in the other, and their sides about the equal angles reciprocally proportional, are equal to one another.

Let ABC, ADE be equal triangles, which have the angle BAC equal to the angle DAE: the sides about the equal angles of the triangles are reciprocally proportional; that is, CA is to AD, as EA to AB.

Let the triangles be placed so that their sides CA, AD be in one straight line; wherefore also EA and AB are in one straight line (14. 1.); join BD. Because the triangle ABC is equal to the triangle ADE, and ABD is another triangle; therefore, triangle CAB : triangle BAD :: triangle EAD : triangle BAD; but CAB : BAD :: CA : AD and EAD : BAD :: EA : AB; therefore CA : AD :: EA : AB (11. 5.), wherefore the sides of the triangles ABC, ADE about the equal angles are reciprocally proportional.

But let the sides of the triangles ABC, ADE, about the equal angles be reciprocally proportional, viz. CA to AD, as EA to AB; the triangle ABC is equal to the triangle ADE.

Having joined BD as before; because CA : AD :: EA : AB; and since CA : AD :: triangle ABC : triangle BAD (1. 6.); and also EA : AB :: triangle EAD : triangle BAD (11. 5.); therefore, triangle ABC : triangle BAD :: triangle EAD : triangle BAD; that is, the triangles ABC, EAD have the same ratio to the triangle BAD; wherefore the triangle ABC is equal (9. 5.) to the triangle EAD. Therefore equal triangles, &c. Q. E. D.

PROP. XVI. THEOR.

If four straight lines be proportionals, the rectangle contained by the extremes is equal to the rectangle contained by the means; And if the rectangle contained by the extremes be equal to the rectangle contained by the means, the four straight lines are proportionals.

Let the four straight lines, AB, CD, E, F be proportionals, viz. as AB to CD, so E to F; the rectangle contained by AB, F is equal to the rectangle contained by CD, E.

From the points A, C draw (11. 1.) AG, CH at right angles to AB, CD; and make AG equal to F, and CH equal to E, and complete the parallelograms BG, DH. Because AB : CD :: E : F; and since E=CH, and F=AG, AB : CD (7. 5.) :: CH : AG; therefore the sides of the parallelograms BG, DH about the equal angles are reciprocally proportional; but parallelograms which have their sides about equal angles reciprocally proportional, are equal to one another (14. 6.); therefore the parallelogram BG is equal to the parallelogram DH: and the parallelogram BG is contained by the straight lines AB, F; because AG is equal to F; and the parallelogram DH is contained by CD and E, because CH is equal to E: therefore the rectangle contained by the straight lines AB, F is equal to that which is contained by CD and E.

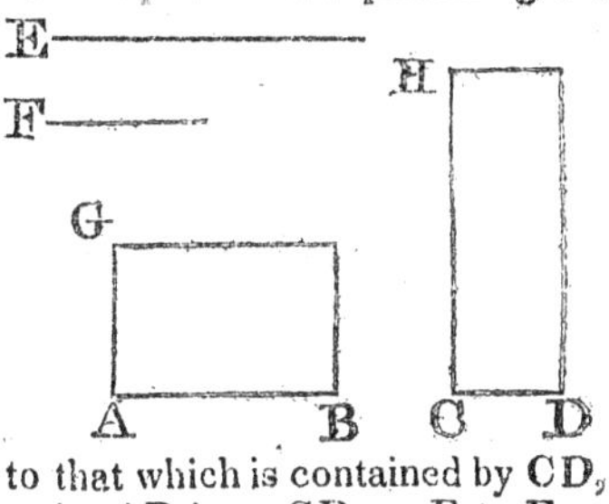

And if the rectangle contained by the straight lines AB, F be equal to that which is contained by CD, F; these four lines are proportionals, viz. AB is to CD, as E to F.

The same construction being made, because the rectangle contained by the straight lines AB, F is equal to that which is contained by CD, E, and the rectangle BG is contained by AB, F, because AG is equal to F; and the rectangle DH, by CD, E, because CH is equal to E; therefore the parallelogram BG is equal to the parallelogram DH, and they are equiangular: but the sides about the equal angles of equal parallelograms are reciprocally proportional (14. 6.): wherefore AB : CD :: CH : AG; but CH=E, and AG=F, therefore AB : CD :: E : F. Wherefore, if four, &c. Q. E. D.

PROP. XVII. THEOR.

If three straight lines be proportionals, the rectangle contained by the extremes is equal to the square of the mean: And if the rectangle contained by the extremes be equal to the square of the mean, the three straight lines are proportionals.

Let the three straight lines, A, B, C be proportionals, viz. as A to B, so B to C; the rectangle contained by A, C is equal to the square of B.

Take D equal to B: and because as A to B, so B to C, and that B is equal to D; A is (7. 5.) to B, as D to C: but if four straight lines be proportionals, the rectangle contained by the extremes is equal to that which is contained by the means (16. 6.); therefore the rectangle A.C = the rectangle B.D; but the rectangle B.D is equal to the square of B, because B=D; therefore the rectangle A.C is equal to the square of B.

A ————————
B ——————
D ——————
C ——————

And if the rectangle contained by A, C be equal to the square of B; A : B :: B : C.

The same construction being made, because the rectangle contained by A, C is equal to the square of B, and the square of B is equal to the rectangle contained by B, D, because B is equal to D; therefore the rectangle contained by A, C is equal to that contained by B, D; but if the rectangle contained by the extremes be equal to that contained by the means, the four straight lines are proportionals (16. 6.): therefore A : B :: D : C, but B=D; wherefore A : B :: B : C: Therefore, if three straight lines, &c. Q. E. D.

PROP. XVIII. PROB.

Upon a given straight line to describe a rectilineal figure similar, and similarly situated to a given rectilineal figure.

Let AB be the given straight line, and CDEF the given rectilineal figure of four sides; it is required upon the given straight line AB to describe a rectilineal figure similar and similarly situated to CDEF.

Join DF, and at the points A, B in the straight line AB, make (23. 1.) the angle BAG equal to the angle at C, and the angle ABG equal to the angle CDF; therefore the remaining angle CFD is equal to the remaining angle AGB (32. 1.): wherefore the triangle FCD is equiangular to the triangle GAB: Again, at the points G, B in the straight line GB, make (23. 1.) the angle BGH equal to the angle DFE, and the angle GBH equal to FDE; therefore the remaining angle FED is equal to the remaining angle GHB, and the triangle FDE equiangular to the triangle GBH: then, because the angle AGB is equal to the angle CFD, and BGH to DFE, the whole angle AGH is equal to the

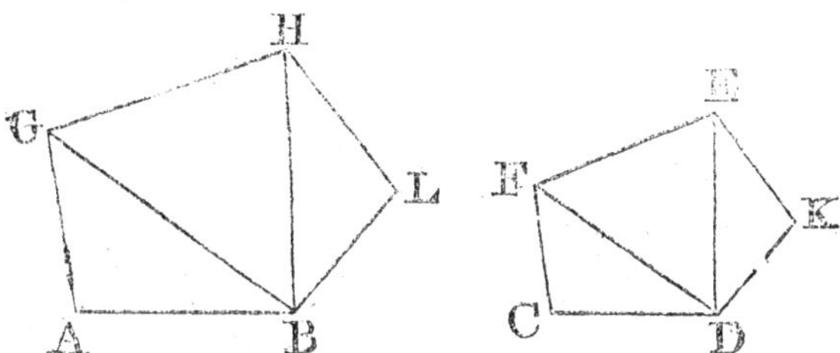

whole CFE: for the same reason, the angle ABH is equal to the angle CDE; also the angle at A is equal to the angle at C, and the angle GHB to FED: Therefore the rectilineal figure ABHG is equiangular to CDEF: but likewise these figures have their sides about the equal angles proportionals: for the triangles GAB, FCD being equiangular,

BA : AG :: DC : CF (4. 6.); for the same reason,
AG : GB :: CF : FD; and because of the equiangular triangles BGH, DFE, GB : GH :: FD : FE; therefore,
ex æquali (22. 5.) AG : GH :: CF : FE.

In the same manner, it may be proved, that

AB : BH :: CD : DE. Also (4. 6.),
GH : HB :: FE : ED. Wherefore, because the rectilineal figures ABHG, CDEF are equiangular, and have their sides about the equal angles proportionals, they are similar to one another (def. 1. 6.).

Next, Let it be required to describe upon a given straight line AB, a rectilineal figure similar, and similarly situated to the rectilineal figure CDKEF.

Join DE, and upon the given straight line AB describe the rectilineal figure ABHG similar, and similarly situated to the quadrilateral figure CDEF, by the former case; and at the points B, H in the straight line BH, make the angle HBL equal to the angle EDK, and the angle BHL equal to the angle DEK; therefore the remaining angle at K is equal to the remaining angle at L; and because the figures ABHG, CDEF are similar. the angle GHB is equal to the angle FED, and BHL is equal to DEK; wherefore the whole angle GHL is equal to the whole angle FEK; for the same reason the angle ABL is equal to the angle CDK: therefore the five-sided figures AGHLB, CFEKD are equiangular; and because the figures AGHB, CFED are similar, GH is to HB as FE to ED; and as HB to HL, so is ED to EK (4. 6.); therefore, ex æquali (22. 5.), GH is to HL, as FE to EK; for the same reason, A is to BL, as CD to DK: and BL is to LH, as (4. 6.) DK to KE, because the triangles BLH, DKE are equiangular: therefore, because the five-sided figures AGHLB, CFEKD are equiangular, and have their sides about the equal angles proportionals, they are similar to one another: and in the same manner a rectilineal figure of six, or more, sides may be described upon a given straight line similar to one given, and so on. Which was to be done.

PROP. XIX. THEOR.

Similar triangles are to one another in the duplicate ratio of the homologous sides.

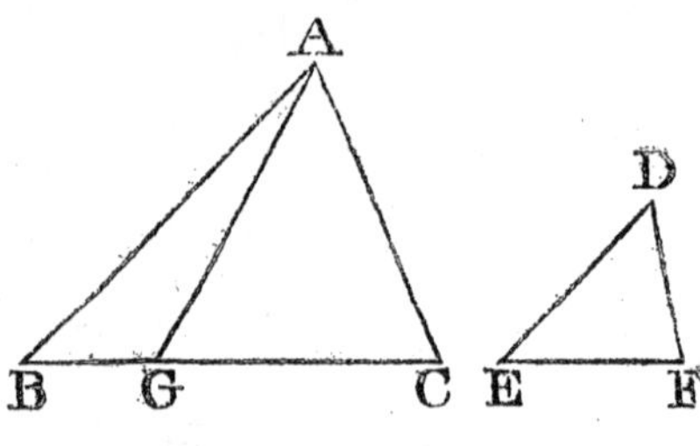

Let ABC, DEF be similar triangles, having the angle B equal to the angle E, and let AB be to BC, as DE to EF, so that the side BC is homologous to EF (def. 13. 5.): the triangle ABC has to the triangle DEF, the duplicate ratio of that which BC has to EF.

Take BG a third proportional to BC and EF (11. 6.), or such that
BC : EF :: EF : BG, and join GA. Then because
AB : BC :: DE : EF, alternately (16. 5.),
AB : DE :: BC : EF; but
BC : EF :: EF : BG; therefore (11. 5.)
AB : DE :: EF : BG: wherefore the sides of the triangles ABG, DEF, which are about the equal angles, are reciprocally proportional: but triangles, which have the sides about two equal angles reciprocally proportional, are equal to one another (15. 6.): therefore the triangle ABG is equal to the triangle DEF; and because that BC is to EF, as EF to BG; and that if three straight lines be proportionals, the first has to the third the duplicate ratio of that which it has to the second; BC therefore has to BG the duplicate ratio of that which BC has to EF. But as BC to BG, so is (1. 6.) the triangle ABC to the triangle ABG: therefore the triangle ABC has to the triangle ABG the duplicate ratio of that which BC has to EF: and the triangle ABG is equal to the triangle DEF; wherefore also the triangle ABC has to the triangle DEF the duplicate ratio of that which BC has to EF. Therefore, similar triangles, &c. Q. E. D.

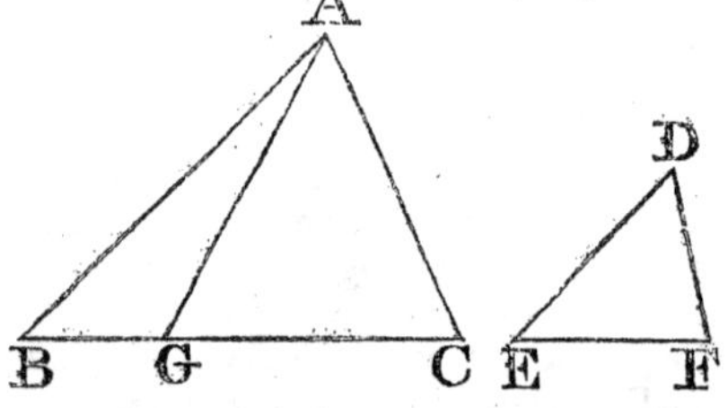

Cor. From this it is manifest, that if three straight lines be proportionals, as the first is to the third, so is any triangle upon the first to a similar, and similarly described triangle upon the second.

PROP. XX. THEOR.

Similar polygons may be divided into the same number of similar triangles, having the same ratio to one another that the polygons have; and the polygons have to one another the duplicate ratio of that which their homologous sides have.

Let ABCDE, FGHKL be similar polygons, and let AB be the homologous side to FG: the polygons ABCDE, FGHKL may be divided into the same number of similar triangles, whereof each has to each the same ratio which the polygons have; and the polygon ABCDE has to the polygon FGHKL a ratio duplicate of that which the side AB has to the side FG.

Join BE, EC, GL, LH: and because the polygon ABCDE is similar to the polygon FGHKL, the angle BAE is equal to the angle GFL (def. 1. 6.), and BA : AE :: GF : FL (def. 1. 6.); wherefore, because the triangles ABE, FGL have an angle in one equal to an angle in the other, and their sides about these equal angles proportionals, the triangle ABE is equianglar (6. 6.), and therefore similar, to the triangle FGL (4. 6.): wherefore the angle ABE is equal to the angle FGL: and, because the polygons are similar, the whole angle ABC is equal (def. 1. 6.) to the whole angle FGH; therefore the remaining angle EBC is equal to the remaining angle LGH: now because the triangles ABE, FGL are similar,

EB : BA :: LG : GF; and also because the polygons are similar, AB : BC :: FG : GH (def. 1. 6.); therefore, ex æquali (22. 5.); EB : BC :: LG : GH; that is, the sides about the equal angles EBC, LGH are proportionals; therefore

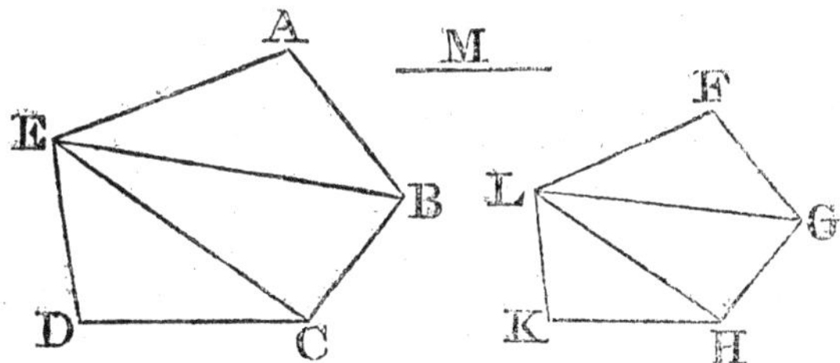

(6. 6.) the triangle EBC is equiangular to the triangle LGH, and similar to it (4. 6.). For the same reason, the triangle ECD is likewise similar to the triangle LHK; therefore the similar polygons ABCDE, FGHKL are divided into the same number of similar triangles.

Also these triangles have, each to each, the same ratio which the polygons have to one another, the antecedents being ABE, EBC, ECD, and the consequents FGL, LGH, LHK: and the polygon ABCDE has to the polygon FGHKL the duplicate ratio of that which the side AB has to the homologous side FG.

Because the triangle ABE is similar to the triangle FGL, ABE has to FGL the duplicate ratio (19. 6.) of that which the side BE has to the side GL: for the same reason, the triangle BEC has to GLH the duplicate ratio of that which BE has to GL: therefore, as the triangle ABE to the triangle FGL, so (11. 5.) is the triangle BEC to the triangle GLH. Again, because the triangle EBC is similar to the triangle LGH, EBC has to LGH the duplicate ratio of that which the side EC has to the side LH: for the same reason, the triangle ECD has to the triangle LHK, the duplicate ratio of that which EC has to LH: therefore, as the triangle EBC to the triangle LGH, so is (11. 5.) the triangle ECD to the triangle LHK: but it has been proved, that the triangle EBC is likewise to the triangle LGH, as the triangle ABE to the triangle FGL. Therefore, as the triangle ABE is to the triangle FGL, so is the triangle EBC to the triangle LGH, and the triangle ECD to the triangle LHK: and therefore, as one of the antecedents to one of the consequents, so are all the antecedents to all the consequents (12. 5.). Wherefore, as the triangle ABE to the tri-

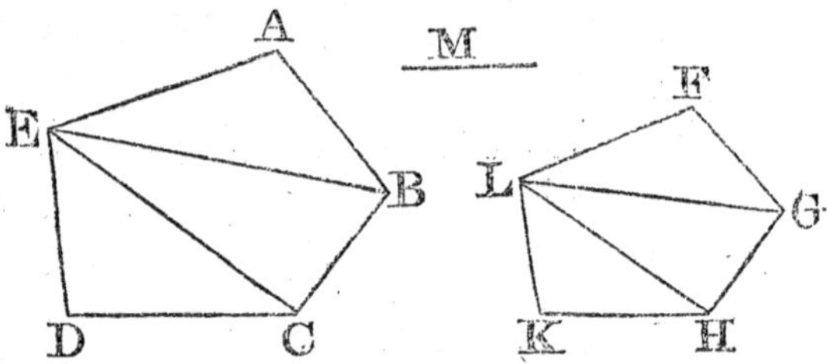

angle FGL, so is the polygon ABCDE to the polygon FGHKL: but the triangle ABE has to the triangle FGL, the duplicate ratio of that which the side AB has to the homologous side FG. Therefore also the polygon ABCDE has to the polygon FGHKL the duplicate ratio of that which AB has to the homologous side FG. Wherefore similar polygons, &c. Q. E. D.

Cor. 1. In like manner it may be proved, that similar figures of four sides, or of any number of sides, are one to another in the duplicate ratio of their homologous sides; and the same has already been proved of triangles: therefore, universally similar rectilineal figures are to one another in the duplicate ratio of their homologous sides.

Cor. 2. And if to AB, FG, two of the homologous sides, a third proportional M be taken, AB has (def. 11. 5.) to M the duplicate ratio of that which AB has to FG: but the four sided figure, or polygon, upon AB has to the four-sided figure, or polygon, upon FG likewise the duplicate ratio of that which AB has to FG: therefore, as AB is to M, so is the figure upon AB to the figure upon FG, which was also proved in triangles (Cor. 19. 6.). Therefore, universally, it is manifest, that if three straight lines be proportionals, as the first is to the third, so is any rectilineal figure upon the first, to a similar, and similarly described rectilineal figure upon the second.

Cor. 3. Because all squares are similar figures, the ratio of any two squares to one another is the same with the duplicate ratio of their sides; and hence, also, any two similar rectilineal figures are to one another as the squares of their homologous sides.

PROP. XXI. THEOR.

Rectilineal figures which are similar to the same rectilineal figure, are also similar to one another.

Let each of the rectilineal figures A, B be similar to the rectilineal figure C: The figure A is similar to the figure B.

Because A is similar to C, they are equiangular, and also have their sides about the equal angles proportionals (def. 1. 6.). Again, because B is similar to C, they are equiangular, and have their sides about the equal angles proportionals (def. 1. 6.): therefore the figures A, B,

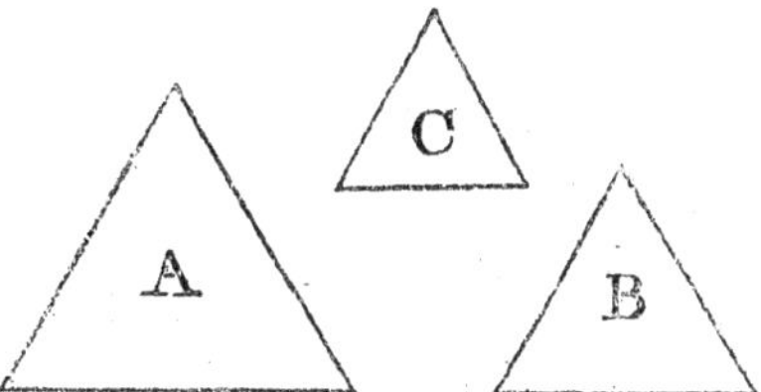

are each of them equiangular to C, and have the sides about the equal angles of each of them, and of C, proportionals. Wherefore the rectilineal figures A and B are equiangular (1. Ax. 1.), and have their sides about the equal angles proportionals (11. 5.). Therefore A is similar (def. 1. 6.) to B. Q. E. D.

PROP. XXII. THEOR.

If four straight lines be proportionals, the similar rectilineal figures similarly described upon them shall also be proportionals; and if the similar rectilineal figures similarly described upon four straight lines be proportionals, those straight lines shall be proportionals.

Let the four straight lines, AB, CD, EF, GH be proportionals, viz. AB to CD, as EF to GH, and upon AB, CD let the similar rectilineal figures KAB, LCD be similarly described; and upon EF, GH the similar rectilineal figures MF, NH, in like manner: the rectilineal figure KAB is to LCD, as MF to NH.

To AB, CD take a third proportional (11. 6.) X; and to EF, GH, a third proportional O; and because

AB : CD :: EF : GH, and
CD : X :: GH : (11. 5.) O, ex æquali (22. 5.)
AB : X :: EF : O. But
AB : X (2. Cor. 20. 6.) :: KAB : LCD; and
EF : O :: (2. Cor. 20. 6.) MF : NH; therefore
KAB : LCD (2. Cor. 20. 6.) :: MF : NH.

And if the figure KAB be to the figure LCD, as the figure MF to the figure NH, AB is to CD, as EF to GH.

Make (12. 6.) as AB to CD, so EF to PR, and upon PR describe (18. 6.) the rectilineal figure SR similar, and similarly situated to ei-

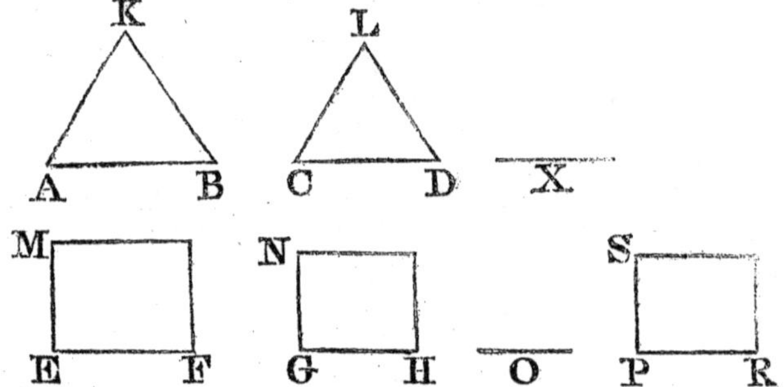

ther of the figures MF, NH: then, because that as AB to CD, so is EF to PR, and upon AB, CD are described the similar and similarly situated rectilineals KAB, LCD, and upon EF, PR, in like manner, the similar rectilineals MF, SR; KAB is to LCD, as MF to SR; but by the hypothesis, KAB is to LCD, as MF to NH; and therefore the rectilineal MF having the same ratio to each of the two NH, SR, these two are equal (9. 5.) to one another: they are also similar, and similarly situated; therefore GH is equal to PR: and because as AB to CD, so is EF to PR, and because PR is equal to GH, AB is to CD, as EF to GH. If therefore four straight lines, &c. Q. E. D.

PROP. XXIII. THEOR.

Equiangular parallelograms have to one another the ratio which is compounded of the ratios of their sides.

Let AC, CF be equiangular parallelograms having the angle BCD equal to the angle ECG; the ratio of the parallelogram AC to the parallelogram CF, is the same with the ratio which is compounded of the ratios of their sides.

Let BC, CG be placed in a straight line; therefore DC and CE are also in a straight line (14. 1.); complete the parallelogram DG; and, taking any straight line K, make (12. 6.) as BC to CG, so K to L; and as DC to CE, so make (12. 6.) L to M: therefore the ratios of K to L, and L to M, are the same with the ratios of the sides, viz. of BC to CG, and of DC to CE. But the ratio of K to M, is that which is said to be compounded (def. 10. 5.) of the ratios of K to L, and L to M; wherefore also K has to M the ratio compounded of the ratios of the sides of the parallelograms. Now, because as BC to CG, so is the parallelogram AC to the parallelogram CH (1. 6.); and as BC to CG, so is K to L; therefore K is (11. 5.) to L, as the parallelogram AC to the parallelogram CH: again, because as DC to CE, so is the parallelogram CH to the parallelogram CF: and as DC to CE, so is L to M; therefore L is (11. 5.) to M, as the parallelogram CH to the parallelogram CF: therefore, since it has been proved, that as K to L, so is the parallelogram AC to the parallelogram CH; and as L to M, so the parallelogram CH to the parallelogram CF; ex æquali (22. 5.), K is to M, as the parallelogram AC to the parallelogram CF; but K has to M the ratio which is compounded of the ratios of the sides; therefore also the parallelogram AC has to the parallelogram CF the ratio which is compounded of the ratios of the sides. Wherefore equiangular parallelograms, &c. Q. E. D.

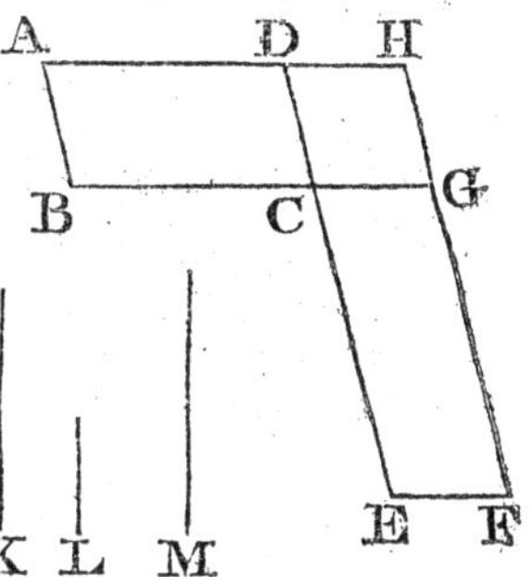

PROP. XXIV. THEOR.

The parallelograms about the diameter of any parallelogram, are similar to the whole, and to one another.

Let ABCD be a parallelogram, of which the diameter is AC; and EG, HK the parallelograms about the diameter: the parallelograms EG, HK are similar, both to the whole parallelogram ABCD, and to one another.

Because DC, GF are parallels, the angle ADC is equal (29. 1.) to the angle AGF: for the same reason, because BC, EF are parallels, the angle ABC is equal to the angle AEF: and each of the angles BCD, EFG is equal to the opposite angle DAB (34. 1.), and therefore are equal to one another, wherefore the parallelograms ABCD, AEFG are equiangular. And because the angle ABC is equal to the angle

AEF, and the angle BAC common to the two triangles BAC, EAF, they are equiangular to one another; therefore (4. 6.) as AB to BC, so is AE to EF; and because the opposite sides of parallelograms are equal to one another (34. 1.), AB is (7. 5.) to AD, as AE to AG; and DC to CB, as GF to FE; and also CD to DA, as FG to GA: therefore the sides of the parallelograms ABCD, AEFG about the equal angles are proportionals; and they are therefore similar to one another (def. 1. 6.): for the same reason, the parallelogram ABCD is similar to the parallelogram FHCK. Wherefore each of the parallelograms, GE, KH is similar to DB: but rectilineal figures which are similar to the same rectilineal figure, are also similar to one another (21. 6.); therefore the parallelogram GE is similar to KH. Wherefore the parallelograms, &c. Q. E. D.

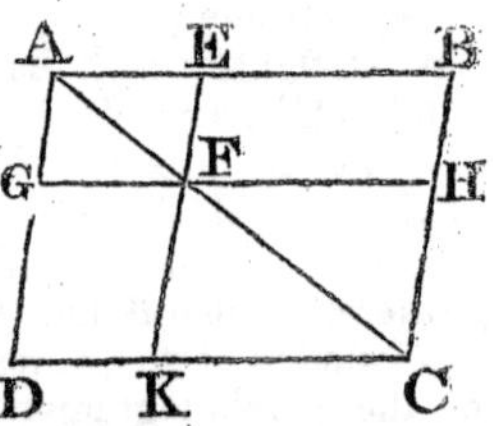

PROP. XXV. PROB.

To describe a rectilineal figure which shall be similar to one, and equal to another given rectilineal figure.

Let ABC be the given rectilineal figure, to which the figure to be described is required to be similar, and D that to which it must be equal. It is required to describe a rectilineal figure similar to ABC, and equal to D.

Upon the straight line BC describe (cor. 45. 1.) the parallelogram BE equal to the figure ABC; also upon CE describe (cor. 45. 1.) the parallelogram CM equal to D, and having the angle FCE equal to the angle CBL: therefore BC and CF are in a straight line (29. 1. 14. 1.), as also LE and EM; between BC and CF find (13. 6.) a mean proportional GH, and upon GH describe (18. 6.) the rectilineal figure KGH similar, and similarly situated, to the figure ABC. And because BC is to GH as GH to CF, and if three straight lines be proportionals, as the first is to the third, so is (2. Cor. 20. 6.) the figure upon the first to the si-

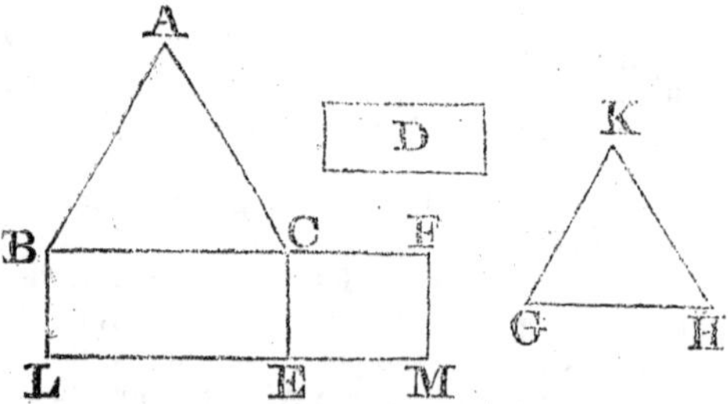

milar and similarly described figure upon the second; therefore as BC to CF, so is the figure ABC to the figure KGH: but as BC to CF, so is (1. 6.) the parallelogram BE to the parallelogram EF: therefore as the figure ABC is to the figure KGH, so is the parallelogram BE to the parallelogram EF (11. 5.): but the rectilineal figure ABC is equal to the parallelogram BE; therefore the rectilineal figure KGH is equal (14. 5.) to the parallelogram EF: but EF is equal to the figure D; wherefore also KGH is equal to D; and it is similar to ABC. Therefore the rectilineal figure KGH has been described similar to the figure ABC, and equal to D. Which was to be done.

PROP. XXVI. THEOR.

If two similar parallelograms have a common angle, and be similarly situated, they are about the same diameter.

Let the parallelograms ABCD, AEFG be similar and similarly situated, and have the angle DAB common; ABCD and AEFG are about the same diameter.

For, if not, let, if possible, the parallelogram BD have its diameter AHC in a different straight line from AF, the diameter of the parallelogram EG, and let GF meet AHC in H; and through H draw HK parallel to AD or BC; therefore the parallelograms ABCD, AKHG being about the same diameter, are similar to one another (24. 6.): wherefore, as DA to AB, so is (def. 1. 6.) GA to AK; but because ABCD and AEFG are similar parallelograms, as DA is to AB, so is GA to AE; therefore (11. 5.) as GA to AE, so GA to AK; wherefore GA has the same ratio to each of the straight lines AE, AK; and consequently AK is equal (9. 5.) to AE, the less to the greater, which is impossible; therefore ABCD and AKHG are not about the same diameter; wherefore ABCD and AEFG must be about the same diameter. Therefore, if two similar, &c. Q. E. D.

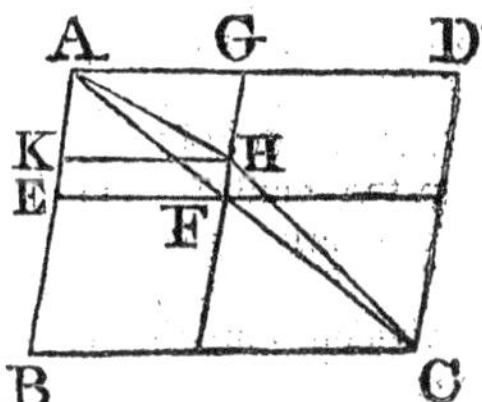

PROP. XXVII. THEOR.

Of all the rectangles contained by the segments of a given straight line, the greatest is the square which is described on half the line.

Let AB be a given straight line, which is bisected in C; and let D be any point in it, the square on AC is greater than the rectangle AD, DB.

For, since the straight line AB is divided in two equal parts in C, and into two unequal parts in D, the rectangle contained by AD and DB, together with the square of CD, is equal to the square of AC

(5. 2.). The square of AC is therefore greater than the rectangle AD.DB. Therefore, &c. Q. E. D.

PROP. XXVIII. PROB.

To divide a given straight line, so that the rectangle contained by its segments may be equal to a given space; but that space must not be greater than the square of half the given line.

Let AB be the given straight line, and let the square upon the given straight line C be the space to which the rectangle contained by the segments of AB must be equal, and this square, by the determination, is not greater than that upon half the straight line AB.

Bisect AB in D, and if the square upon AD be equal to the square upon C, the thing required is done: But if it be not equal to it, AD must be greater than C, according to the determination: Draw DE at right angles to AB, and make it equal to C: produce ED to F, so that EF be equal to AD or DB, and from the centre E, at the distance EF, describe a circle meeting AB in G. Join EG; and because AB is divided equally in D, and unequally in G, $AG.GB + DG^2 =$ (5. 2) $DB^2 = EG^2$. But (47. 1.) $ED^2+DG^2=EG^2$; therefote $AG.GB+DG^2=ED^2+DG^2$, and taking away DG^2, $AG.GB=ED^2$. Now ED=C, therefore the rectangle AG.GB is equal to the square of C: and the given line AB is divided in G, so that the rectangle contained by the segments AG, GB is equal to the square upon the given straight line C. Which was to be done.

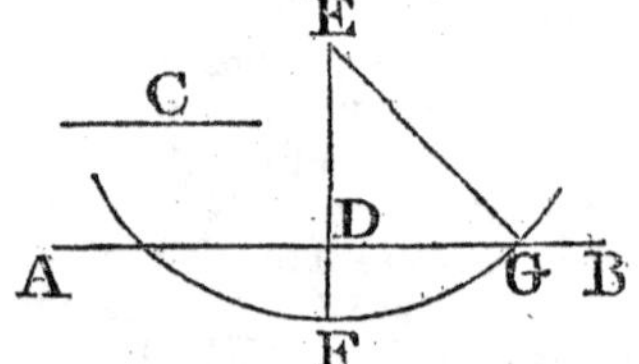

PROP. XXIX. PROB.

To produce a given straight line, so that the rectangle contained by the segments between the extremities of the given line, and the point to which it is produced, may be equal to a given space.

Let AB be the given straight line, and let the square upon the given straight line C be the space to which the rectangle under the segments of AB produced, must be equal.

Bisect AB in D, and draw BE at right angles to it, so that BE be equal to C; and having joined DE, from the centre D at the distance

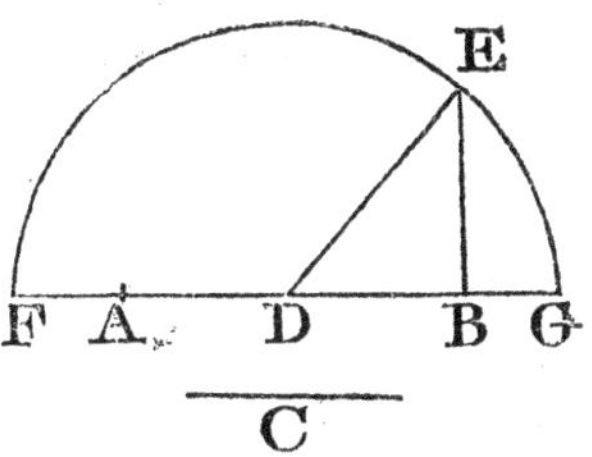

DE describe a circle meeting AB produced in G. And because AB is bisected in D, and produced to G, (6. 2.) $AG.GB+DB^2=DG^2=DE^2$.

But (47. 1.) $DE^2 = DB^2 + BE^2$, therefore $AG.GB + DB^2 = DB^2 + BE^2$ and $AG.GB=BE^2$. Now, BE =C; wherefore the straight line AB is produced to G, so that the rectangle contained by the segments AG, GB of the line produced, is equal to the square of C. Which was to be done.

PROP. XXX. PROB.

To cut a given straight line in extreme and mean ratio.

Let AB be the given straight line; it is required to cut it in extreme and mean ratio.

Upon AB describe (46. 1.) the square BC, and produce CA to D, so that the rectangle CD.DA may be equal to the square CB (29. 6.). Take AE equal to AD, and complete the rectangle DF under DC and AE, or under DC and DA. Then, because the rectangle CD.DA is equal to the square CB, the rectangle DF is equal to CB. Take away the common part CE from each, and the remainder FB is equal to the remainder DE. But FB is the rectangle contained by FE and EB, that is, by AB and BE; and DE is the square upon AE; therefore AE is a mean proportional between AB and BE (17. 6.), or AB is to AE as AE to EB. But AB is greater than AE; wherefore AE is greater than EB (14. 5.): Therefore the straight line AB is cut in extreme and mean ratio in E (def. 3. 6.). Which was to be done.

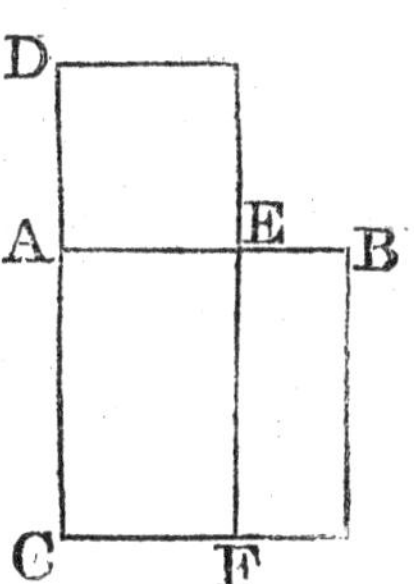

Otherwise.

Let AB be the given straight line; it is required to cut it in extreme and mean ratio.

Divide AB in the point C, so that the rectangle contained by AB, BC be equal to the square of AC (11. 2.): Then because the rectangle AB.BC is equal to the square of AC, as BA to AC, so is AC

A C B

to CB (17. 6.); Therefore AB is cut in extreme and mean ratio in C (def. 3. 6.). Which was to be done.

PROP. XXXI. THEOR.

In right angled triangles, the rectilineal figure described upon the side opposite to the right angle, is equal to the similar, and similarly described figures upon the sides containing the right angle.

Let ABC be a right angled triangle, having the right angle BAC: The rectilineal figure described upon BC is equal to the similar, and similarly described figures upon BA, AC.

Draw the perpendicular AD; therefore, because in the right angled triangle ABC, AD is drawn from the right angle at A perpendicular to the base BC, the triangles ABD, ADC are similar to the whole triangle ABC, and to one another (8. 6.), and because the triangle ABC is similar to ADB, as CB to BA, so is BA to BD (4. 6.); and because these three straight lines are proportionals, as the first to the third, so is the figure upon the first to the similar, and similarly described figure upon the second (2. Cor. 20. 6.): Therefore, as CB to BD, so is the figure upon CB to the similar and similarly described figure upon BA: and inversely (B. 5.), as DB to BC, so is the figure upon BA to that upon BC; for the same reason as DC to CB, so is the figure upon CA to that upon CB. Wherefore, as BD and DC together to BC, so are the figures upon BA and on AC, together, to the figure upon BC (24. 5.); therefore the figures on BA, and on AC, are together equal to that on BC; and they are similar figures. Wherefore, in right angled triangles, &c. Q. E. D.

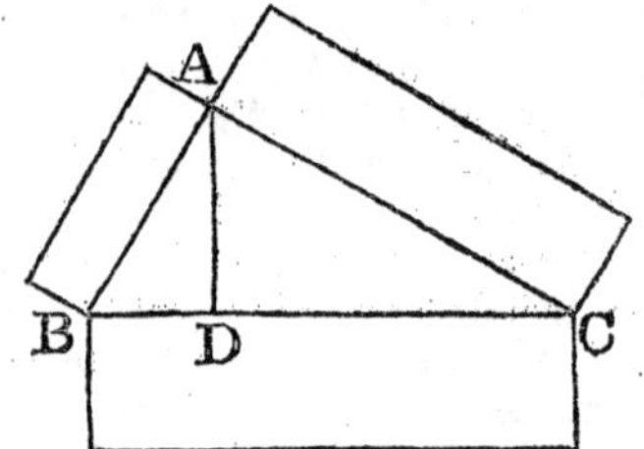

PROP. XXXII. THEOR.

If two triangles, which have two sides of the one proportional to two sides of the other, be joined at one angle, so as to have their homologous sides parallel to one another; their remaining sides shall be in a straight line.

Let ABC, DCE be two triangles which have two sides BA, AC proportional to the two CD, DE, viz. BA to AC, as CD to DE; and let AB be parallel to DC, and AC to DE; BC and CE are in a straight line.

Because AB is parallel to DC, and the straight line AC meets them, the alternate angles BAC, ACD are equal (29. 1.); for the same reason, the angle CDE is equal to the angle ACD; wherefore also BAC is equal to CDE: And because the triangles ABC, DCE have one angle at A equal to one at D, and the sides about these angles proportionals, viz. BA to AC, as CD to DE, the triangle ABC is equiangular (6. 6.) to DCE: Therefore the angle ABC is equal to the angle DCE: And the angle BAC was proved to be equal to ACD: Therefore the whole angle ACE is equal to the two angles ABC, BAC; add the common angle ACB, then the angles ACE, ACB are equal to the angles ABC, BAC, ACB: But ABC, BAC, ACB are equal to two right angles (32. 1.); therefore also the angles ACE, ACB are equal to two right angles: And since at the point C, in the straight line AC, the two straight lines BC, CE, which are on the opposite sides of it, make the adjacent angles ACE, ACB equal to two right angles; therefore (14. 1.) BC and CE are in a straight line. Wherefore, if two triangles, &c. Q. E. D.

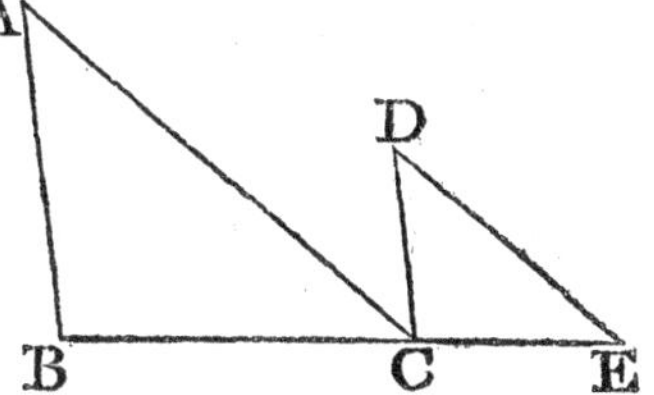

PROP. XXXIII. THEOR.

In equal circles, angles, whether at the centres or circumferences, have the same ratio which the arches, on which they stand, have to one another: So also have the sectors.

Let ABC, DEF be equal circles; and at their centres the angles BGC, EHF, and the angles BAC, EDF at their circumferences; as the arch BC to the arch EF, so is the angle BGC to the angle EHF, and the angle BAC to the angle EDF: and also the sector BGC to the sector EHF.

Take any number of arches CK, KL, each equal to BC, and any number whatever FM, MN each equal to EF; and join GK, GL, HM, HN. Because the arches BC, CK, KL are all equal, the angles BGC, CGK, KGL are also all equal (27. 3.): Therefore, what multiple soever the arch BL is of the arch BC, the same multiple is the angle BGL of the angle BGC: For the same reason, whatever multiple the arch EN is of the arch EF, the same multiple is the angle EHN of the angle EHF. But if the arch BL, be equal to the arch EN, the angle BGL is also equal (27. 3.) to the angle EHN; or if the arch BL be greater than EN, likewise the angle BGL is greater than EHN: and if less, less: There being then four magnitudes, the two arches, BC, EF, and the two angles BGC, EHF, and of the arch BC, and of the angle BGC, have been taken any equimultiples whatever, viz. the arch BL, and the angle BGL; and of the arch EF, and of the angle EHF, any equimultiples whatever, viz. the arch EN, and the angle EHN: And it has been proved, that if the arch BL be greater than EN, the angle BGL is greater than EHN; and if equal, equal; and if less, less: As therefore, the arch BC to the arch EF, so (def. 5. 5.) is the angle BGC to the angle EHF: But as the angle BGC is to the angle EHF, so is (15. 5.) the angle BAC to the angle EDF, for each is double of each (20. 3.): Therefore, as the circumference BC is to EF, so is the angle BGC to the angle EHF, and the angle BAC to the angle EDF.

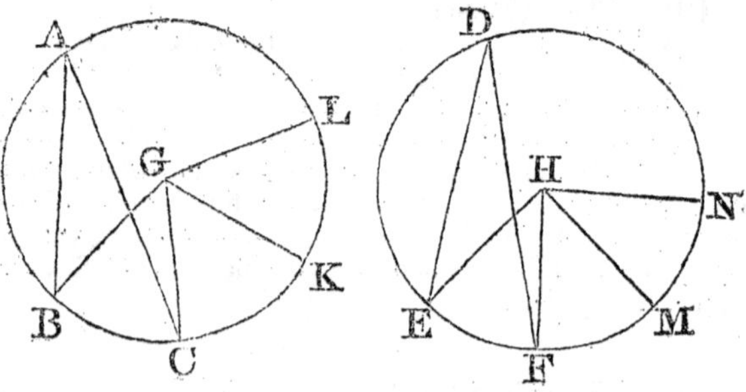

Also, as the arch BC to EF, so is the sector BGC to the sector EHF. Join BC, CK, and in the arches BC, CK take any points X, O, and join BX, XC, CO, OK: Then, because in the triangles GBC, GCK, the two sides BG, GC are equal to the two CG, GK, and also contain equal angles; the base BC is equal (4. 1.) to the base CK, and the triangle GBC to the triangle GCK: And because the arch BC is equal to the arch CK, the remaining part of the whole circumference of the circle ABC is equal to the remaining part of the whole circumference of the same circle: Wherefore the angle BXC is equal to the angle COK (27. 3.); and the segment BXC is therefore similar to the segment COK (def. 9. 3.); and they are upon equal straight lines BC, CK: But similar segments of circles upon equal straight lines are equal (24. 3.) to one another: Therefore the segment BXC is equal to the segment COK: And the triangle BGC is equal to the triangle CGK; therefore the whole, the sector BGC is equal to the whole, the sector CGK: For the same reason, the sector KGL is equal to each of the sectors BGC, CGK; and in the same manner, the sectors EHF, FHM, MHN may be proved equal to one another: Therefore, what multiple soever the arch BL is of the arch BC, the same multiple is the sector BGL of the sector BGC. For the same reason, whatever multiple the arch EN is of EF, the same multiple is the sector EHN of the sector EHF: Now if the arch BL be equal to EN, the sector BGL is equal to the sector EHN;

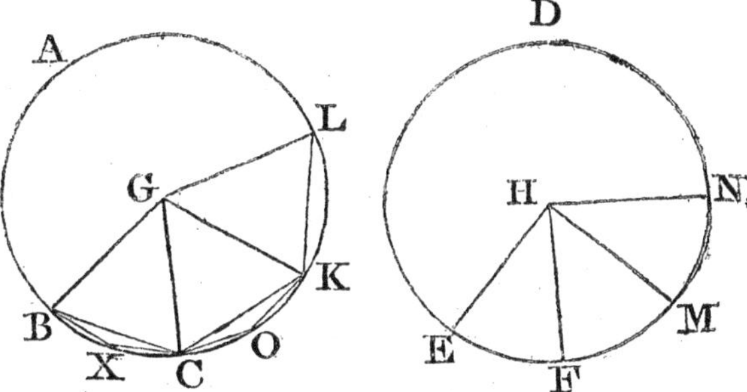

and if the arch BL be greater than EN, the sector BGL is greater than the sector EHN; and if less, less: Since, then, there are four magnitudes, the two arches BC, EF, and the two sectors BGC, EHF, and of the arch BC, and sector BGC, the arch BL and the sector BGL are any equimultiples whatever; and of the arch EF, and sector EHF, the arch EN and sector EHN, are any equimultiples whatever; and it has been proved, that if the arch BL be greater than EN, the sector BGL is greater than the sector EHN; if equal, equal; and if less, less; therefore (def. 5. 5.), as the arch BC is to the arch EF, so is the sector BGC to the sector EHF. Wherefore, in equal circles, &c. Q. E. D.

PROP. B. THEOR.

If an angle of a triangle be bisected by a straight line, which likewise cuts the base; the rectangle contained by the sides of the triangle is equal to the rectangle contained by the segments of the base, together with the square of the straight line bisecting the angle.

Let ABC be a triangle, and let the angle BAC be bisected by the straight line AD; the rectangle BA.AC is equal to the rectangle BD.DC, together with the square of AD.

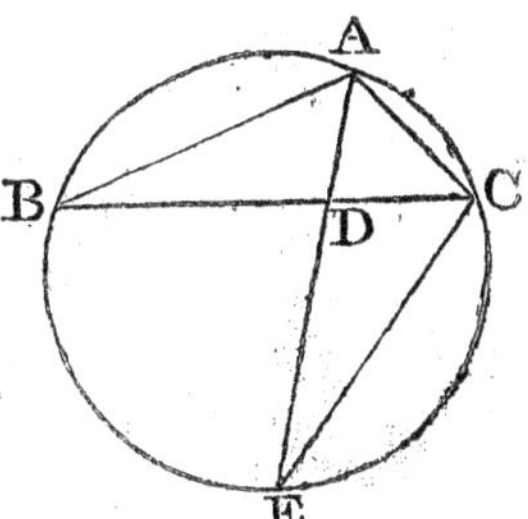

Describe the circle (5. 4.) ACB about the triangle, and produce AD to the circumference in E, and join EC. Then, because the angle BAD is equal to the angle CAE, and the angle ABD to the angle (21. 3.) AEC, for they are in the same segment; the triangles ABD, AEC are equiangular to one another: Therefore BA : AD : : EA : (4. 6.) AC, and consequently, BA.AC = (16. 6.) AD.AE = ED.DA (3. 2.)+DA^2. But ED.DA= BD.DC, therefore BA.AC=BD.DC+ DA^2. Wherefore, if an angle, &c. Q. E. D.

PROP. C. THEOR.

If from any angle of a triangle a straight line be drawn perpendicular to the base; the rectangle contained by the sides of the triangle is equal to the rectangle contained by the perpendicular, and the diameter of the circle described about the triangle.

Let ABC be a triangle, and AD the perpendicular from the angle A to the base BC; the rectangle BA.AC is equal to the rectangle contained by AD and the diameter of the circle described about the triangle.

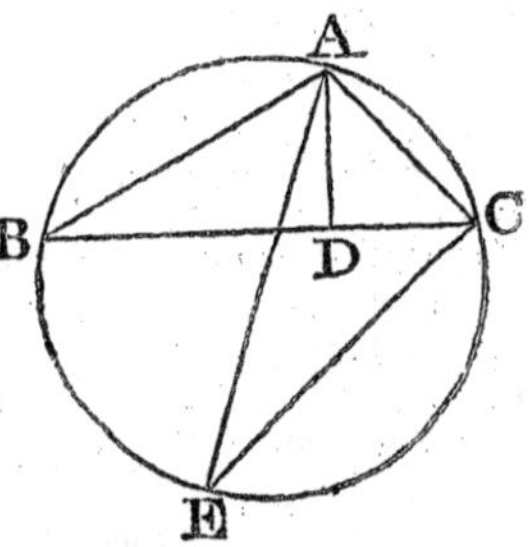

Describe (5. 4.) the circle ACB about the triangle, and draw its diameter AE, and join EC: Because the right angle BDA is equal (3. 3.) to the angle ECA in a semicircle, and the angle ABD to the angle AEC, in the same segment (21. 3.); the triangles ABD, AEC are equiangular: Therefore, as (4. 6.) BA to AD, so is EA to AC: and consequently the rectangle BA.AC is equal (16. 6.) to the rectangle EA.AD. If, therefore, from an angle, &c. Q. E. D.

PROP. D. THEOR.

The rectangle contained by the diagonals of a quadrilateral inscribed in a circle, is equal to both the rectangles, contained by its opposite sides.

Let ABCD be any quadrilateral inscribed in a circle, and let AC, BD be drawn; the rectangle AC.BD is equal to the two rectangles AB.CD, and AD.BC.

Make the angle ABE equal to the angle DBC; add to each of these the common angle EBD, then the angle ABD is equal to the angle EBC: And the angle BDA is equal to (21. 3.) the angle BCE, because they are in the same segment; therefore the triangle ABD is equiangular to the triangle BCE. Wherefore (4. 6.), BC : CE :: BD : DA, and consequently (16. 6.) BC.DA=BD.CE. Again, because the angle ABE is equal to the angle DBC, and the angle (21. 3.) BAE to the angle BDC, the triangle ABE is equiangular to the triangle BCD; therefore BA : AE :: BD : DC, and BA.DC=BD.AE: But it was shewn that BC.DA=BD.CE; wherefore BC.

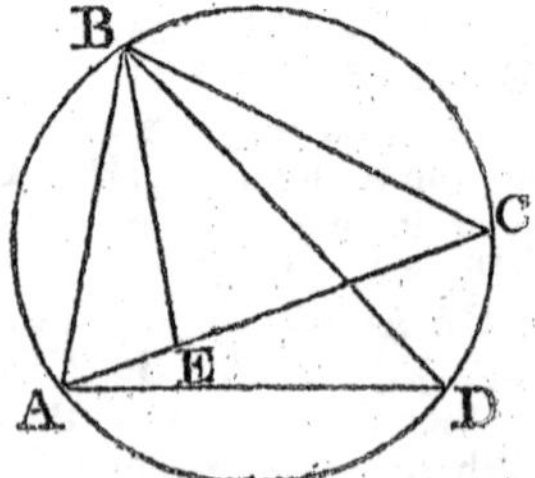

DA+BA.DC=BD.CE+BD.AE=BD.AC (1. 2.). That is, the rectangle contained by BD and AC, is equal to the rectangles contained by AB, CD, and AD, BC. Therefore the rectangle, &c. Q. E. D.

PROP. E. THEOR.

If an arch of a circle be bisected, and from the extremities of the arch, and from the point of bisection, straight lines be drawn to any point in the circumference, the sum of the two lines drawn from the extremities of the arch will have to the line drawn from the point of bisection, the same ratio which the straight line subtending the arch has to the straight line subtending half the arch.

Let ABD be a circle, of which AB is an arch bisected in C, and from A, C, and B to D, any point whatever in the circumference, let AD, CD, BD be drawn; the sum of the two lines AD and DB has to DC the same ratio that BA has to AC.

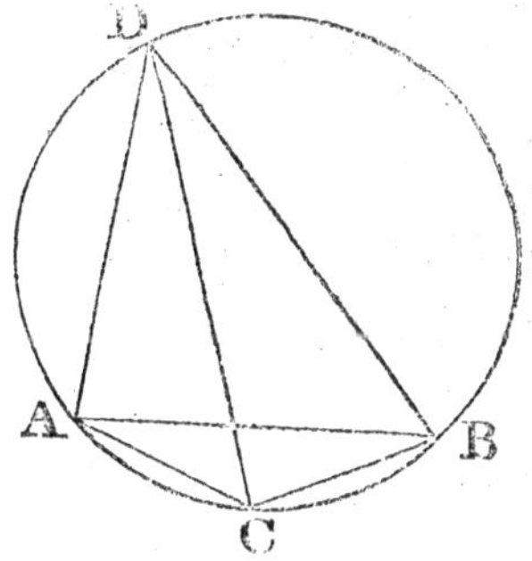

For since ACBD is a quadrilateral inscribed in a circle, of which the diagonals are AB and CD, AD.CB+DB.AC (D. 6.)=AB.CD: but AD.CB + DB.AC=AD.AC + DB.AC, because CB=AC. Therefore AD.AC+DB.AC, that is (1. 2.), (AD+DB) AC=AB.CD. And because the sides of equal rectangles are reciprocally proportional (14. 6.), AD+DB : DC : : AB : AC. Wherefore, &c. Q. E. D.

PROP. F. THEOR.

If two points be taken in the diameter of a circle, such that the rectangle contained by the segments intercepted between them and the centre of the circle be equal to the square of the radius: and if from these points two straight lines be drawn to any point whatsoever in the circumference of the circle, the ratio of these lines will be the same with the ratio of the segments intercepted between the two first mentioned points and the circumference of the circle.

Let ABC be a circle, of which the centre is D, and in DA produced, let the points E and F be such that the rectangle ED, DF is equal to the square of AD; from E and F to any point B in the circumference, let EB, FB be drawn; FB : BE : : FA : AE.

Join BD, and because the rectangle FD, DE is equal to the square of AD, that is, of DB, FD : DB : : DB : DE (17. 6.).

The two triangles, FDB, BDE have therefore the sides proportional that are about the common angle D; therefore they are equiangular (6. 6.), the angle DEB being equal to the angle DBF, and DBE to DFB. Now since the sides about these equal angles are also propor-

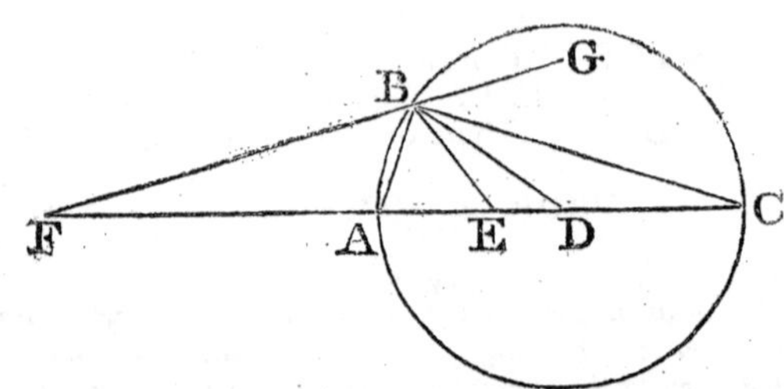

tional (4. 6.), FB : BD :: BE : ED, and alternately (16. 5.), FB : BE :: BD : ED, or FB : BE :: AD : DE. But because FD : DA :: DA : DE, by division (17. 5.), FA : DA :: AE : ED, and alternately (11. 5.) FA : AE :: DA : ED. Now it has been shewn that FB : BE :: AD : DE, therefore FB : BE :: FA : AE. Therefore, &c. Q. E. D.

COR. If AB be drawn, because FB : BE :: FA : AE, the angle FBE is bisected (3. 6.) by AB. Also, since FD : DC :: DC : DE, by composition (18. 5.), FC : DC :: CE : ED, and since it has been shewn that FA : AD (DC) :: AE : ED, therefore, ex æquo, FA : AE :: FC : CE. But FB : BE :: FA : AE, therefore, FB : BE :: FC : CE (11. 5.); so that if FB be produced to G, and if BC be drawn, the angle EBG is bisected by the line BC (A 6.).

PROP. G. THEOR.

If from the extremity of the diameter of a circle a straight line be drawn in the circle, and if either within the circle or produced without it, it meet a line perpendicular to the same diameter, the rectangle contained by the straight line drawn in the circle, and the segment of it, intercepted between the extremity of the diameter and the perpendicular, is equal to the rectangle contained by the diameter, and the segment of it cut off by the perpendicular.

Let ABC be a circle, of which AC is a diameter, let DE be per-

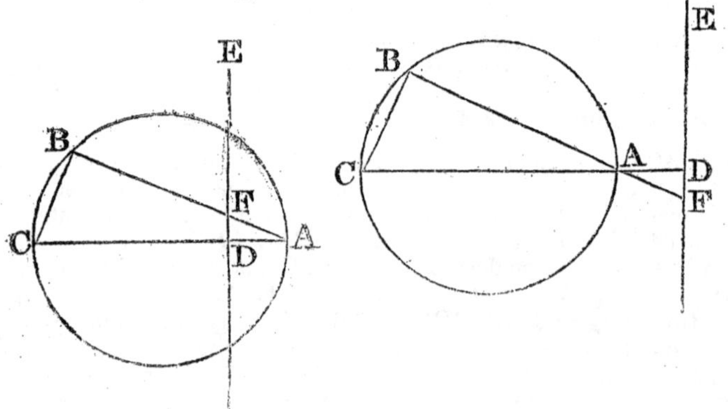

pendicular to the diameter AC, and let AB meet DE in F; the rectangle BA.AF is equal to the rectangle CA.AD. Join BC, and because ABC is an angle in a semicircle, it is a right angle (31. 3.): Now, the angle ADF is also a right angle (Hyp.); and the angle BAC is either the same with DAF, or vertical to it; therefore the triangles ABC, ADF are equiangular, and BA : AC :: AD : AF (4. 6.); therefore also the rectangle BA.AF, contained by the extremes, is equal to the rectangle AC.AD contained by the means (16. 6.). If therefore, &c. Q. E. D.

PROP. H. THEOR.

The perpendiculars drawn from the three angles of any triangle to the opposite sides intersect one another in the same point.

Let ABC be a triangle, BD and CE two perpendiculars intersecting one another in F: Let AF be joined, and produced if necessary, let it meet BC in G, AG is perpendicular to BC.

Join DE, and about the triangle AEF let a circle be described, AEF: then, because AEF is a right angle, the circle described about the triangle AEF will have AF for its diameter (31. 3.). In the same manner, the circle described about the triangle ADF has AF for its diameter; therefore the points A, E, F and D are in the circumference of the same circle. But because the angle EFB is equal to the angle DFC (15. 1.), and also the angle BEF to the angle CDF, being both right angles, the triangles BEF, and CDF are equiangular, and therefore BF : EF :: CF : FD (4. 6.), or alternately (16. 5.) BF : FC :: EF : FD. Since, then, the sides about the equal angles BFC, EFD are proportionals, the triangles BFC, EFD are also equiangular (6. 6.); wherefore the angle FCB is equal to the angle EDF. But EDF is equal to EAF, because they are angles in the same segment (21. 3.); therefore the angle EAF is equal to the angle FCG: Now, the angles AFE, CFG are also equal, because they are vertical angles; therefore the remaining angles AEF, FGC are also equal (32. 1.): But AEF is a right angle, therefore FGC is a right angle, and AG is perpendicular to BC. Q. E. D.

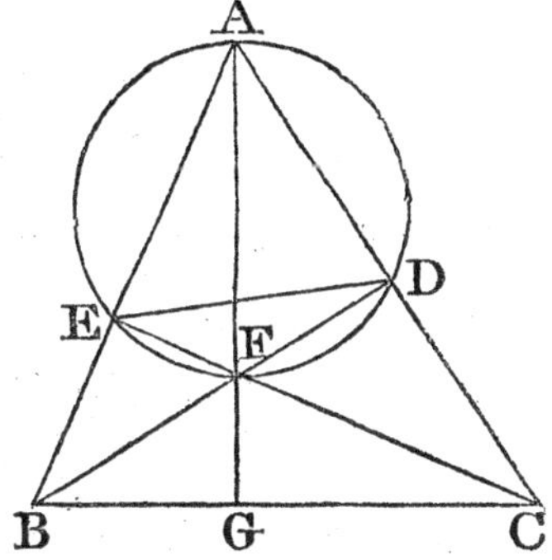

COR. The triangle ADE is similar to the triangle ABC. For the two triangles BAD, CAE having the angles at D and E right angles, and the angle at A common, are equiangular, and therefore BA : AD :: CA : AE, and alternately BA : CA :: AD : AE; therefore the two triangles BAC, DAE, have the angle at A common, and the sides about that angle proportionals, therefore they are equiangular (6. 6.) and similar.

Hence the rectangles BA.AE, CA.AD are equal.

PROP. K. THEOR.

If from any angle of a triangle a perpendicular be drawn to the opposite side or base: the rectangle contained by the sum and difference of the other two sides, is equal to the rectangle contained by the sum and difference of the segments, into which the base is divided by the perpendicular.

Let ABC be a triangle, AD a perpendicular drawn from the angle A on the base BC, so that BD, DC are the segments of the base; (AC+AB) (AC−AB)=(CD+DB) (CD−DB).

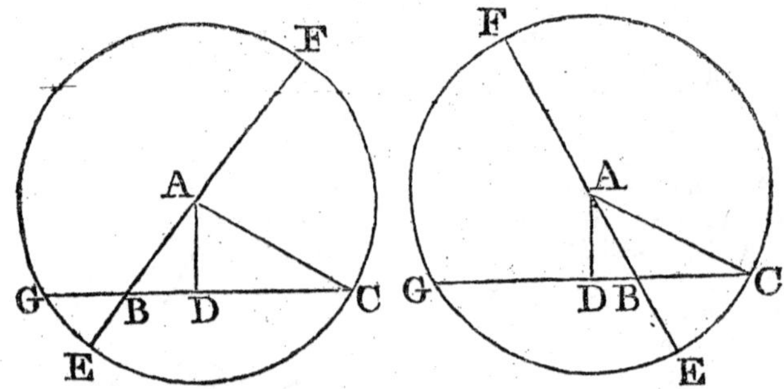

From A as a centre with the radius AC, the greater of the two sides, describe the circle CFG: produce AB to meet the circumference in E and F, and CB to meet it in G. Then because AF=AC, BF=AB+AC, the sum of the sides; and since AE=AC, BE=AC, −AB= the difference of the sides. Also, because AD drawn from the centre cuts GC at right angles, it bisects it; therefore, when the perpendicular falls within the triangle, BG=DG—DB=DC−DB= the difference of the segments of the base, and BC=BD+DC= the sum of the segments. But when AD falls without the triangle, BG= DG+DB=CD+DB= the sum of the segments of the base, and BC =CD−DB= the difference of the segments of the base. Now, in both cases, because B is the intersection of the two lines FE, GC, drawn in the circle, FB.BE=CB.BG; that is, as has been shewn, (AC+AB) (AC−AB)=(CD+DB) (CD−DB). Therefore, &c.

SUPPLEMENT

TO THE

ELEMENTS

OF

GEOMETRY.

ELEMENTS

OF

GEOMETRY.

SUPPLEMENT.

BOOK I.

OF THE QUADRATURE OF THE CIRCLE.

DEFINITIONS.

I.

A CHORD of an arch of a circle is the straight line joining the extremities of the arch; or the straight line which subtends the arch.

II.

The perimeter of any figure is the length of the line or lines, by which it is bounded.

III.

The area of any figure is the space contained within it.

AXIOM.

The least line that can be drawn between two points, is a straight line; and if two figures have the same straight line for their base, that which is contained within the other, if its bounding line or lines be not any where convex toward the base, has the least perimeter.

COR. 1. Hence the perimeter of any polygon inscribed in a circle is less than the circumference of the circle.

COR. 2. If from a point two straight lines be drawn touching a circle, these two lines are together greater than the arch intercepted between them; and hence the perimeter of any polygon described about a circle is greater than the circumference of the circle.

X

PROP. I. THEOR.

If from the greater of two unequal magnitudes there be taken away its half, and from the remainder its half; and so on; There will at length remain a magnitude less than the least of the proposed magnitudes.

Let AB and C be two unequal magnitudes, of which AB is the greater. If from AB there be taken away its half, and from the remainder its half, and so on; there shall at length remain a magnitude less than C.

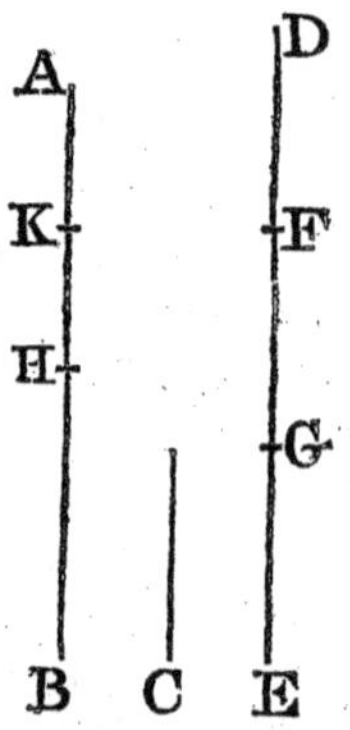

For C may be multipled so as, at length, to become greater than AB. Let DE, therefore, be a multiple of C, which is greater than AB, and let it contain the parts DF, FG, GE, each equal to C. From AB take BH equal to its half, and from the remainder AH, take HK equal to its half, and so on, until there be as many divisions in AB as there are in DE; And let the divisions in AB be AK, KH, HB. And because DE is greater than AB, and EG taken from DE is not greater than its half, but BH taken from AB is equal to its half; therefore the remainder GD is greater than the remainder HA. Again, because GD is greater than HA, and GF is not greater than the half of GD, but HK is equal to the half of HA; therefore, the remainder FD is greater than the remainder AK. And FD is equal to C, therefore C is greater than AK; that is, AK is less than C. Q. E. D.

PROP. II. THEOR.

Equilateral polygons, of the same number of sides, inscribed in circles, are similar, and are to one another as the squares of the diameters of the circles.

Let ABCDEF and GHIKLM be two equilateral polygons of the same number of sides inscribed in the circles ABD, and GHK; ABCDEF and GHIKLM are similar, and are to one another as the squares of the diameters of the circles ABD, GHK.

Find N and O the centres of the circles, join AN and BN, as also GO and HO, and produce AN and GO till they meet the circumferences in D and K.

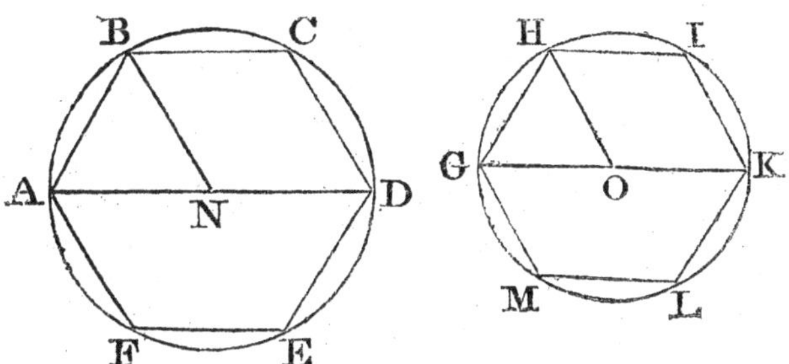

Because the straight lines AB, BC, CD, DE, EF, FA, are all equal, the arches AB, BC, CD, DE, EF, FA are also equal (28. 3.). For the same reason, the arches GH, HI, IK, KL, LM, MG are all equal, and they are equal in number to the others; therefore, whatever part the arch AB is of the whole circumference, ABD, the same is the arch GH of the circumference GHK. But the angle ANB is the same part of four right angles, that the arch AB is of the circumference ABD (33. 6.); and the angle GOH is the same part of four right angles that the arch GH is of the circumference GHK (33. 6.) therefore the angles ANB, GOH are each of them the same part of four right angles, and therefore they are equal to one another. The isosceles triangles ANB, GOH are therefore equiangular (6. 6.), and the angle ABN equal to the angle GHO; in the same manner, by joining NC, OI, it may be proved that the angles NBC, OHI are

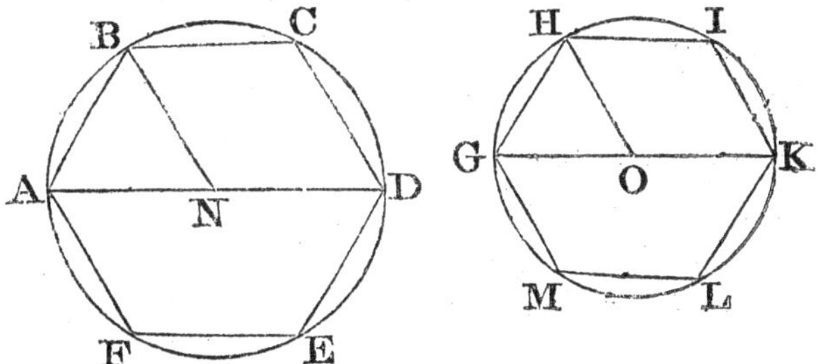

equal to one another, and to the angle ABN. Therefore the whole angle ABC is equal to the whole GHI; and the same may be proved of the angles BCD, HIK, and of the rest. Therefore, the polygons ABCDEF and GHIKLM are equiangular to one another; and since they are equilateral, the sides about the equal angles are proportionals; the polygon ABCD is therefore similar to the polygon GHIKLM (def. 1. 6.). And because similar polygons are as the squares of their homologous sides (20. 6.), the polygon ABCDEF is to the polygon GHIKLM as the square of AB to the square of GH; but because the triangles ANB, GOH are equiangular, the square of AB is to the square of GH as the square of AN to the square of GO (4. 6.), or as four times the square of AN to four times the square (15. 5.) of GO, that is, as the square of AD to the square of GK (2. Cor. 8. 2.). Therefore also, the polygon ABCDEF is to the polygon GHIKLM as the square of AD to the square of GK; and they have also been shewn to be similar. Therefore, &c. Q. E. D.

COR. Every equilateral polygon inscribed in a circle is also equiangular: For the isosceles triangles, which have their common vertex in the centre, are all equal and similar; therefore, the angles at their bases are all equal, and the angles of the polygon are therefore also equal.

PROP. III. THEOR.

The side of any equilateral polygon inscribed in a circle being given, to find the side of a polygon of the same number of sides described about the circle.

Let ABCDEF be an equilateral polygon inscribed in the circle ABD; it is required to find the side of an equilateral polygon of the same number of sides described about the circle.

Find G the centre of the circle; join GA, GB, bisect the arch AB in H; and through H draw KHL touching the circle in H, and meeting GA and GB produced in K and L; KL is the side of the polygon required.

Produce GF to N, so that GN may be equal to GL; join KN, and from G draw GM at right angles to KN, join also HG.

Because the arch AB is bisected in H, the angle AGH is equal to the angle BGH (27. 3.); and because KL touches the circle in H, the angles LHG, KHG are right angles (16. 3.); therefore, there are two angles of the triangle HGK, equal to two angles of the triangle HGL, each to each. But the side GH is common to these triangles; therefore they are equal (26. 1.), and GL is equal to GK. Again, in the triangles KGL, KGN, because GN is equal to GL; and GK common, and also the angle LGK equal to the angle KGN; therefore the base KL is equal to the base KN (4. 1.). But because the triangle KGN is isosceles, the angle GKN is equal

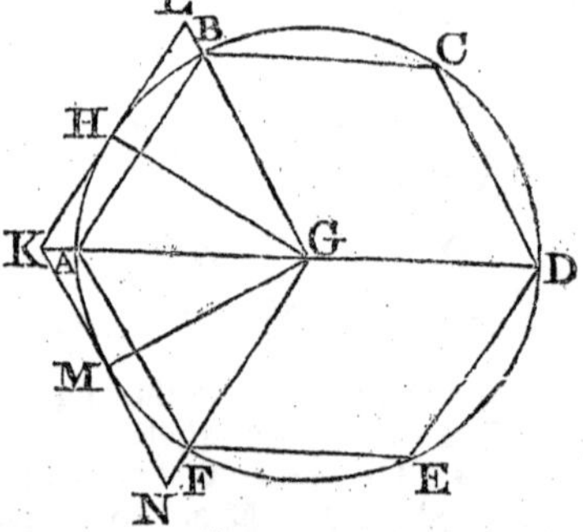

to the angle GNK, and the angles GMK, GMN are both right angles by construction; wherefore, the triangles GMK, GMN have two angles of the one equal to two angles of the other, and they have also the side GM common, therefore they are equal (26. 1.), and the side KM is equal to the side MN, so that KN is bisected in M. But KN is equal to KL, and therefore their halves KM and KH are also equal. Wherefore, in the triangles GKH, GKM, the two sides GK and KH are equal to the two GK and KM, each to each; and the angles GKH, GKM, are also equal, therefore GM is equal to GH (4. 1.); wherefore, the point M is in the circumference of the circle; and because KMG is a right angle, KM touches the circle. And in the same manner, by joining the centre and the other angular points of the inscribed polygon, an equilateral polygon may be described about the circle, the sides of which will each be equal to KL, and will be equal in number to the sides of the inscribed polygon. Therefore, KL is the side of an equilateral polygon, described about the circle, of the same number of sides with the inscribed polygon ABCDEF; which was to be found.

Cor. 1. Because GL, GK, GN, and the other straight lines drawn from the centre G to the angular points of the polygon described about the circle ABD are all equal; if a circle be described from the centre G, with the distance GK, the polygon will be inscribed in that circle; and therefore, it is similar to the polygon ABCDEF (2. 1.).

Cor. 2. It is evident that AB, a side of the inscribed polygon is to KL, a side of the circumscribed, as the perpendicular from G upon AB, to the perpendicular from G upon KL, that is, to the radius of the circle; therefore also, because magnitudes have the same ratio with their equimultiples (15. 5.), the perimeter of the inscribed polygon is to the perimeter of the circumscribed, as the perpendicular from the centre, on a side of the inscribed polygon, to the radius of the circle.

PROP. IV. THEOR.

A circle being given, two similar polygons may be found, the one described about the circle, and the other inscribed in it, which shall differ from one another by a space less than any given space.

Let ABC be the given circle, and the square of D any given space; a polygon may be inscribed in the circle ABC, and a similar polygon described about it, so that the difference between them shall be less than the square of D.

In the circle ABC apply the straight line AE equal to D, and let AB be a fourth part of the circumference of the circle. From the circumference AB take away its half, and from the remainder its half, and so on till the circumference AF is found less than the circumference AE (1. 1. Sup.). Find the centre G; draw the diameter AC, as also the straight lines AF and FG; and having bisected the circumference

AF in K, join KG, and draw HL touching the circle in K, and meeting GA and GF produced in H and L; join CF.

Because the isosceles triangles HGL and AGF have the common angle AGF, they are equiangular (6. 6.), and the angles GHK, GAF are therefore equal to one another. But the angle GKH, CFA are also equal, for they are right angles; therefore the triangles HGK, ACF, are likewise equiangular (32. 1.)

And because the arch AF was found by taking from the arch AB its half, and from that remainder its half, and so on, AF will be contained a certain number of times, exactly, in the arch AB, and therefore it will also be contained a certain number of times, exactly, in the whole circumference, ABC; and the straight line AF is therefore the side of an equilateral polygon inscribed in the circle ABC. Wherefore also, HL is the side of an equilateral polygon, of the same number of sides, described about ABC (3. 1. Sup.). Let the polygon described about the circle be called M, and the polygon inscribed be called N; then,

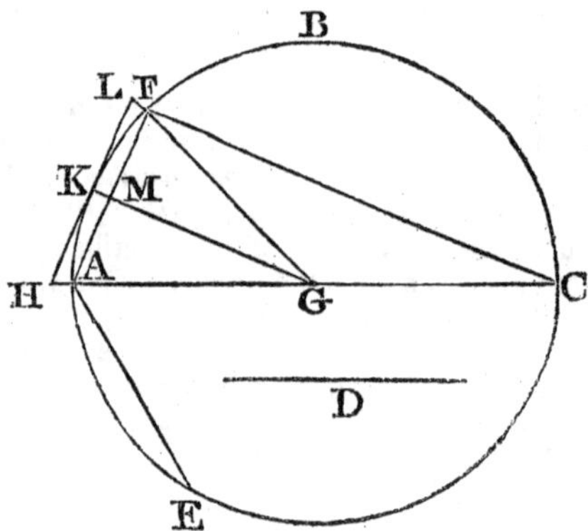

because these polygons are similar (Cor. 3. 1.) they are as the squares of the homologous sides HL and AF (Sup. 3. Cor. 20. 6.), that is, because the triangles HLG, AFG are similar, as the square of HG to the square of AG, that is of GK. But the triangles HGK, ACF have been proved to be similar, and therefore the square of AC is to the square of CF as the polygon M to the polygon N; and, by conversion, the square of AC is to its excess above the squares of CF, that is, to the square of AF (47. 1.), as the polygon M to its excess above the polygon N. But the square of AC, that is, the square described about the circle ABC is greater than the equilateral polygon of eight sides described about the circle, because it contains that polygon; and, for the same reason, the polygon of eight sides is greater than the polygon of sixteen, and so on; therefore, the square of AC is greater than any polygon described about the circle by the continual bisection of the arch AB; it is therefore greater than the polygon M. Now, it has been demonstrated, that the square of AC is to the square of AF as the polygon M to the difference of the polygons; therefore, since the square of AC is greater than M, the square of AF is greater than the difference of the polygons (14. 5.). The difference of the polygons is therefore less than the square of AF; but AF is less than D; therefore, the difference of the polygons is less than the square of D; that is, than the given space. Therefore, &c. Q. E. D.

COR. 1. Because the polygons M and N differ from one another more than either of them differs from the circle, the difference between each of them and the circle is less than the given space, viz. the square of D. And therefore, however small any given space may be, a polygon may be inscribed in the circle, and another described about it, each of which shall differ from the circle by a space less than the given space.

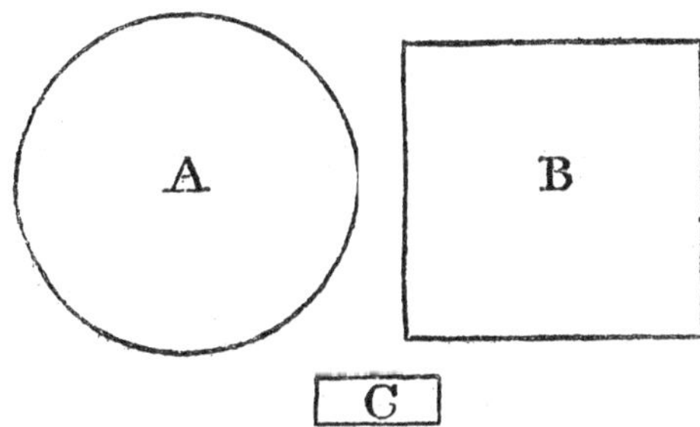

COR. 2. The space B which is greater than any polygon that can be inscribed in the circle A, and less than any polygon that can be described about it, is equal to the circle A. If not, let them be unequal; and first, let B exceed A by the space C. Then, because the polygons described about the circle A are all greater than B, by hypothesis; and because B is greater than A by the space C, therefore no polygon can be described about the circle A, but what must exceed it by a space greater than C, which is absurd. In the same manner, if B be less than A by the space C, it is shewn that no polygon can be inscribed in the circle A, but what is less than A by a space greater than C, which is also absurd. Therefore, A and B are not unequal, that is, they are equal to one another.

PROP. V. THEOR.

The area of any circle is equal to the rectangle contained by the semidiameter, and a straight line equal to half the circumference.

Let ABC be a circle of which the centre is D, and the diameter AC; if in AC produced there be taken AH equal to half the circumference, the area of the circle is equal to the rectangle contained by DA and AH.

Let AB be the side of any equilateral polygon inscribed in the circle ABC; bisect the circumference AB in G, and through G draw EGF touching the circle, and meeting DA produced in E, and DB

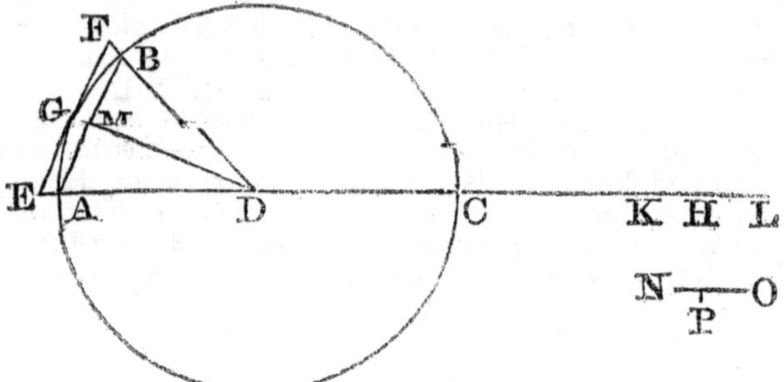

produced in F; EF will be the side of an equilateral polygon described about the circle ABC (3. 1. Sup.). In AC produced take AK equal to half the perimeter of the polygon whose side is AB; and AL equal to half the perimeter of the polygon whose side is EF. Then AK will be less, and AL greater than the straight line AH (Ax. 1. Sup.). Now, because in the triangle EDF, DG is drawn perpendicular to the base, the triangle EDF is equal to the rectangle contained by DG and the half of EF (41. 1.); and as the same is true of all the other equal triangles having their vertices in D, which make up the polygon described about the circle; therefore, the whole polygon is equal to the rectangle contained by DG and AL, half the perimeter of the polygon (1. 2.), or by DA and AL. But AL is greater than AH, therefore the rectangle DA.AL is greater than the rectangle DA.AH; the rectangle DA.AH is therefore less than the rectangle DA.AL, that is, than any polygon described about the circle ABC.

Again, the triangle ADB is equal to the rectangle contained by DM the perpendicular, and one half of the base AB, and it is therefore less than the rectangle contained by DG, or DA, and the half of AB.

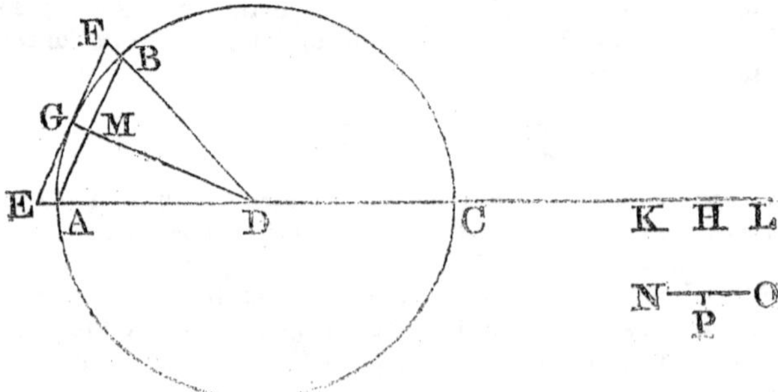

And as the same is true of all the other triangles having their vertices in D, which make up the inscribed polygon, therefore the whole of the inscribed polygon is less than the rectangle contained by DA, and AK half the perimeter of the polygon. Now, the rectangle DA.AK is less than DA.AH; much more, therefore, is the polygon whose side is AB less than DA.AH; and the rectangle DA.AH is therefore greater than any polygon inscribed in the circle ABC. But the same rectangle DA.AH has been proved to be less than any polygon described about the circle ABC; therefore, the rectangle DA.AH is equal to the circle ABC (2. Cor. 4. 1. Sup.) Now, DA is the semidiameter of the circle ABC, and AH the half of its circumference. Therefore, &c. Q. E. D.

Cor. 1. Because DA : AH :: DA^2 : DA.AH (1. 6.), and because by this proposition, DA.AH=the area of the circle, of which DA is the radius: therefore, as the radius of any circle to the semicircumference, or as the diameter to the whole circumference, so is the square of the radius to the area of the circle.

Cor. 2. Hence a polygon may be described about a circle, the perimeter of which shall exceed the circumference of the circle by a line that is less than any given line. Let NO be the given line. Take in NO the part NP less than its half, and less also than AD, and let a polygon be described about the circle ABC, so that its excess above ABC may be less than the square of NP (1. Cor. 4. 1. Sup.). Let the side of this polygon be EF. And since, as has been proved, the circle is equal to the rectangle DA.AH, and the polygon to the rectangle DA.AL, the excess of the polygon above the circle is equal to the rectangle DA.HL; therefore the rectangle DA.HL is less than the square of NP; and therefore, since DA is greater than NP, HL is less than NP, and twice HL less than twice NP, wherefore, much more is twice HL less than NO. But HL is the difference between half the perimeter of the polygon whose side is EF, and half the circumference of the circle; therefore, twice HL is the difference between the whole perimeter of the polygon and the whole circumference of the circle (5. 5.). The difference, therefore, between the perimeter of the polygon and the circumference of the circle is less than the given line NO.

Cor. 3. Hence also, a polygon may be inscribed in a circle, such that the excess of the circumference above the perimeter of the polygon may be less than any given line. This is proved like the preceding.

PROP. VI. THEOR.

The areas of circles are to one another in the duplicate ratio, or as the squares of their diameters.

Let ABD and GHL be two circles, of which the diameters are AD and GL; the circle ABD is to the circle GHL as the square of AD to the square of GL.

Let ABCDEF and GHKLMN be two equilateral polygons of the same number of sides inscribed in the circles ABD, GHL; and let Q

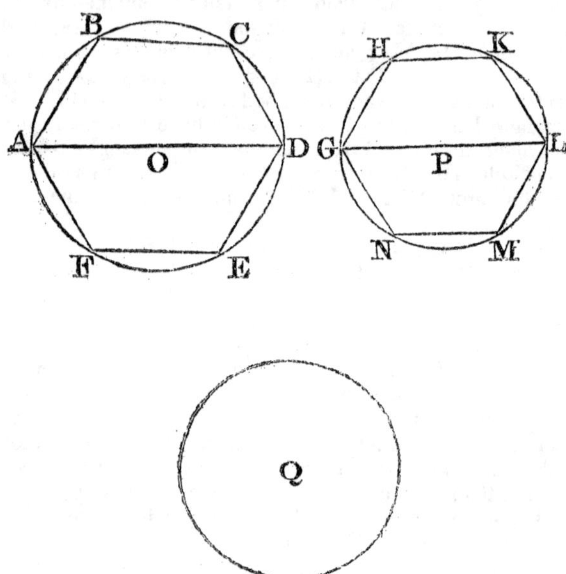

be such a space that the square of AD is to the square of GL as the circle ABD to the space Q. Because the polygons ABCDEF and GHKLMN are equilateral and of the same number of sides, they are similar (2. 1. Sup.), and their areas are as the squares of the diameters of the circles in which they are inscribed. Therefore AD^2 : GL^2 : : polygon ABCDEF : polygon GHKLMN ; but AD^2 : GL^2 : : circle ABD : Q ; and therefore, ABCDEF : GHKLM : : circle ABD : Q. Now, circle ABD $>$ ABCDEF ; therefore Q $>$ GHKLMN (14. 5.), that is, Q is greater than any polygon inscribed in the circle GHL.

In the same manner it is demonstrated, that Q is less than any polygon described about the circle GHL ; wherefore the space Q is equal to the circle GHL (2. Cor. 4. 1. Sup.). Now, by hypothesis, the circle ABD is to the space Q as the square of AD to the square of GL ; therefore the circle ABD is to the circle GHL as the square of AD to the square of GL. Therefore, &c. Q. E. D.

Cor. 1. Hence the circumferences of circles are to one another as their diameters.

Let the straight line X be equal to half the circumference of the circle ABD, and the straight line Y to half the circumference of the

X ————————————————

Y ——————————————

circle GHL: And because the rectangles AO.X and GP.Y are equal to the circles ABD and GHL (5. 1. Sup.); therefore AO.X : GP.Y :: $AD^2 : GL^2$ :: $AO^2 : GP^2$; and alternately, AO.X : AO^2 :: GP.Y : GP^2; whence, because rectangles that have equal altitudes are as their bases (1. 6.), X : AO :: Y : GP, and again alternately, X : Y :: AO : GP; wherefore, taking the doubles of each, the circumference ABD is to the circumference GHL as the diameter AD to the diameter GL.

Cor. 2. The circle that is described upon the side of a right angled triangle opposite to the right angle, is equal to the two circles described on the other two sides. For the circle described upon SR is to the circle described upon RT as the square of SR to the square of RT; and the circle described upon TS is to the circle described upon RT as the square of ST to the square of RT. Wherefore, the circles described on SR and on ST are to the circle described on RT as the squares of SR and of ST to the square of RT (24. 5.). But the squares of RS and of ST are equal to the square of RT (47. 1.); therefore the circles described on RS and ST are equal to the circle described on RT.

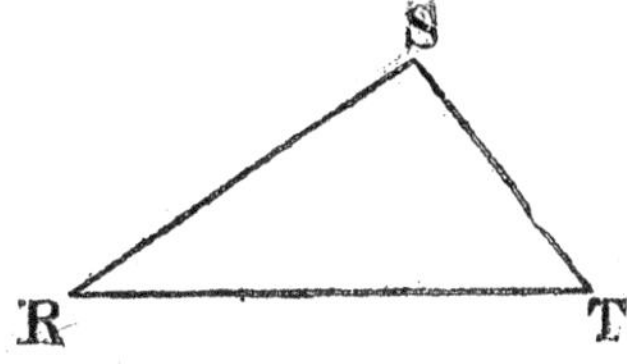

PROP. VII. THEOR.

Equiangular parallelograms are to one another as the products of the numbers proportional to their sides.

Let AC and DF be two equiangular parallelograms, and let M, N, P and Q be four numbers, such that AB : BC :: M : N; AB : DE :: M : P; and AB : EF :: M : Q, and therefore ex æquali, BC : EF :: N : Q. The parallelogram AC is to the parallelogram DF as MN to PQ.

Let NP be the product of N into P, and the ratio of MN to PQ will be compounded of the ratios (def. 10. 5.) of MN to NP, and of NP to PQ. But the ratio of MN to NP is the same with that of M

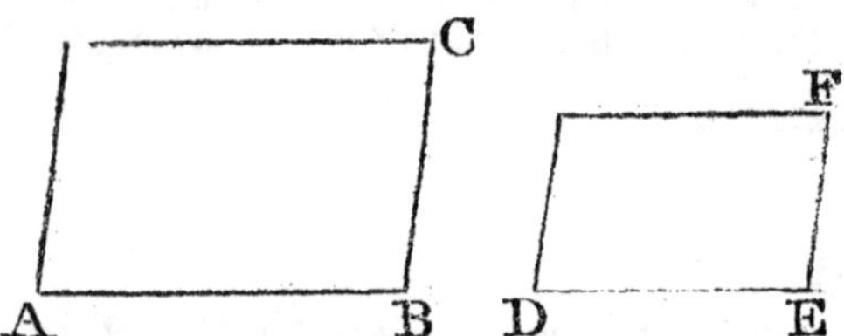

to P (15. 5.), because MN and NP are equimultiples of M and P; and for the same reason, the ratio of NP to PQ is the same with that of N to Q; therefore the ratio of MN to PQ is compounded of the ratios of M to P, and of N to Q. Now, the ratio of M to P is the same with that of the side AB to the side DE (by Hyp.); and the ratio of N to Q the same with that of the side BC to the side EF. Therefore, the ratio of MN to PQ is compounded of the ratios of AB to DE, and of BC to EF. And the ratio of the parallelogram AC to the parallelogram DF is compounded of the same ratios (23. 6.); therefore, the parallelogram AC is to the parallelogram DF as MN, the product of the numbers M and N, to PQ, the product of the numbers P and Q. Therefore, &c. Q. E. D.

COR. 1. Hence, if GH be to KL as the number M to the number F; the square described on GH will be to the square described on KL as MM, the square of the number M to NN, the square of the number N.

G H K L

COR. 2. If A, B, C, D, &c, are any lines, and m, n, r, s, &c. numbers proportional to them; viz. $A : B :: m : n$, $A : C :: m : r$, $A : D :: m : s$, &c.; and if the rectangle contained by any two of the lines be equal to the square of a third line, the product of the numbers, proportional to the first two, will be equal to the square of the numbers proportional to the third; that is, if $A.C = B^2$, $m \times r = n \times n$, or $= n^2$.

For by this Prop. $A.C : B^2 :: m \times r : n^2$; but $A.C = B^2$, therefore $m \times r = n^2$. Nearly in the same way it may be demonstrated, that whatever is the relation between the rectangles contained by these lines, there is the same between the products of the numbers proportional to them.

So also conversely if m and r be numbers proportional to the lines A and C; if also $A.C = B^2$, and if a number n be found such, that $n^2 = mr$, then $A : B :: m : n$. For let $A : B :: m : q$, then since, m, q, r are proportional to A, B, and C, and $A.C = B^2$; therefore, as has just been proved, $q^2 = m \times r$; but $n^2 = q \times r$, by hypothesis, therefore $n^2 = q^2$, and $n = q$; wherefore $A : B :: m : n$.

SCHOLIUM.

In order to have numbers proportional to any set of magnitudes of the same kind, suppose one of them to be divided into any number, m of equal parts, and let H be one of those parts. Let H be found in times in the magnitude B, r times in C, s times in D, &c., then it is evident that the numbers m, n, r, s are proportional to the magnitudes A, B, C and D. When therefore it is said in any of the following propositions, that a line as A = a number m, it is understood that $A=m \times H$, or that A is equal to the given magnitude H multipled by m, and the same is understood of the other magnitudes, B, C, D, and their proportional numbers, H being the common measure of all the magnitudes. This common measure is omitted for the sake of brevity in the arithmetical expression; but is always implied, when a line, or other geometrical magnitude, is said to be equal to a number. Also, when there are fractions in the number to which the magnitude is called equal, it is meant that the common measure H is farther subdivided into such parts as the numerical fraction indicates. Thus, if A=360.375, it is meant that there is a certain magnitude H, such that $A=360\times H+\frac{375}{1000}\times H$, or that A is equal to 360 times H, together with 375 of the thousandth parts of H. And the same is true in all other cases, where numbers are used to express the relations of geometrical magnitudes.

PROP. VIII. THEOR.

The perpendicular drawn from the centre of a circle on the chord of any arch is a mean proportional between half the radius and the line made up of the radius and the perpendicular drawn from the centre on the chord of double that arch: And the chord of the arch is a mean proportional between the diameter and a line which is the difference between the radius and the aforesaid perpendicular from the centre.

Let ADB be a circle, of which the centre is C; DBE any arch, and DB the half of it; let the chords DE, DB be drawn; as also CF and CG at right angles to DE and DB: if CF be produced it will meet the circumference in B: let it meet it again in A, and let AC be bisected in H; CG is a mean proportional between AH and AF; and

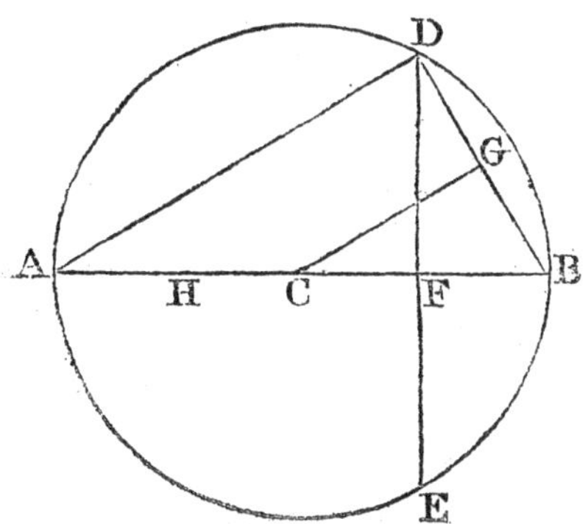

BD a mean proportional between AB, and BF, the excess of the radius above CF.

Join AD; and because ADB is a right angle, being an angle in a semicircle; and because CGB is also a right angle, the triangles ABD, CBG are equiangular, and, AB : AD :: BC : CG (4. 6.), or alternately, AB : BC :: AD : CG; and therefore, because AB is double of BC, AD is double of CG, and the square of AD therefore equal to four times the square of CG.

But, because ADB is a right angled triangle, and DE a perpendicular on AB, AD is a mean proportional between AB and AF (8. 6.), and $AD^2=AB.AF$ (17. 6.), or since AB is $=4AH$, $AD^2=4AH.AF$. Therefore also, because $4CG^2=AD^2$, $4CG^2=4AH.AF$, and $CG^2=AH.AF$; wherefore CG is a mean proportional between AH and AF, that is, between half the radius and the line made up of the radius, and the perpendicular on the chord of twice the arch BD.

Again, it is evident, that BD is a mean proportional between AB and BF (8. 6.), that is, between the diameter and the excess of the radius above the perpendicular, on the chord of twice the arch DB. Therefore, &c. Q. E. D.

PROP. IX. THEOR.*

The circumference of a circle exceeds three times the diameter, by a line less than ten of the parts, of which the diameter contains seventy, but greater than ten of the parts whereof the diameter contains seventy-one.

Let ABD be a circle, of which the centre is C, and the diameter AB; the circumference is greater than three times AB, by a line less than $\frac{10}{70}$, or $\frac{1}{7}$, of AC, but greater than $\frac{10}{71}$ of AC.

In the circle ABD apply the straight line BD equal to the radius

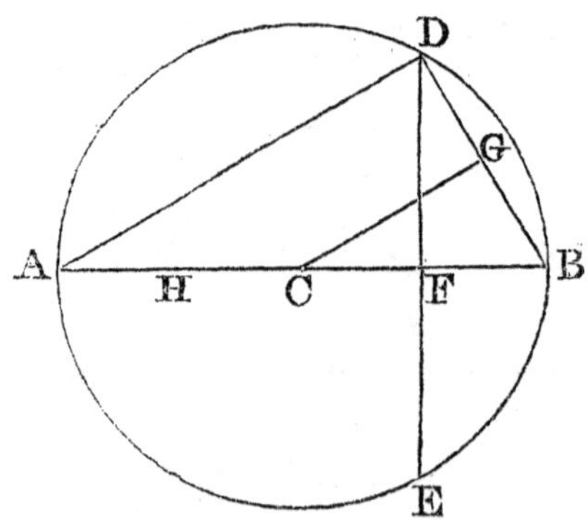

* In this proposition, the character +, placed after a number, signifies that something is to be added to it; and the character —, on the other hand, signifies that something is to be taken away from it.

BC: Draw DF perpendicular to BC, and let it meet the circumference again in E; draw also CG perpendicular to BD: produce BC to A, bisect AC in H, and join CD.

It is evident, that the arches BD, BE are each of them one-sixth of the circumference (Cor. 15. 4.), and that therefore the arch DBE is one third of the circumference. Wherefore, the line (8. 1. Sup.) CG is a mean proportional between AH, half the radius, and the line AF. Now because the sides BD, DC, of the triangle BDC are equal, the angles DCF, DBF are also equal; and the angles DFC, DFB being equal, and the side DF common to the triangles DBF, DCF, the base BF is equal to the base CF, and BC is bisected in F.

Therefore, if AC or BC=1000, AH=500, CF=500, AF=1500, and CG being a mean proportional between AH and AF, CG^2=(17. 6.) AH.AF=500×1500=750000; wherefore CG=866.0254+, because $(866.0254)^2$ is less than 750000. Hence also, AC+CG=1866. 0254+.

Now, as CG is the perpendicular drawn from the centre C, on the chord of one-sixth of the circumference, if P = the perpendicular from C on the chord of one-twelfth of the circumference, P will be a mean proportional between AH (8. 1. Sup.) and AC+CG, and P^2= AH (AC + CG)=500×(1866.0254+)=933012.7 +. Therefore, P=965.9258+, because $(965.9258)^2$ is less than 933012.7. Hence also, AC+P=1965.9258+.

Again, if Q = the perpendicular drawn from C on the chord of one twenty-fourth of the circumference, Q will be a mean proportional between AH and AC+P, and Q^2=AH (AC+P)=500(1965.9258+) =982962.9 +; and therefore Q=991.4449+, because $(991.4449)^2$ is less than 982962.9. Therefore also AC+Q=1991.4449+.

In like manner, if S be the perpendicular from C on the chord of one forty-eighth of the circumference, S^2 = AH (AC + Q) = 500 (1991.4449+)=995722.45 + ; and S=997.8589+, because $(997. 8589)^2$ is less than 995722.45. Hence also, AC+S=1997.8589+.

Lastly, if T be the perpendicular from C on the chord of one ninety-sixth of the circumference, T^2= AH (AC+S)=500 (1997.8589 +) = 998929.45 +, and T = 999.46458+. Thus T, the perpendicular on the chord of one ninety-sixth of the circumference, is greater than 999.46458 of those parts of which the radius contains 1000.

But by the last proposition, the chord of one ninety-sixth part of the circumference is a mean proportional between the diameter and the excess of the radius above S, the perpendicular from the centre on the chord of one forty-eighth of the circumference. Therefore, the square of the chord of one ninety-sixth of the circumference = AB (AC - S) = 2000 × (2.1411 —.) = 4282.2—; and therefore the chord itself =65.4386—, because $(65.4386)^2$ is greater than 4282.2. Now, the chord of one ninty-sixth of the circumference, or the side of an equilateral polygon of ninety-six sides inscribed in the circle, being 65.4386—, the perimeter of that polygon will be = (65.4386 —)96 = 6282.1056—.

Let the perimeter of the circumscribed polygon of the same number of sides, be M, then (2 Cor. 2. 1. Sup.) T : AC :: 6282.1056— : M, that is, (since T=999.46458+, as already shewn),

999.46458 + : 1000 :: 6282.1056 – : M ; if then N be such, that 999.46458 : 1000 :: 6282.1056— : N ; ex æquo perturb. 999.46458+ : 999.46458 :: N : M ; and, since the first is greater than the second, the third is greater than the fourth, or N is greater than M.

Now, if a fourth proportional be found to 999.46458, 1000 and 6282.1056 viz. 6285.461—, then,

because, 999.46458 : 1000 :: 6282.1056 : 6285.461 –,

and as before, 999.46458 : 1000 :: 6282.1056— : N;

therefore, 6282.1056 : 6282.1056— :: 6285.461 – N, and as the first of these proportionals is greater than the second, the third, viz. 6285.

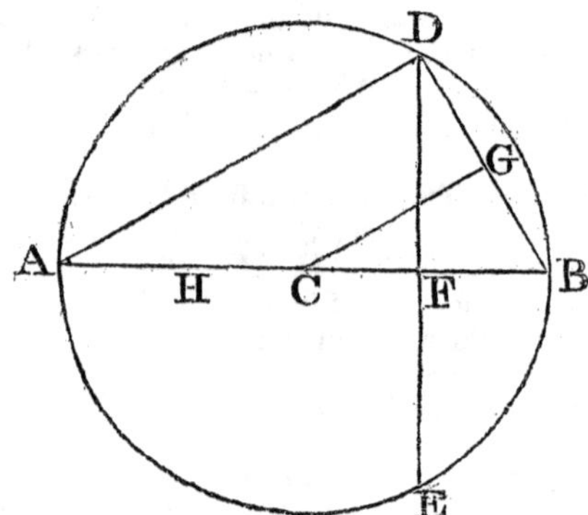

461— is greater than N, the fourth. But N was proved to be greater than M; much more, therefore, is 6285.461 greater than M, the perimeter of a polygon of ninety-six sides circumscribed about the circle; that is, the perimeter of that polygon is less than 6285.461; now, the circumference of the circle is less than the perimeter of the polygon; much more, therefore, is it less than 6285.461; wherefore the circumference of a circle is less than 6285.461 of those parts of which the radius contains 1000. The circumference, therefore, has to the diameter a less ratio (8. 5.) than 6285.461 has to 2000, or than 3142.7305 has to 1000: but the ratio of 22 to 7 is greater than the ratio of 3142.7305 to 1000, therefore the circumference has a less ratio to the diameter than 22 has to 7, or the circumference is less than 22 of the parts of which the diameter contains 7.

It remains to demonstrate, that the part by which the circumference exceeds the diameter is greater than $\frac{10}{71}$ of the diameter.

It was before shewn, that $CG^2 = 750000$; wherefore $CG = 866.02545-$, because $(866.02545)^2$ is greater than 750000; therefore $AC + CG = 1866.02545-$.

Now, P being, as before, the perpendicular from the centre on the chord of one twelfth of the circumference, $P^2 = AH\ (AC+CG) = 500 \times (1866.02545) - = 933012.73-$; and $P = 965.92585-$, because $(965.92585)^2$ is greater than 633012.73. Hence also, $AC+P = 1965.92585-$.

Next, as Q = the perpendicular drawn from the centre on the chord of one twenty-fourth of the circumference, $Q^2 = AH\ (AC+P) = 500 \times (1965.92585-) = 982962.93-$; and $Q = 991.44495-$, because $(991.44495)^2$ is greater than 982962.93. Hence also, $AC+Q = 1991.44495-$.

In like manner, as S is the perpendicular from C on the chord of one forty-eighth of the circumference, $S^2 = AH\ (AC+Q) = 500(1991.44495-) = 995722.475-$, and $S = (997.85895-)$ because $(997.85895)^2$ is greater than 995722.475.

But the square of the chord of the ninety-sixth part of the circumference $= AB\ (AC-S) = 2000\ (2.14105+) = 4282.1+$, and the chord itself $= 65.4377+$ because $(65.4377)^2$ is less than 4282.1. Now the chord of one ninety-sixth part of the circumference being $= 65.4377+$, the perimeter of a polygon of ninety-six sides inscribed in the circle $= (65.4377+)\ 96 = 6282.019+$. But the circumference of the circle is greater than the perimeter of the inscribed polygon; therefore the circumference is greater than 6282.019, of those parts of which the radius contains 1000; or than 3141.009 of the parts of which the radius contains 500, or the diameter contains 1000. Now, 3141.009 has to 1000 a greater ratio than $3+\frac{10}{71}$ to 1; therefore the circumference of the circle has a greater ratio to the diameter than $3+\frac{10}{71}$ has to 1; that is, the excess of the circumference above three times the diameter is greater than ten of those parts of which the diameter contains 71; and it has already been shewn to be less than ten of those of which the diameter contains 70. Therefore, &c. Q. E. D.

Cor. 1. Hence the diameter of a circle being given, the circumference may be found nearly, by making as 7 to 22, so the given diameter to a fourth proportional, which will be greater than the circumference. And if as 1 to $3+\frac{10}{71}$, or as 71 to 223, so the given diameter to a fourth proportional, this will be nearly equal to the circumference, but will be less than it.

COR. 2. *Because the difference between* $\frac{1}{7}$ and $\frac{10}{71}$ is $\frac{1}{497}$, therefore the lines found by these proportions differ by $\frac{1}{497}$ of the diameter. Therefore the difference of either of them from the circumference must be less than the 497th part of the diameter.

COR. 3. As 7 to 22, so the square of the radius to the area of the circle nearly.

For it has been shewn, that (1. Cor. 5. 1. Sup.) the diameter of a circle is to its circumference as the square of the radius to the area of the circle; but the diameter is to the circumference nearly as 7 to 22, therefore the square of the radius is to the area of the circle nearly in that same ratio.

SCHOLIUM.

It is evident that the method employed in this proposition, for finding the limits of the ratio of the circumference of the diameter, may be carried to a greater degree of exactness, by finding the perimeter of an inscribed and of a circumscribed polygon of a greater number of sides than 96. The manner in which the perimeters of such polygons approach nearer to one another, as the number of their sides increases, may be seen from the following Table, which is constructed on the principles explained in the foregoing Proposition, and in which the radius is supposed = 1.

NO. of Sides of the Polygon.	Perimeter of the inscribed Polygon.	Perimeter of the circumscribed Polygon.
6	6.000000	6.822033—
12	6.211657+	6.430781—
24	6.265257+	6.319320—
48	6.278700+	6.292173—
96	6.282063+	6.285430—
192	6.282904+	6.283747—
384	6.283115+	6.283327—
768	6.283167+	6.283221—
1536	6.283180+	6.283195—
3072	6.283184+	6.283188—
6144	6.283185+	6.283186—

The part that is wanting in the numbers of the second column, to make up the entire perimeter of any of the inscribed polygons, is less than unit in the sixth decimal place; and in like manner, the part by which the numbers in the last column exceed the perimeter of any of the circumscribed polygons is less than a unit in the sixth decimal place, that is than $\frac{1}{1000000}$ of the radius. Also, as the numbers in the second column are less than the perimeters of the inscribed polygons, they are each of them less than the circumference, of the circle; and for the same reason, each of those in the third column is greater than the circumference. But when the arch of $\frac{1}{6}$ of the circumference is bisected ten times, the number of sides in the polygon is 6144, and the numbers in the Table differ from one another only by $\frac{1}{1000000}$ part of the radius, and therefore the perimeters of the polygons differ by less than that quantity; and consequently the circumference of the circle, which is greater than the least, and less than the greatest of these numbers, is determined within less than the millioneth part of the radius.

Hence also, if R be the radius of any circle, the circumference is greater than $R \times 6.283185$, or than $2R \times 3.141592$, but less than $2R \times 3.141593$; and these numbers differ from one another only by a millioneth part of the radius. So also $R^2 + 3.141592$ is less, and $R^2 \times 3.141593$ greater than the area of the circle; and these numbers differ from one another only by a millioneth part of the square of the radius.

In this way, also, the circumference and the area of the circle may be found still nearer to the truth; but neither by this, nor by any other method yet known to geometers, can they be exactly determined, though the errors of both may be reduced to a less quantity than any that can be assigned.

ELEMENTS

OF

GEOMETRY.

SUPPLEMENT.

BOOK II.

OF THE INTERSECTION OF PLANES.

DEFINITIONS.

I.

A STRAIGHT line is perpendicular or at right angles to a plane, when it makes right angles with every straight line which it meets in that plane.

II.

A plane is perpendicular to a plane, when the straight lines drawn in one of the planes perpendicular to the common section of the two planes, are perpendicular to the other plane.

III.

The inclination of a straight line to a plane is the acute angle contained by that straight line, and another drawn from the point in which the first line meets the plane, to the point in which a perpendicular to the plane, drawn from any point of the first line, meets the same plane.

IV.

The angle made by two planes which cut one another, is the angle contained by two straight lines drawn from any, the same point in the line of their common section, at right angles to that line, the one, in the one plane, and the other, in the other. Of the two adjacent angles made by two lines drawn in this manner, that which is acute is also called the inclination of the planes to one another.

V.

Two planes are said to have the same, or a like inclination to one another, which two other planes have, when the angles of inclination above defined are equal to one another.

VI.

A straight line is said to be parallel to a plane, when it does not meet the plane, though produced ever so far.

VII.

Planes are said to be parallel to one another, which do not meet, though produced ever so far.

VIII.

A solid angle is an angle made by the meeting of more than two plane angles, which are not in the same plane in one point.

PROP. I. THEOR.

One part of a straight line cannot be in a plane and another part above it.

If it be possible, let AB, part of the straight line ABC, be in the plane, and the part BC above it: and since the straight line AB is in the plane, it can be produced in that plane (2. Post. 1.); let it be produced to D: Then ABC and ABD are two straight lines, and they have the common segment AB, which is impossible (Cor. def. 3. 1.). Therefore ABC is not a straight line. Wherefore one part, &c. Q. E. D.

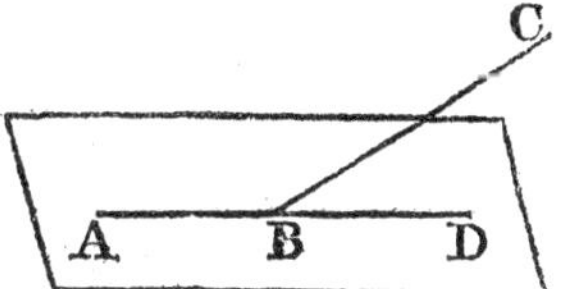

PROP. II. THEOR.

Any three straight lines which meet one another, not in the same point, are in one plane.

Let the three straight lines AB, CD, CB meet one another in the points B, C and E; AB, CD, CB are in one plane.

Let any plane pass through the straight line EB, and let the plane be turned about EB, produced, if necessary, until it pass through the point C: Then, because the points E, C are in this plane, the straight line EC is in it (def. 5. 1.): for the same reason, the straight line BC is in the same; and, by the hypothesis, EB is in it; therefore the three straight lines EC, CB, BE are in one plane: but the whole of the lines DC,

AB, and BC produced, are in the same plane with the parts of them EC, EB, BC (1. 2. Sup.). Therefore AB, CD, CB, are all in one plane. Wherefore, &c. Q. E. D.

COR. It is manifest, that any two straight lines which cut one another are in one plane: Also, that any three points whatever are in one plane.

PROP. III. THEOR.

If two planes cut one another, their common section is a straight line.

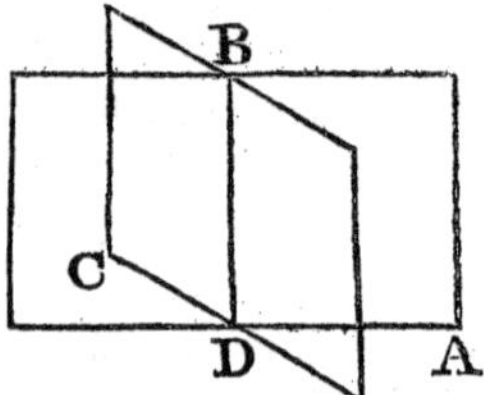

Let two planes AB, BC cut one another, and let B and D be two points in the line of their common section. From B to D draw the straight line BD; and because the points B and D are in the plane AB, the straight line BD is in that plane (def. 5. 1.): for the same reason it is in the plane CB; the straight line BD is therefore common to the planes AB and BC, or it is the common section of these planes. Therefore, &c. Q. E. D.

PROP. IV. THEOR.

If a straight line stand at right angles to each of two straight lines in the point of their intersection, it will also be at right angles to the plane in which these lines are.

Let the straight line AB stand at right angles to each of the straight lines EF, CD in A, the point of their intersection: AB is also at right angles to the plane passing through EF, CD.

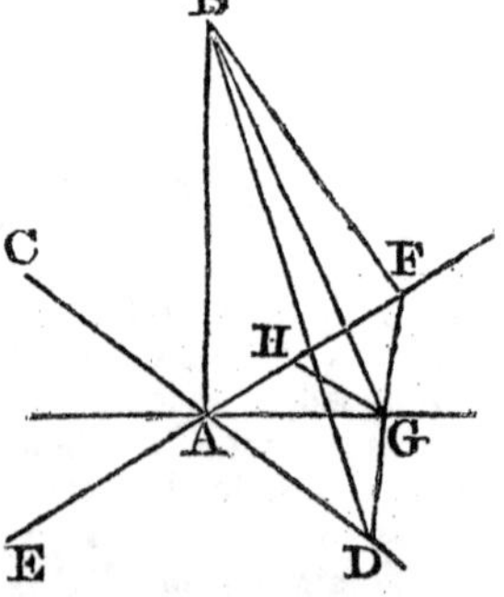

Through A draw any line AG in the plane in which are EF and CD; let G be any point in that line; draw GH parallel to AD; and make HF=HA, join FG; and when produced let it meet CA in D; join BD, BG, BF. Because GH is parallel to AD, and FH=HA: therefore FG=GD, so that the line DF is bisected in G. And because BAD is a right angle, $BD^2=AB^2+AD^2$(47. 1.); and for the same reason, $BF^2=AB^2+AF^2$, therefore $BD^2+BF^2=2AB^2+AD^2+AF^2$; and because DF is bisected in G (A. 2.), $AD^2+AF^2=2AG^2+2GF^2$, therefore $BD^2+BF^2=2AB^2+2AG^2+2GF^2$. But $BD^2+BF^2=$ (A. 2.) $2BG^2+2GF^2$, therefore $2BG^2+2GF^2=2AB^2+$

$2AG^2 + 2GF^2$; and taken $2GF^2$ from both, $2BG^2 = 2AB^2 + 2AG^2$, or $BG^2 = AB^2 + AG^2$; whence BAG (48. 1.) is a right angle. Now AG is any straight line drawn in the plane of the lines AD, AF; and when a straight line is at right angles to any straight line which it meets with in a plane, it is at right angles to the plane itself (def. 1. 2. Sup.). AB is therefore at right angles to the plane of the lines AF, AD. Therefore, &c. Q. E. D.

PROP. V. THEOR.

If three straight lines meet all in one point, and a straight line stand at right angles to each of them in that point: these three straight lines are in one and the same plane.

Let the straight line AB stand at right angles to each of the straight lines BC, BD, BE, in B, the point where they meet; BC, BD, BE are in one and the same plane.

If not, let BD and BE, if possible, be in one plane, and BC be above it; and let a plane pass through AB, BC, the common section of which with the plane, in which BD and BE are, shall be a straight (3. 2. Sup.) line; let this be BF: therefore the three straight lines AB, BC, BF are all in one plane, viz. that which passes through AD, BC; and because AB stands at right angles to each of the straight lines BD, BE, it is also at right angles (4. 2. Sup.) to the plane passing through them; and therefore makes right angles with every straight line meeting it in that plane; but BF which is in that plane meets it; therefore the angle ABF is a right angle; but the angle ABC, by the hypothesis is also a right angle; therefore the angle ABF is equal to the angle ABC, and they are both in the same plane, which is impossible: therefore the straight line BC is not above the plane in which are BD and BE: Wherefore the three straight lines BC, BD, BE are in one and the same plane. Therefore, if three straight lines, &c. Q. E. D.

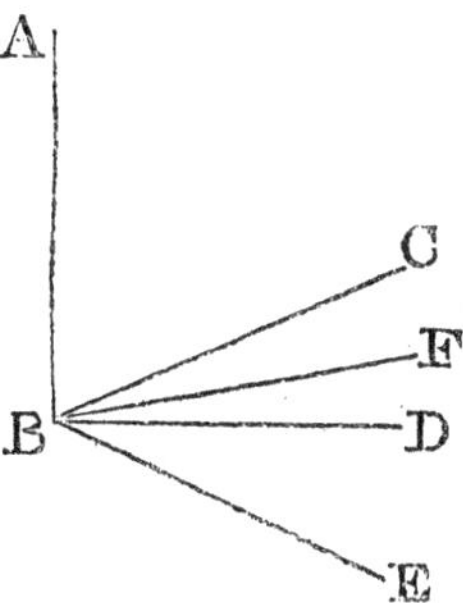

PROP. VI. THEOR.

Two straight lines which are at right angles to the same plane, are parallel to one another.

Let the straight lines AB, CD be at right angles to the same plane BDE; AB is parallel to CD.

Let them meet the plane in the points B, D. Draw DE at right angles to DB, in the plane BDE, and let E be any point in it: Join AE, AD, EB. Because ABE is a right angle, $AB^2 + BE^2 =$ (47. 1.) AE^2, and because BDE is a right angle, $BE^2 = BD^2 + DE^2$; therefore $AB^2 + BD^2 + DE^2 = AE^2$; now, $AB^2 + BD^2 = AD^2$, because ABD is a right angle, therefore $AD^2 + DE^2 = AE^2$, and ADE is therefore a (48. 1.) right angle. Therefore ED is perpendicular to the three lines BD, DA, DC, whence these lines are in one plane (5. 2. Sup.). But AB is in the plane in which are BD, DA, because any three straight lines, which meet one another, are in one plane (2. 2. Sup.): Therefore AB, BD, DC are in one plane; and each of the angles ABD, BDC is a right angle; therefore AB is parallel (28. 1.) to CD. Wherefore, if two straight lines, &c. Q. E. D.

PROP. VII. THEOR.

If two straight lines be parallel, and one of them at right angles to a plane; the other is also at right angles to the same plane.

Let AB, CD be two parallel straight lines, and let one of them AB be at right angles to a plane; the other CD is at right angles to the same plane.

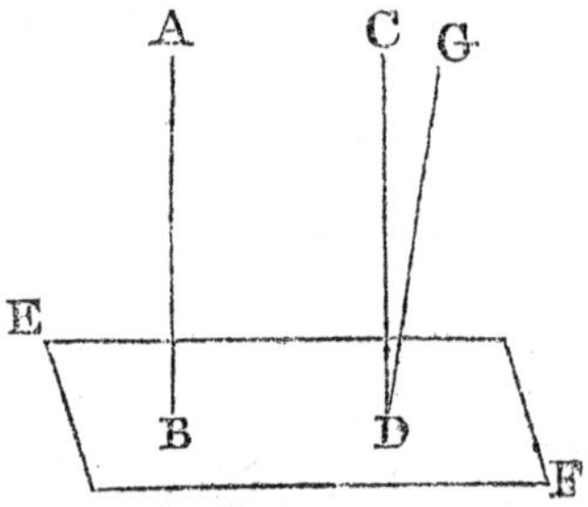

For, if CD be not perpendicular to the plane to which AB is perpendicular, let DG be perpendicular to it. Then (6. 2. Sup.) DG is parallel to AB: DG and DC therefore are both parallel to AB, and are drawn through the same point D, which is impossible (11. Ax. 1.) Therefore, &c. Q. E. D.

PROP. VIII. THEOR.

Two straight lines which are each of them parallel to the same straight line, though not both in the same plane with it, are parallel to one another.

Let AB, CD be each of them parallel to EF, and not in the same plane with it; AB shall be parallel to CD.

In EF take any point G, from which draw, in the plane passing through EF, AB, the straight line GH at right angles to EF; and in the plane passing through EF, CD, draw GK at right angles to the same EF. And because EF is perpendicular both to GH and GK, it is perpendicular (4. 2. Sup.) to the plane HGK passing through them: and EF is parallel to AB; therefore AB is at right angles (7. 2. Sup.) to the plane HGK. For the same reason, CD is likewise at right angles to the plane HGK. Therefore AB, CD are each of them at right angles to the plane HGK. But if two straight lines are at right angles to the same plane, they are parallel (6. 2. Sup.) to one another. Therefore AB is parallel to CD. Wherefore two straight lines, &c. Q. E. D.

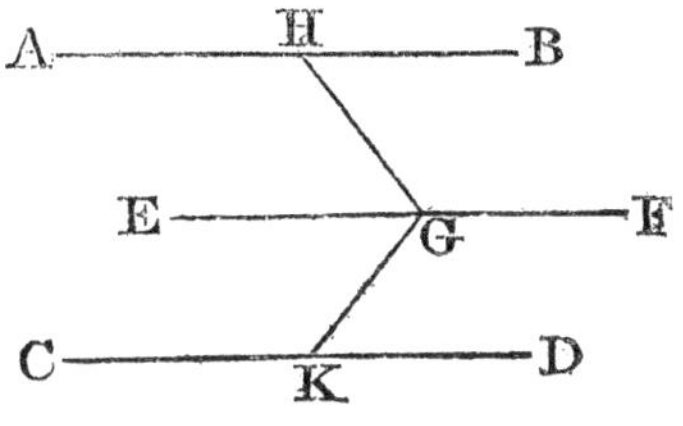

PROP. IX. THEOR.

If two straight lines meeting one another be parallel to two others that meet one another, though not in the same plane with the first two; the first two and the other two shall contain equal angles.

Let the two straight lines AB, BC which meet one another be parallel to the two straight lines DE, EF that meet one another, and are not in the same plane with AB, BC. The angle ABC is equal to the angle DEF.

Take BA, BC, ED, EF all equal to one another; and join AD, CF, BE, AC, DF: Because BA is equal and parallel to ED, therefore AD is (33. 1.) both equal and parallel to BE. For the same reason, CF is equal and parallel to BE. Therefore AD and CF are each of them equal and parallel to BE. But straight lines that are parallel to the same straight line, though not in the same plane with it, are parallel (8. 2. Sup.) to one another. Therefore AD is parallel to CF; and it is equal to it, and AC, DF join them towards the same parts; and therefore (33 1) AC is equal and parallel to DF. And because AB, BC are equal to DE, EF, and the base AC to the base DF; the angle ABC is equal (8. .) to the angle DEF. Therefore, if two straight lines, &c. Q. E. D.

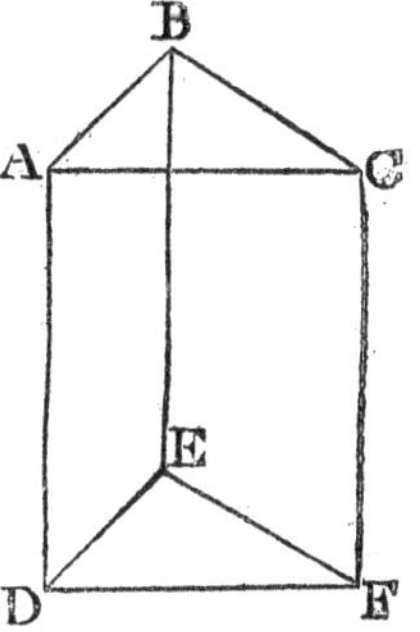

PROP. X. PROB.

To draw a straight line perpendicular to a plane, from a given point above it.

Let A be the given point above the plane BH, it is required to draw from the point A a straight line perpendicular to the plane BH.

In the plane draw any straight line BC, and from the point A draw (12. 1.) AD perpendicular to BC. If then AD be also perpendicular to the plane BH, the thing required is already done; but if it be not, from the point D draw (11. 1.), in the plane BH, the straight line DE at right angles to BC; and from the point A draw AF perpendicular to DE; and through F draw (31. 1.) GH parallel to BC: and because BC is at right angles to ED, and DA, BC is at right angles (4. 2. Sup.) to the plane passing through ED, DA. And GH is parallel to BC; but if two straight lines be parallel, one of which is at right angles to a plane, the other shall be at right (7. 2. Sup.) angles to the same plane; wherefore GH is at right angles to the plane through ED, DA, and is perpendicular (def. 1. 2. Sup.) to every straight line meeting it in that plane. But AF, which is in the plane through ED, DA, meets it: Therefore GH is perpendicular to AF, and consequently AF is perpendicular to GH; and AF is also perpendicular to DE: Therefore AF is perpendicular to each of the straight lines GH, DE. But if a straight line stands at right angles to each of two straight lines in the point of their intersection, it is also at right angles to the plane passing through them (4. 2. Sup.). And the plane passing through ED, GH is the plane BH; therefore AF is perpendicular to the plane BH; so that, from the given point A, above the plane BH, the straight line AF is drawn perpendicular to that plane. Which was to be done.

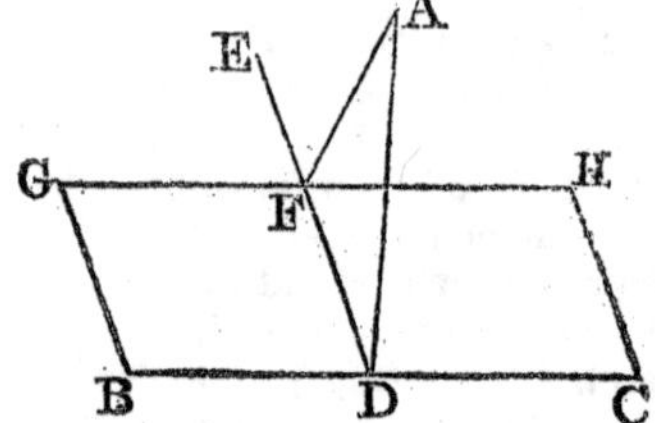

Cor. If it be required from a point C in a plane to erect a perpendicular to that plane, take a point A above the plane, and draw AF perpendicular to the plane; then, if from C a line be drawn parallel to AF, it will be the perpendicular required; for being parallel to AF it will be perpendicular to the same plane to which AF is perpendicular (7. 2. Sup.).

PROP. XI. THEOR.

From the same point in a plane, there cannot be two straight lines at right angles to the plane, upon the same side of it: And there can be but one perpendicular to a plane from a point above it.

For, if it be possible, let the two straight lines AC, AB be at right angles to a given plane from the same point A in the plane, and upon the same side of it; and let a plane pass through BA, AC; the common section of this plane with the given plane is a straight (3. 2. Sup.) line passing through A: Let DAE be their common section: Therefore the straight lines AB, AC, DAE are in one plane: And because CA is at right angles to the given plane, it makes right angles with every straight line meeting it in that plane. But DAE, which is in that plane, meets CA: therefore CAE is a right angle. For the same reason BAE is a right angle. Wherefore the angle CAE is equal to the angle BAE; and they are in one plane, which is impossible. Also, from a point above a plane, there can be but one perpendicular to that plane; for if there could be two, they would be parallel (6. 2. Sup.) to one another, which is absurd. Therefore, from the same point, &c. Q. E. D.

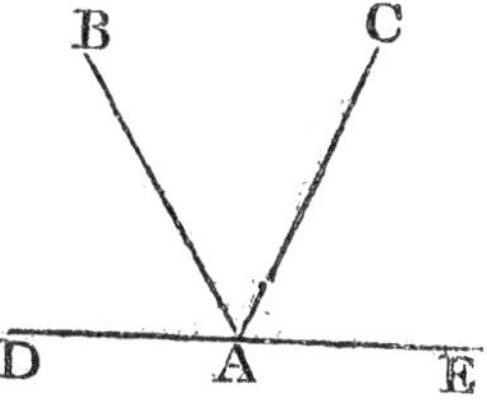

PROP. XII. THEOR.

Planes to which the same straight line is perpendicular, are parallel to one another.

Let the straight line AB be perpendicular to each of the planes CD, EF: these planes are parallel to one another.

If not, they must meet one another when produced, and their common section must be a straight line GH, in which take any point K, and join AK, BK: Then, because AB is perpendicular to the plane EF, it is perpendicular (def. 1. 2. Sup.) to the straight line BK which is in that plane, and therefore ABK is a right angle. For the same reason, BAK is a right angle; wherefore the two angles ABK, BAK of the triangle ABK are equal to two right angles, which is impossible (17. 1.): Therefore the planes CD, EF, though produced, do not meet one another; that is, they are parallel (def. 7. 2. Sup.). Therefore planes, &c. Q. E. D.

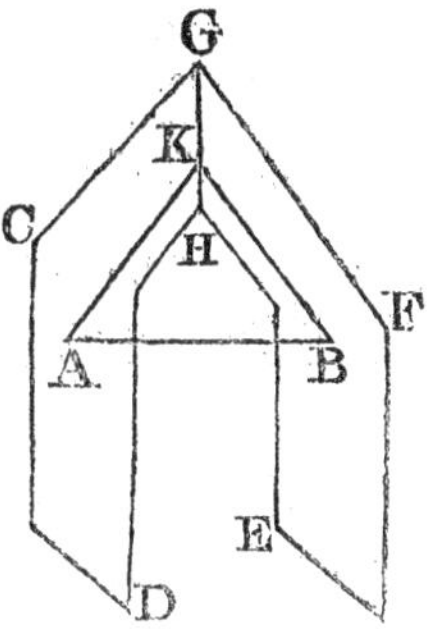

PROP. XIII. THEOR.

If two straight lines meeting one another, be parallel to two straight lines which also meet one another, but are not in the same plane with the first two: the plane which passes through the first two is parallel to the plane passing through the others.

Let AB, BC, two straight lines meeting one another, be parallel to DE, EF that meet one another, but are not in the same plane with AB, BC: The planes through AB, BC, and DE, EF shall not meet, though produced.

From the point B draw BG perpendicular (10. 2. Sup.) to the plane which passes through DE, EF, and let it meet that plane in G; and through G draw GH parallel to ED (31. 1.), and GK parallel to EF:

And because BG is perpendicular to the plane through DE, EF, it must make right angles with every straight line meeting it in that plane (1. def. 2. Sup.). But the straight lines GH, GK, in that plane meet it: Therefore each of the angles BGH, BGK is a right angle: And because BA is parallel (8. 2. Sup.) to GH (for each of them is parallel to DE), the angles GBA, BGH are together equal (29. 1.) to two right angles: And BGH is a right angle; therefore also GBA is a right angle, and GB perpendicular to BA: For the same reason, GB is perpendicular to BC: Since, therefore, the straight line GB stands at right angles to the two straight lines BA, BC, that cut one another in B; GB is perpendicular (4. 2. Sup.) to the plane through BA, BC: And it is perpendicular to the plane through DE, EF; therefore BG is perpendicular to each of the planes through AB, BC, and DE, EF: But planes to which the same straight line is perpendicular, are parallel (12. 2. Sup.) to one another: Therefore the plane through AB, BC, is parallel to the plane through DE, EF. Wherefore, if two straight lines, &c. Q. E. D.

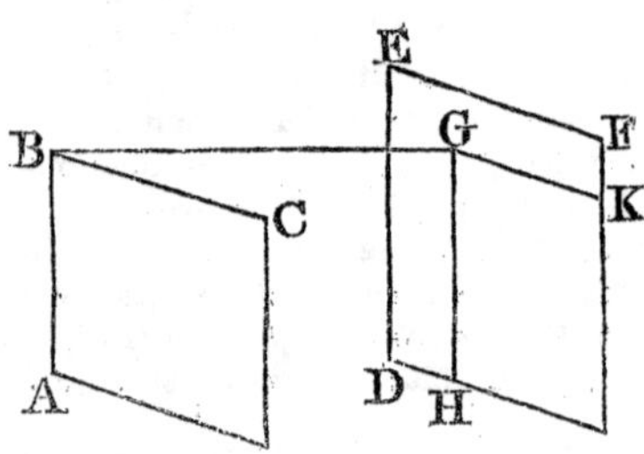

Cor. It follows from this demonstration, that if a straight line meet two parallel planes, and be perpendicular to one of them, it must be perpendicular to the other also.

PROP. XIV. THEOR.

If two parallel planes be cut by another plane, their common sections with it are parallels.

Let the parallel planes AB, CD be cut by the plane EFHG, and let their common sections with it be EF, GH; EF is parallel to GH.

For the straight lines EF and GH are in the same plane, viz. EFHG which cuts the planes AB and CD; and they do not meet though produced; for the planes in which they are do not meet; therefore EF and GH are parallel (def. 30. 1.). Q. E. D.

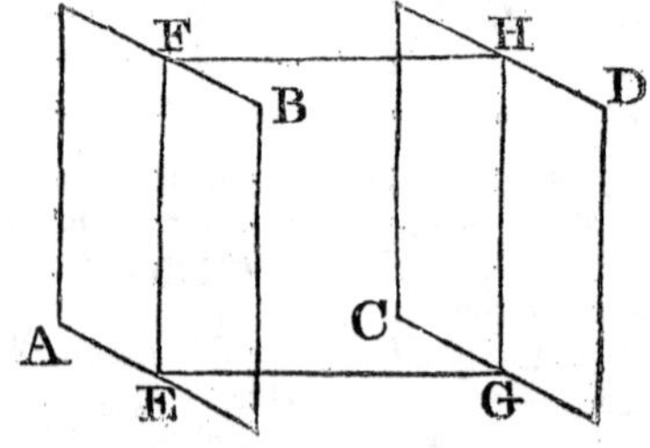

PROP. XV. THEOR.

If two parallel planes be cut by a third plane, they have the same inclination to that plane.

Let AB and CD be two parallel planes, and EH a third plane cutting them: The planes AB and CD are equally inclined to EH.

Let the straight lines EF and GH be the common section of the plane EH with the two planes AB and CD; and from K, any point in EF, draw in the plane EH the straight line KM at right angles to EF, and let it meet GH in L; draw also KN at right angles to EF in the plane AB: and through the straight lines KM, KN, let a plane be made to pass, cutting the plane CD in the line LO. And because EF and GH are the common sections of the plane EH with the two parallel planes AB and CD, EF is parallel to GH (14. 2. Sup.). But EF is at right angles to the plane that passes through KN and KM (4. 2. Sup.), because it is at right angles to the lines KM and KN:

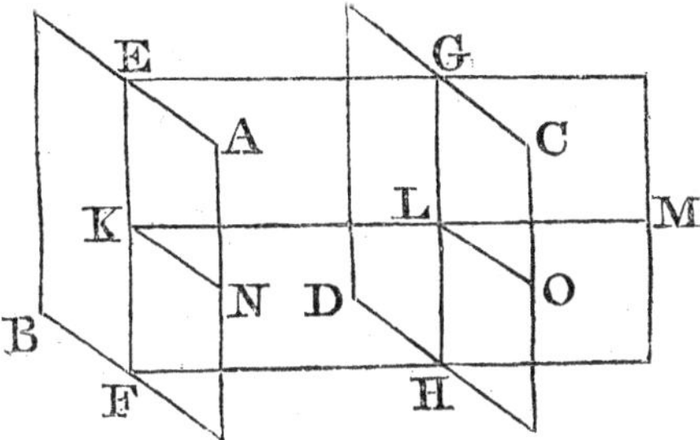

therefore GH is also at right angles to the same plane (7. 2. Sup.), and it is therefore at right angles to the lines LM, LO which it meets in that plane. Therefore, since LM and LO are at right angles to LG, the common section of the two planes CD and EH, the angle OLM is the inclination of the plane CD to the plane EH (4. def. 2. Sup.). For the same reason the angle MKN is the inclination of the plane AB to the plane EH. But because KN and LO are parallel, being the common sections of the parallel planes AB and CD with a third plane, the interior angle NKM is equal to the exterior angle OLM (29. 1.); that is, the inclination of the plane AB to the plane EH, is equal to the inclination of the plane CD to the same plane EH. Therefore, &c. Q. E. D.

PROP. XVI. THEOR.

If two straight lines be cut by parallel planes, they must be cut in the same ratio.

Let the straight lines AB, CD be cut by the parallel planes GH, KL, MN, in the points A, E, B; C, F, D: As AE is to EB, so is CF to FD.

Join AC, BD, AD, and let AD meet the plane KL in the point X; and join EX, XF: Because the two parallel planes KL, MN are cut by the plane EBDX, the common sections EX, BD, are parallel (14. 2. Sup.). For the same reason, because the two parallel planes GH, KL are cut by the plane AXFC, the common sections AC, XF are parallel: And because EX is parallel to BD, a side of the triangle ABD, as AE to EB, so is (2. 6.) AX to XD. Again, because XF is parallel AC, a side of the triangle ADC, as AX to XD, so is CF to FD: and it was proved that AX is to XD, as AE to EB: Therefore (11. 5.), as AE to EB, so is CF to FD. Wherefore, if two straight lines, &c. Q. E. D.

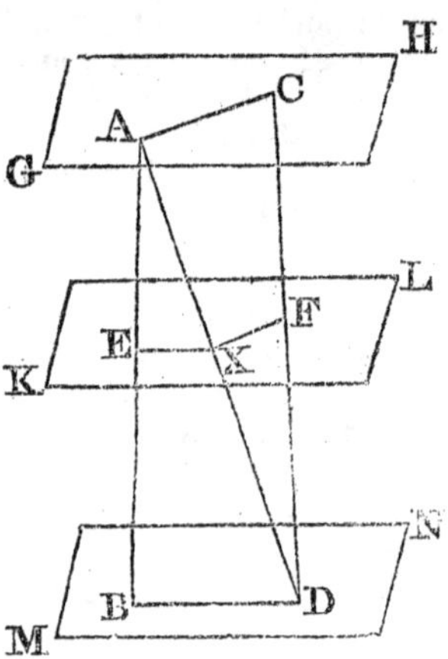

PROP. XVII. THEOR.

If a straight line be at right angles to a plane, every plane which passes through that line is at right angles to the first mentioned plane.

Let the straight line AB be at right angles to a plane CK; every plane which passes through AB is at right angles to the plane CK.

Let any plane DE pass through AB, and let CE be the common section of the planes DE, CK; take any point F in CE, from which draw FG in the plane DE at right angles to CE: And because AB is perpendicular to the plane CK, therefore it is also perpendicular to every straight line meeting it in that plane (1. def. 2. Sup.); and consequently it is perpendicular to CE: Wherefore ABF is a right angle; But GFB is likewise a right angle; therefore AB is parallel (28. 1.) to FG. And AB is at right angles to the plane CK: therefore FG is

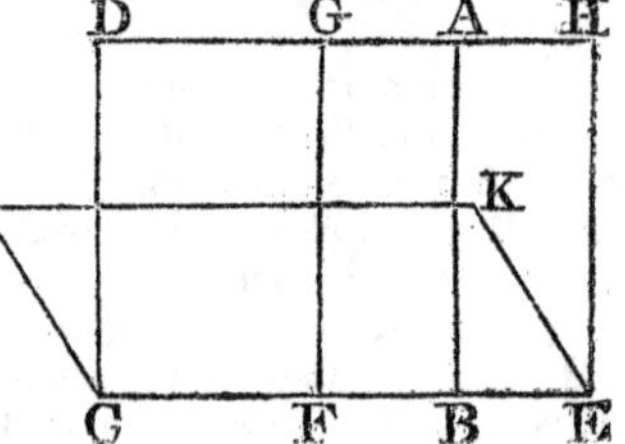

also at right angles to the same plane (7. 2. Sup.) But one plane is at right angles to another plane when the straight lines drawn in one of the planes, at right angles to their common section, are also at right angles to the other plane (def. 2. 2.); and any straight line FG in the plane DE, which is at right angles to CE, the common section of the planes, has been proved to be perpendicular to the other plane CK; therefore the plane DE is at right angles to the plane CK. In like manner, it may be proved that all the planes which pass through AC are at right angles to the plane CK. Therefore, if a straight line, &c. Q. E. D.

PROP. XVIII. THEOR.

If two planes cutting one another be each of them perpendicular to a third plane, their common section is perpendicular to the same plane.

Let the two planes AB, BC be each of them perpendicular to a third plane, and BD be the common section of the first two; BD is perpendicular to the plane ADC.

From D in the plane ADC, draw DE perpendicular to AD, and DF to DC. Because DE is perpendicular to AD, the common section of the planes AB and ADC; and because the plane AB is at right angles to ADC, DE is at right angles to the plane AB (def. 2. 2. Sup.), and therefore also to the straight line BD in that plane (def. 1. 2. Sup.). For the same reason, DF is at right angles to DB. Since BD is therefore at right angles to both the lines DE and DF, it is at right angles to the plane in which DE and DF are, that is, to the plane ADC (4. 2. Sup.). Wherefore, &c. Q. E. D.

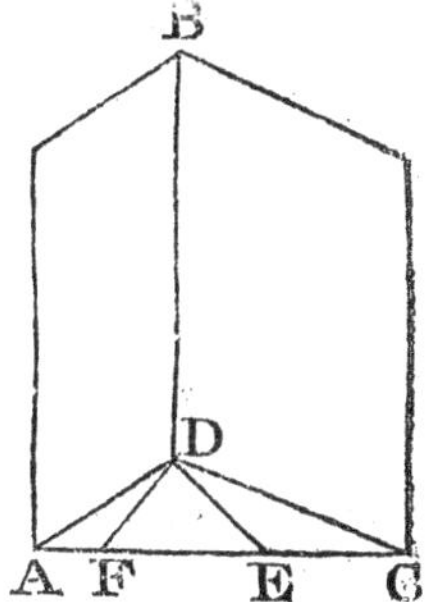

PROP. XIX. THEOR.

Two straight lines not in the same plane being given in position, to draw a straight line perpendicular to them both.

Let AB and CD be the given lines, which are not in the same plane; it is required to draw a straight line which shall be perpendicular both to AB and CD.

In AB take any point E, and through E draw EF parallel to CD, and let EG be drawn perpendicular to the plane which passes through EB, BF (10. 2. Sup.). Through AB and EG let a plane pass, viz. GK.

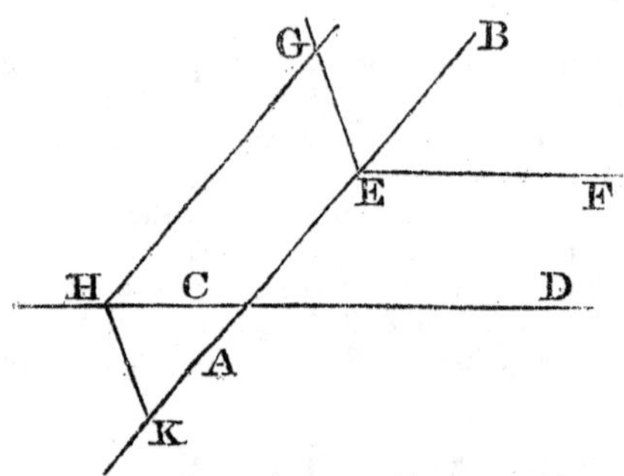

and let this plane meet CD in H; from H draw HK perpendicular to AB; and HK is the line required. Through H, draw HG parallel to AB.

Then, since HK and GE, which are in the same plane, are both at right angles to the straight line AB, they are parallel to one another. And because the lines HG, HD are parallel to the lines EB, EF, each to each, the plane GHD is parallel to the plane (13. 2. Sup.) BEF; and therefore EG, which is perpendicular to the plane BEF, is perpendicular also to the plane (Cor. 13. 2. Sup.) GHD. Therefore HK, which is parallel to GE, is also perpendicular to the plane GHD (7. 2. Sup.), and it is therefore perpendicular to HD (def. 1. 2. Sup.), which is in that plane, and it is also perpendicular to AB; therefore HK is drawn perpendicular to the two given lines, AB and CD. Which was to be done.

PROP. XX. THEOR.

If a solid angle be contained by three plane angles, any two of these angles are greater than the third.

Let the solid angle at A be contained by the three plane angles BAC, CAD, DAB. Any two of them are greater than the third.

If the angles BAC, CAD, DAB be all equal, it is evident that any two of them are greater than the third. But if they are not, let BAC be that angle which is not less than either of the other two, and is greater than one of them, DAB; and at the point A in the straight line AB, make in the plane which passes through BA, AC, the angle BAE equal (23. 1.) to the angle DAB; and make AE equal to AD, and through E draw BEC cutting AB, AC in the points B, C, and join DB, DC. And because DA is equal to AE, and AB is common to the two triangles ABD, ABE, and also the angle DAB equal to the angle EAB; therefore the base DB is equal

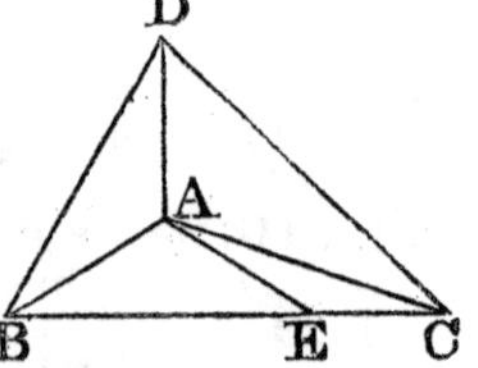

(4. 1.) to the base BE. And because BD, DC are greater (20. 1.) than CB, and one of them BD has been proved equal to BE, a part of CB, therefore the other DC is greater than the remaining part EC. And because DA is equal to AE, and AC common, but the base DC greater than the base EC; therefore the angle DAC is greater (25. 1.) than the angle EAC; and, by the construction, the angle DAB is equal to the angle BAE; wherefore the angles DAB, DAC are together greater than BAE, EAC, that is, than the angle BAC. But BAC is not less than either of the angles DAB, DAC; therefore BAC, with either of them, is greater than the other. Wherefore, if a solid angle, &c. Q. E. D.

PROP. XXI. THEOR.

The plane angles which contain any solid angle are together less than four right angles.

Let A be a solid angle contained by any number of plane angles BAC, CAD, DAE, EAF, FAB; these together are less than four right angles.

Let the planes which contain the solid angle at A be cut by another plane, and let the section of them by that plane be the rectilineal figure BCDEF. And because the solid angle at B is contained by three plane angles CBA, ABF, FBC, of which any two are greater (20. 2. Sup.) than the third, the angles CBA, ABF are greater than the angle FBC: For the same reason, the two plane angles at each of the points C, D, E, F, viz. the angles which are at the bases of the triangles having the common vertex A, are greater than the third angle at the same point, which is one of the angles of the figure BCDEF: therefore all the angles at the bases of the triangles are together greater than all the angles of the figure: and because all the angles of the triangles are together equal to twice as many right angles as there are triangles (32. 1); that is, as there are sides in the figure BCDEF; and because all the angles of the figure, together with four right angles, are likewise equal to twice as many right angles as there are sides in the figure (cor. 1. 32. 1.); therefore all the angles of the triangles are equal to all the angles of the rectilineal figure, together with four right angles. But all the angles at the bases of the triangles are greater than all the angles of the rectilineal, as has been proved. Wherefore, the remaining angles of the triangles, viz. those at the vertex, which contain the solid angle at A, are less than four right angles. Therefore every solid angle, &c. Q. E. D.

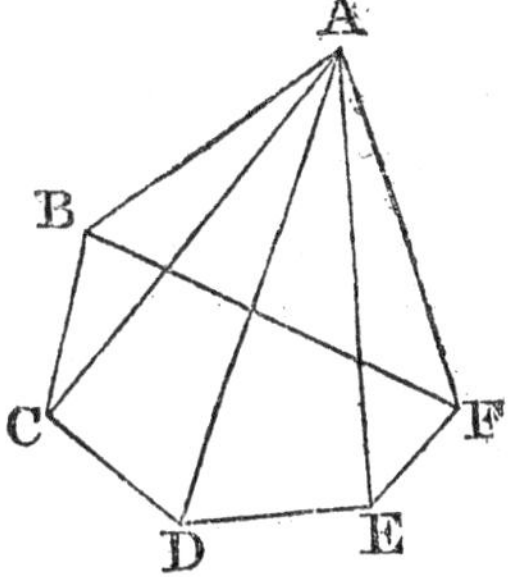

Otherwise.

Let the sum of all the angles at the bases of the triangles $=S$; the sum of all the angles of the rectilineal figure BCDEF$=\Sigma$; the sum of the plane angles at $A=X$, and let $R=$ a right angle.

Then, because $S+X=$twice (32. 1.) as many right angles as there are triangles, or as there are sides of the rectilineal figure BCDEF, and as $\Sigma+4R$ is also equal to twice as many right angles as there are sides of the same figure; therefore $S+X=\Sigma+4R$. But because of the three plane angles which contain a solid angle, any two are greater than the third, $S > \Sigma$; and therefore $X < 4R$; that is, the sum of the plane angles which contain the solid angle at A is less than four right angles. Q. E. D.

SCHOLIUM.

It is evident, that when any of the angles of the figure BCDEF is exterior, like the angle at D, in the annexed figure, the reasoning in the above proposition does not hold, because the solid angles at the base are not all contained by plane angles, of which two belong to the triangular planes, having their common vertex in A, and the third is an interior angle of the rectilineal figure, or base. Therefore, it cannot be concluded that S is necessarily greater than Σ. This proposition, therefore, is subject to a limitation, which is farther explained in the notes on this book.

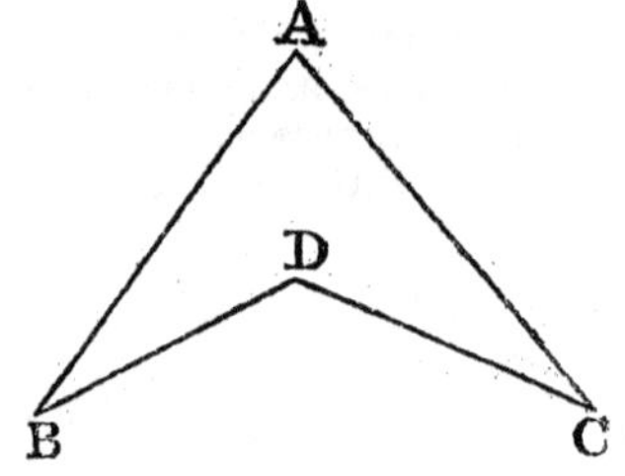

ELEMENTS

OF

GEOMETRY.

SUPPLEMENT.

BOOK III.

OF THE COMPARISON OF SOLIDS.

DEFINITIONS.

I.

A Solid is that which has length, breadth, and thickness.

II.

Similar solid figures are such as are contained by the same number of similar planes similarly situated, and having like inclinations to one another.

III.

A pyramid is a solid figure contained by planes that are constituted betwixt one plane and a point above it in which they meet.

IV.

A prism is a solid figure contained by plane figures, of which two that are opposite are equal, similar, and parallel to one another; and the others are parallelograms.

V.

A parallelopiped is a solid figure contained by six quadrilateral figures, whereof every opposite two are parallel.

VI.

A cube is a solid figure contained by six equal squares.

VII.

A sphere is a solid figure described by the revolution of a semicircle about a diameter, which remains unmoved.

VIII.

The axis of a sphere is the fixed straight line about which the semicircle revolves.

IX.

The centre of a sphere is the same with that of the semicircle.

X.

The diameter of a sphere is any straight line which passes through the centre, and is terminated both ways by the superficies of the sphere.

XI.

A cone is a solid figure described by the revolution of a right angled triangle about one of the sides containing the right angle, which side remains fixed.

XII.

The axis of a cone is the fixed straight line about which the triangle revolves.

XIII.

The base of a cone is the circle described by that side, containing the right angle, which revolves.

XIV.

A cylinder is a solid figure described by the revolution of a right angled parallelogram about one of its sides, which remains fixed.

XV.

The axis of a cylinder is the fixed straight line about which the parallelogram revolves.

XVI.

The bases of a cylinder are the circles described by the two revolving opposite sides of the parallelogram.

XVII.

Similar cones and cylinders are those which have their axes, and the diameters of their bases proportionals.

PROP. I. THEOR.

If two solids be contained by the same number of equal and similar planes similarly situated, and if the inclination of any two contiguous planes in the one solid be the same with the inclination of the two equal, and similarly situated planes in the other, the solids themselves are equal and similar.

Let AG and KQ be two solids contained by the same number of equal and similar planes, similarly situated so that the plane AC is similar and equal to the plane KM, the plane AF to the plane KP; BG to LQ, GD to QN, DE to NO, and FH to PR Let also the inclination of the plane AF to the plane AC be the same with that of the plane KP to the plane KM, and so of the rest; the solid KQ is equal and similar to the solid AG.

Let the solid KQ be applied to the solid AG, so that the bases KM

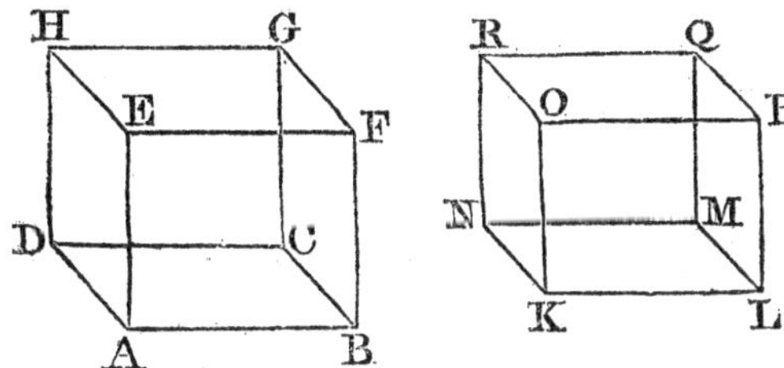

and AC, which are equal and similar, may coincide (8. Ax. 1.), the point N coinciding with the point D, K with A, L with B, and so on. And because the plane KM coincides with the plane AC, and, by hypothesis, the inclination of KR to KM is the same with the inclination of AH to AC, the plane KR will be upon the plane AH, and will coincide with it, because they are similar and equal (8. Ax. 1.), and because their equal sides KN and AD coincide. And in the same manner it is shewn that the other planes of the solid KQ coincide with the other planes of the solid AG, each with each: wherefore the solids KQ and AG do wholly coincide, and are equal and similar to one another. Therefore, &c. Q. E. D.

PROP. II. THEOR.

If a solid be contained by six planes, two and two of which are parallel, the opposite planes are similar and equal parallelograms.

Let the solid CDGH be contained by the parallel planes AC, GF; BG, CE; FB, AE: its opposite planes are similar and equal parallelograms.

Because the two parallel planes BG, CF, are cut by the plane AC their common sections AB, CD are parallel (14. 2. Sup.). Again, because the two parallel planes BF, AE are cut by the plane AC, their common sections AD, BC are parallel (14. 2. Sup.): and AB is parallel to CD; therefore AC is a parallelogram. In like manner, it may be proved that each of the figures CE, FG, GB, BF, AE is a parallelogram; join AH, DF; and because AB is parallel to DC, and BH to CF; the two straight lines AB, BH, which meet one another, are parallel to DC and CF, which meet one another; wherefore, though the first two are not in the same plane with the other two, they contain equal angles (9. 2. Sup.); the angle ABH is therefore equal to the angle DCF. And because AB, BH, are equal to DC, CF, and the angle ABH equal to the angle DCF; therefore the base AH is equal (4. 1.) to the base DF, and the triangle ABH to the triangle DCF: For the same reason, the triangle AGH is equal to the triangle DEF: and therefore the parallelogram BG is equal and similar to the parallelogram CE. In the same manner, it may be proved, that the parallelogram AC is equal and similar to the parallelogram GF, and the parallelogram AE to BF. Therefore, if a solid, &c. Q. E. D.

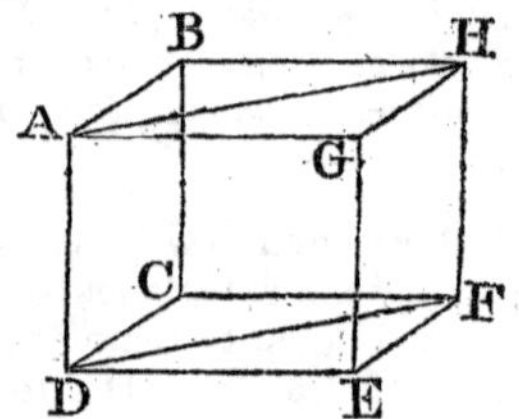

PROP. III. THEOR.

If a solid parallelopiped be cut by a plane parallel to two of its opposite planes, it will be divided into two solids, which will be to one another as the bases.

Let the solid parallelopiped ABCD be cut by the plane EV, which is parallel to the opposite planes AR, HD, and divides the whole into the solids ABFV, EGCD; as the base AEFY to the base EHCF, so is the solid ABFV to the solid EGCD.

Produce AH both ways, and take any number of straight lines HM, MN, each equal to EH, and any number AK, KL each equal to EA, and complete the parallelograms LO, KY, HQ, MS, and the solids LP,

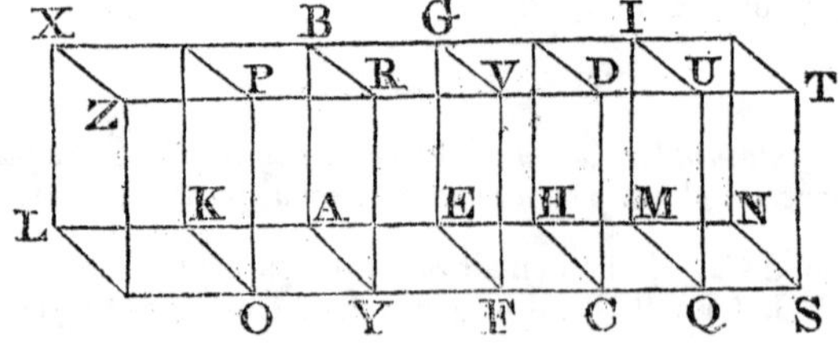

KR, HU, MT; then, because the straight lines LK, KA, AE are all

equal, and also the straight lines KO, AY, EF which make equal angles with LK, KA, AE, the parallelograms LO, KY, AF are equal and similar (36. 1. & def. 1. 6.): and likewise the parallelograms KX, KB, AG; as also (2. 3. Sup.) the parallelograms LZ, KP, AR, because they are opposite planes. For the same reason, the parallelograms EC, HQ, MS are equal (36. 1. & def. 1. 6.); and the parallelograms HG, HI, IN, as also (2. 3. Sup.) HD, MU, NT; therefore three planes of the solid LP, are equal and similar to three planes of the solid KR, as also to three planes of the solid AV: but the three planes opposite to these three are equal and similar to them (2. 3. Sup.) in the several solids; therefore the solids LP, KR, AV are contained by equal and similar planes. And because the planes LZ, KP, AR are parallel, and are cut by the plane XV, the inclination of LZ to XP is equal to that of KP to PB; or of AR to BV (15. 2. Sup.) and the same is true of the other contiguous planes, therefore the solids LP, KR, and AV, are equal to one another (1. 3. Sup.). For the same reason, the three solids ED, HU, MT are equal to one another; therefore what multiple soever the base LF is of the base AF, the same multiple is the solid LV of the solid AV; for the same reason, whatever multiple the base NF is of the base HF, the same multiple is the solid NV of the solid ED: And if the base LF be equal to the base NF, the solid LV is equal (1. 3. Sup.) to the solid NV; and if the base LF be greater than the base NF, the solid LV is greater than the solid NV: and if less, less. Since then there are four magnitudes, viz. the two bases AF, FH, and the two solids AV, ED, and of the base AF and solid AV, the base LF and solid LV are any equimultiples whatever; and of the base FH and solid ED, the base FN and solid NV are any equimultiples whatever; and it has been proved, that if the base LF is greater than the base FN, the solid LV is greater than the solid NV; and if equal, equal; and if less, less: Therefore (def. 5. 5.), as the base AF is to the base FH, so is the solid AV to the solid ED. Wherefore, if a solid, &c. Q. E. D.

Cor. Because the parallelogram AF is to the parallelogram FH as YF to FC (1. 6.), therefore the solid AV is to the solid ED as YF to FC.

PROP. IV. THEOR.

If a solid parallelopiped be cut by a plane passing through the diagonals of two of the opposite planes, it will be cut into two equal prisms.

Let AB be a solid parallelopiped, and DE, CF the diagonals of the opposite parallelograms AH, GB, viz. those which are drawn betwixt the equal angles in each; and because CD, FE are each of them parallel to GA, though not in the same plane with it, CD, FE are parallel (8. 2. Sup.); wherefore the diagonals CF, DE are in the plane in which the parallels are, and are themselves parallels (14. 2. Sup.): the plane CDEF cuts the solid AB into two equal parts.

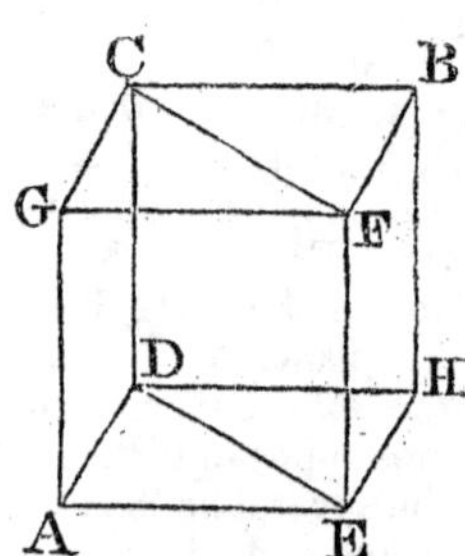

Because the triangle CGF is equal (34. 1.) to the triangle CBF, and the triangle DAE to DHE; and since the parallelogram CA is equal (2. 3. Sup.) and similar to the opposite one BE; and the parallelogram GE to CH: therefore the planes which contain the prisms CAE, CBE, are equal and similar, each to each; and they are also equally inclined to one another, because the planes AC, EB are parallel, as also AF and BD, and they are cut by the plane CE (15. 2. Sup.). Therefore the prism CAE is equal to the prism CBE (1. 3. Sup.), and the solid AB is cut into two equal prisms by the plane CDEF. Q. E. D.

N. B. The insisting straight lines of a parallelopiped, mentioned in the following propositions, are the sides of the parallelograms betwixt the base and the plane parallel to it.

PROP. V. THEOR.

Solid parallelopipeds upon the same base, and of the same altitude, the insisting straight lines of which are terminated in the same straight lines in the plane opposite to the base, are equal to one another.

Let the solid parallelopipeds AH, AK, be upon the same base AB, and of the same altitude, and let their insisting straight lines AF, AG, LM, LN, be terminated in the same straight line FN, and let the insisting lines CD, CE, BH, BK be terminated in the same straight line DK; the solid AH is equal to the solid AK.

Because CH, CK are parallelograms, CB is equal (34. 1.) to each of the opposite sides DH, EK: wherefore DH is equal to EK: add, or take away the common part HE; then DE is equal to HK: Wherefore also the triangle CDE is equal (38. 1.) to the triangle BHK: and the parallelogram DG is equal (36. 1.) to the parallelogram HN. For the same reason, the triangle AFG is equal to the triangle LMN, and the parallelogram CF is equal (2. 3. Sup.) to the parallelogram BM, and CG to BN; for they are opposite. Therefore the planes which

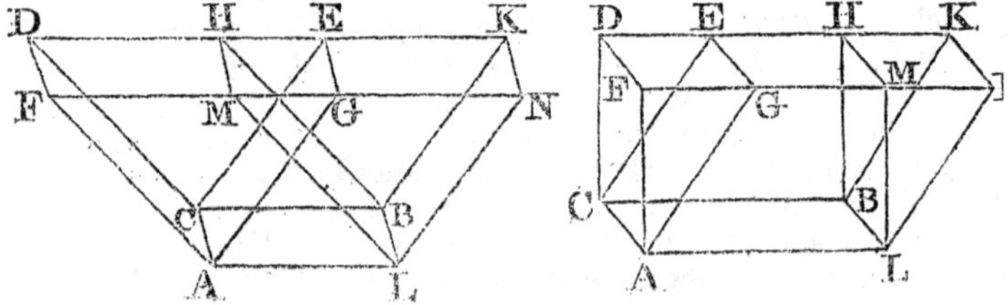

contain the prism DAG are similar and equal to those which contain the prism HLN, each to each; and the contiguous planes are also equally inclined to one another (15. 2. Sup.), because that the parallel planes AD and LH, as also AE and LK, are cut by the same plane DN: therefore the prisms DAG, HLN are equal (1. 3. Sup.). If therefore the prism LNH be taken from the solid, of which the base is the parallelogram AB, and FDKN the plane opposite to the base; and if from this same solid there be taken the prism AGD, the remaining solid, viz. the parallelopiped AH is equal to the remaining parallelopiped AK. Therefore solid parallelopipeds, &c. Q. E. D.

PROP. VI. THEOR.

Solid parallelopipeds upon the same base, and of the same altitude, the insisting straight lines of which are not terminated in the same straight lines in the plane opposite to the base, are equal to one another.

Let the parallelopipeds CM, CN, be upon the same base AB, and of the same altitude, but their insisting straight lines AF, AG, LM, LN, CD, CE, BH, BK, not terminated in the same straight lines; the solids CM, CN are equal to one another.

Produce FD, MH, and NG, KE, and let them meet one another in the points O, P, Q, R; and join AO, LP, BQ, CR. Because the planes (def. 5. 3. Sup.), LBHM and ACDF are parallel, and because the

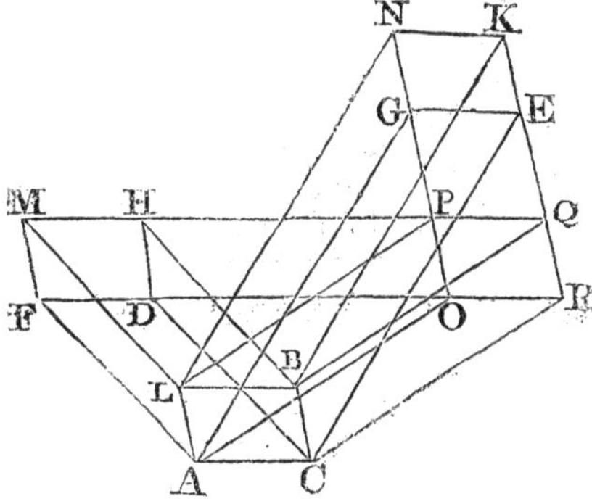

plane LBHM is that in which are the parallels LB, MHPQ (def. 5. 3. Sup.), and in which also is the figure BLPQ; and because the plane ACDF is that in which are the parallels AC, FDOR, and in which also is the figures CAOR; therefore the figures BLPQ, CAOR, are in parallel planes. In like manner, because the planes ALNG and CBKE are parallel, and the plane ALNG is that in which are the parallels AL, OPGN, and in which also is the figure ALPO; and the plane CBKE is that in which are the parallels CB, RQEK, and in which also is the figure CBQR; therefore the figures ALPO, CBQR are in parallel planes. But the planes ACBL, ORQP are also parallel; therefore the solid CP is a parallelopiped. Now the solid parallelopiped CM is equal (5. 2. Sup.) to the solid parallelopiped CP; because they are upon the same base, and their insisting straight lines AF, AO, CD, CR; LM, LP, BH, BQ are terminated in the same straight lines FR, MQ: and the solid CP is equal (5. 2. Sup.) to the solid CN; for they are upon the same base ACBL, and their insisting straight lines AO, AG, LP, LN; CR, CE, BQ, BK are terminated in the same straight lines ON, RK; Therefore the solid CM is equal to the solid CN. Wherefore solid parallelopipeds, &c. Q. E. D.

PROP. VII. THEOR.

Solid parallelopipeds which are upon equal bases, and of the same altitude, are equal to one another.

Let the solid parallelopipeds, AE, CF, be upon equal bases AB, CD, and be of the same altitude; the solid AE is equal to the solid CF.

Case 1. Let the insisting straight lines be at right angles to the bases AB, CD, and let the bases be placed in the same plane, and so as that the sides CL, LB, be in a straight line; therefore the straight line LM, which is at right angles to the plane in which the bases are, in the point L, is common (11. 2. Sup.) to the two solids AE, CF; let the other insisting lines of the solids be AG, HK, BE; DF, OP, CN: and first, let the angle ALB be equal to the angle CLD; then AL, LD are in a straight line (14. 1.). Produce OD, HB, and let them meet in Q and complete the solid parallelopiped LR, the base of which is the parallelogram LQ, and of which LM is one of its insisting straight lines: therefore, because the parallelogram AB is equal to CD, as the base AB is to the base LQ, so is (7. 5.) the base CD to the same LQ: and because the solid parallelopiped AR is cut by the plane LMEB, which is parallel to the opposite planes AK, DR; as the base AB is to the

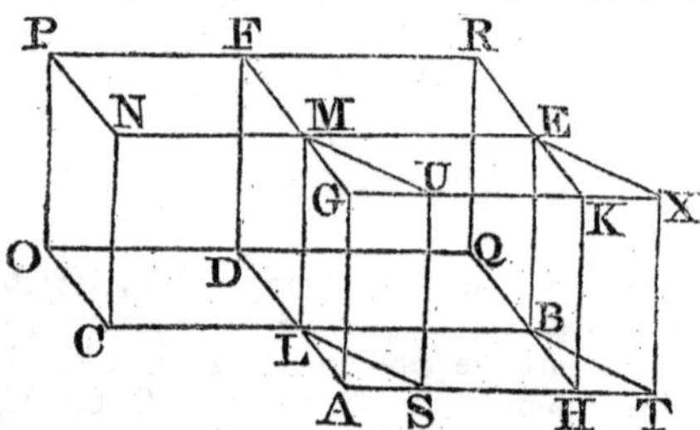

base LQ, so is (3. 3. Sup.) the solid AE to the solid LR: for the same reason because the solid parallelopiped CR is cut by the plane LMFD, which is parallel to the opposite planes CP, BR; as the base CD to the base LQ; so is the solid CF to the solid LR; but as the base AB to the base LQ, so the base CD to the base LQ, as has been proved: therefore as the solid AE to the solid LR, so is the solid CF to the solid LR; and therefore the solid AE is equal (9. 5.) to the solid CF.

But let the solid parallelopipeds, SE, CF be upon equal bases SB, CD, and be of the same altitude, and let their insisting straight lines be at right angles to the bases; and place the bases SB, CD in the same plane, so that CL, LB be in a straight line; and let the angles SLB, CLD be unequal; the solid SE is also in this case equal to the solid CF. Produce DL, TS until they meet in A, and from B draw BH parallel to DA; and let HB, OD produced meet in Q, and complete the solids AE, LR: therefore the solid AE, of which the base is the parallelogram LE, and AK the plane opposite to it, is equal (5. 3. Sup.) to the solid SE, of which the base is LE, and SX the plane opposite; for they are upon the same base LE, and of the same altitude, and their insisting straight lines, viz. LA, LS, BH, BT; MG, MU, EK, EX, are in the same straight lines AT, GX: and because the parallelogram AB is equal (35. 1.) to SB, for they are upon the same base LB, and between the same parallels LB, AT; and because the base SB is equal to the base CD; therefore the base AB is equal to the base CD; but the angle ALB is equal to the angle CLD: therefore, by the first case, the solid AE is equal to the solid CF; but the solid AE is equal to the solid SE, as was demonstrated; therefore the solid SE is equal to the solid CF.

Case 2. If the insisting straight lines AG, HK, BE, LM; CN, RS,

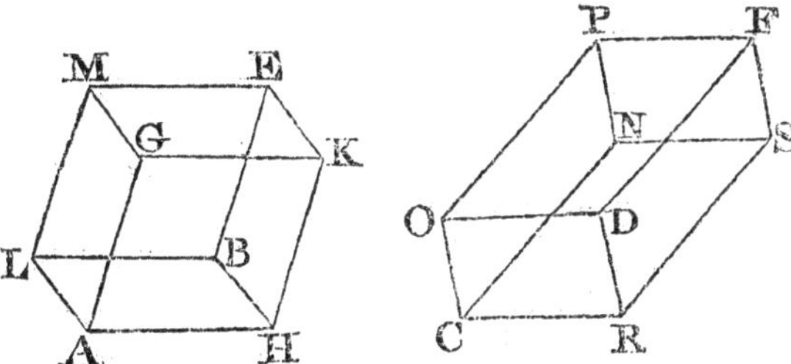

DF, OP, be not at right angles to the bases AB, CD; in this case likewise the solid AE is equal to the solid CF. Because solid parallelopipeds on the same base, and of the same altitude, are equal (6. 3. Sup.), if two solid parallelopipeds be constituted on the bases AB and CD of the same altitude with the solids AE and CF, and with their insisting lines perpendicular to their bases, they will be equal to the solids AE and CF; and, by the first case of this proposition, they will be equal to one another; wherefore, the solids AE and CF are also equal. Wherefore, solid parallelopipeds, &c. Q. E. D.

PROP. VIII. THEOR.

Solid parallelopipeds which have the same altitude, are to one another as their bases.

Let AB, CD be solid parallelopipeds of the same altitude: they are to one another as their bases; that is, as the base AE to the base CF, so is the solid AB to the solid CD.

To the straight line FG apply the parallelogram FH equal (Cor. 45. 1.) to AE, so that the angle FGH be equal to the angle LCG;

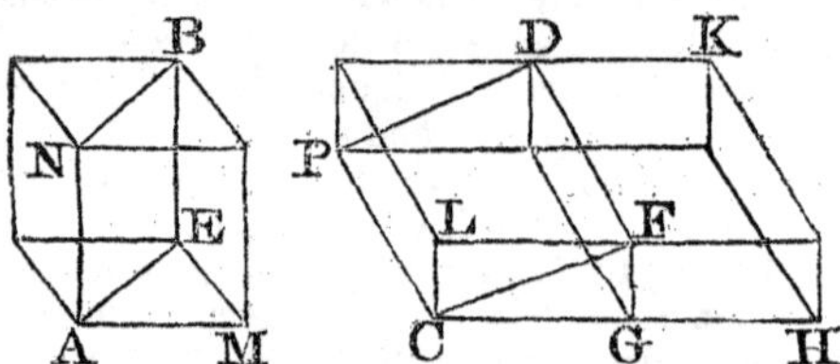

and complete the solid parallelopiped GK upon the base FH, one of whose insisting lines is FD, whereby the solids CD, GK must be of the same altitude. Therefore the solid AB is equal (7. 3. Sup.) to the solid GK, because they are upon equal bases AE, FH, and are of the same altitude: and because the solid parallelopiped CK is cut by the plane DG which is parallel to its opposite planes, the base HF is (3. 3. Sup.) to the base FC, as the solid HD to the solid DC: But the base HF is equal to the base AE, and the solid GK to the solid AB: therefore, as the base AE to the base CF, so is the solid AB to the solid CD. Wherefore solid parallelopipeds, &c. Q. E. D.

Cor. 1. From this it is manifest, that prisms upon triangular bases, and of the same altitude, are to one another as their bases. Let the prisms BNM, DPG, the bases of which are the triangles AEM, CFG, have the same altitude; complete the parallelograms AE, CF, and the solid parallelopipeds AB, CD, in the first of which let AN, and in the other let CP be one of the insisting lines. And because the solid parallelopipeds AB, CD have the same altitude, they are to one another as the base AE is to the base CF; wherefore the prisms, which are their halves (4. 3. Sup.) are to one another, as the base AE to the base CF; that is, as the triangle AEM to the triangle CFG.

Cor. 2. Also a prism and a parallelopiped, which have the same altitude, are to one another as their bases; that is, the prism BNM is to the parallelopiped CD as the triangle AEM to the parallelogram LG. For by the last Cor. the prism BNM is to the prism DPG as the triangle AME to the triangle CGF, and therefore the prism BNM is to twice the prism DPG as the triangle AME to twice the triangle CGF

(4. 5.); that is, the prism BNM is to the parallelopiped CD as the triangle AME to the parallelogram LG.

PROP. IX. THEOR.

Solid parallelopipeds are to one another in the ratio that is compounded of the ratios of the areas of their bases, and of their altitudes.

Let AF and GO be two solid parallelopipeds, of which the bases are the parallelograms AC and GK, and the altitudes, the perpendiculars let fall on the planes of these bases from any point in the opposite planes EF and MO; the solid AF is to the solid GO in a ratio compounded of the ratios of the base AC to the base GK, and of the perpendicular on AC, to the perpendicular on GK.

Case 1. When the insisting lines are perpendicular to the bases AC and GK, or when the solids are upright.

In GM, one of the insisting lines of the solid GO, take GQ equal to AE, one of the insisting lines of the solid AF, and through Q let a plane pass parallel to the plane GK, meeting the other insisting lines

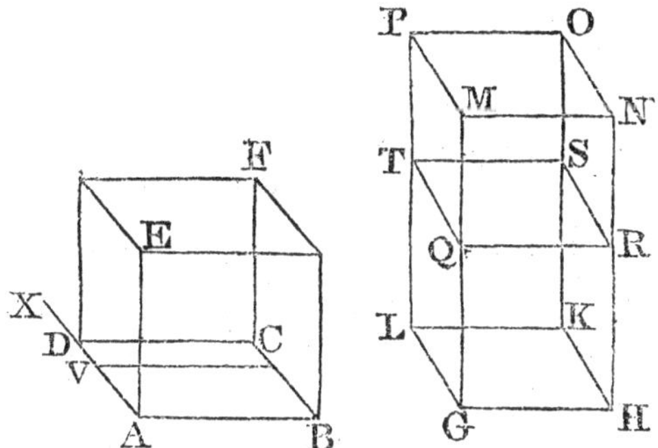

of the solid GO in the points R, S and T. It is evident that GS is a solid parallelopiped (def. 5. 3. Sup.), and that it has the same altitude with AF, viz. GQ or AE. Now the solid AF is to the solid GO in a ratio compounded of the ratios of the solid AF to the solid GS (def. 10. 5.), and of the solid GS to the solid GO; but the ratio of the solid AF to the solid GS, is the same with that of the base AC to the base GK (8. 3. Sup.), because their altitudes AE and GQ are equal; and the ratio of the solid GS to the solid GO, is the same with that of GQ to GM (3. 2. Sup.); therefore, the ratio which is compounded of the ratios of the solid AF to the solid GS, and of the solid GS to the solid GO, is the same with the ratio which is compounded of the ratios of the base AC to the base GK, and of the altitude AE to the altitude GM (F. 5.). But the ratio of the solid AF to the solid GO, is that which is compounded of the ratios of AF to GS, and of GS to GO; therefore the ratio of the solid AF to the solid GO is compounded of the ratios of the base AC to the base GK, and of the altitude AE to the altitude GM.

Case 2. When the insisting lines are not perpendicular to the bases.

Let the parallelograms AC and GK be the bases as before, and let AE and GM be the altitudes of two parallelopipeds Y and Z on these bases. Then, if the upright parallelopipeds AF and GO be constituted on the bases AC and GK, with the altitudes AE and GM, they will be equal to the parallelopipeds Y and Z (7. 3. Sup.). Now, the solids AF and GO, by the first case, are in the ratio compounded of the ratios of the bases AC and GK, and of the altitudes AE and GM; therefore also the solids Y and Z have to one another a ratio that is compounded of the same ratios. Therefore, &c. Q. E. D.

Cor. 1. Hence, two straight lines may be found having the same ratio with the two parallelopipeds AF and GO. To AB, one of the sides of the parallelogram AC, apply the parallelogram BV equal to GK, having an angle equal to the angle BAD (44. 1.); and as AE to GM, so let AV be to AX (12. 6.), then AD is to AX as the solid AF to the solid GO. For the ratio of AD to AX is compounded of the ratios (def. 10. 5.) of AD to AV, and of AV to AX; but the ratio of AD to AV is the same with that of the parallelogram AC to the parallelogram BV (1. 6.) or GK; and the ratio of AV to AX is the same with that of AE to GM; therefore the ratio of AD to AX is compounded of the ratios of AC to GK, and of AE to GM (E. 5.). But the ratio of the solid AF to the solid GO is compounded of the same ratios; therefore, as AD to AX, so is the solid AF to the solid GO.

Cor. 2. If AF and GO are two parallelopipeds, and if to AB, to the perpendicular from A upon DC, and to the altitude of the parallelopiped AF, the numbers L, M, N be proportional: and if to AB, to GH, to the perpendicular from G on LK, and to the altitude of the parallelopiped GO, the numbers L, l, m, n be proportional; the solid AF is to the solid GO as $L\times M\times N$ to $l\times m\times n$.

For it may be proved, as in the 7th of the 1st of the Sup. that $L\times M\times N$ is to $l\times m\times n$ in the ratio compounded of the ratio of $L\times M$ to $l\times m$, and of the ratio of N to n. Now the ratio of $L\times M$ to $l\times m$ is that of the area of the parallelogram AC to that of the parallelogram GK; and the ratio of N to n is the ratio of the altitudes of the parallelopipeds, by hypothesis, therefore, the ratio of $L\times M\times N$ to $l\times m\times n$ is compounded of the ratio of the areas of the bases, and of the ratio of the altitudes of the parallelopipeds AF and GO; and the ratio of the parallelopipeds themselves is shewn, in this proposition, to be compounded of the same ratios; therefore it is the same with that of the product $L\times M\times N$ to the product $l\times m\times n$.

Cor. 3. Hence all prisms are to one another in the ratio compounded of the ratios of their bases, and of their altitudes. For every prism is equal to a parallelopiped of the same altitude with it, and of an equal base (2. Cor. 8. 3. Sup.).

PROP. X. THEOR.

Solid parallelopipeds, which have their bases and altitudes reciprocally proportional, are equal; and parallelopipeds which are equal, have their bases and altitudes reciprocally proportional.

Let AG and KQ be two solid parallelopipeds, of which the bases are AC and KM, and the altitudes AE and KO. and let AC be to KM as KO to AE; the solids AG and KQ are equal.

As the base AC to the base KM, so let the straight line KO be to the straight line S. Then, since AC is to KM as KO to S, and also

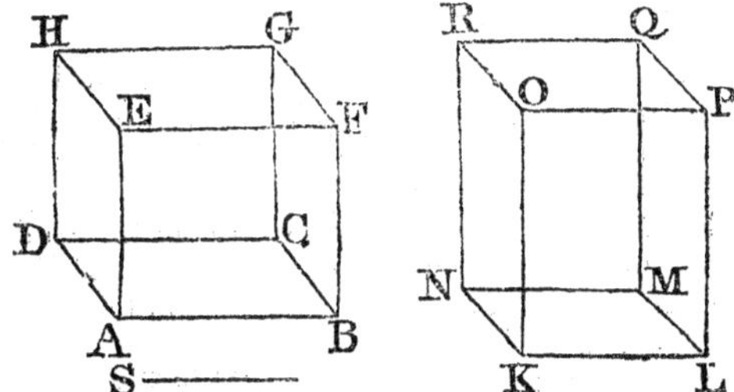

by hypothesis, AC to KM as KO to AE, KO has the same ratio to S that it has to AE (11. 5.); wherefore AF is equal to S (9. 5.). But the solid AG is to the solid KQ, in the ratio compounded of the ratios of AE to KO, and of AC to KM (9. 3. Sup.), that is, in the ratio compounded of the ratios of AE to KO, and of KO to S. And the ratio of AE to S is also compounded of the same ratios (def. 10. 5.); therefore, the solid AG has to the solid KQ the same ratio that AE has to S. But AE was proved to be equal to S, therefore AG is equal to KQ.

Again, if the solids AG and KQ be equal, the base AC is to the base KM as the altitude KO to the altitude AE. Take S, so that AC may be to KM as KO to S, and it will be shewn, as was done above, that the solid AG is to the solid KQ as AE to S; now, the solid AG is, by hypothesis, equal to the solid KQ: therefore, AE is equal to S; but, by construction, AC is to KM, as KO is to S; therefore, AC is to KM as KO to AE. Therefore, &c. Q. E. D.

Cor. In the same manner, it may be demonstrated, that equal prisms have their bases and altitudes reciprocally proportional, and conversely.

PROP. IX. THEOR.

Similar solid parallelopipeds are to one another in the triplicate ratio of their homologous sides.

Let AG, KQ be two similar parallelopipeds, of which AB and KL are two homologous sides; the ratio of the solid AG to the solid KQ is triplicate of the ratio of AB to KL.

Because the solids are similar, the parallelograms AF, KB are similar (def. 2. 3. Sup.), as also the parallelograms AH, KR; therefore,

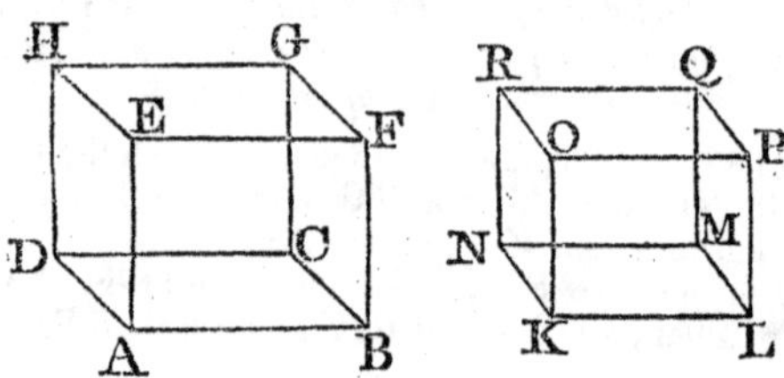

the ratios of AB to KL, of AE to KO, and of AD to KN are all equal (def. 1. 6.). But the ratio of the solid AG to the solid KQ is compounded of the ratios of AC to KM, and of AE to KO. Now, the ratio of AC to KM, because they are equiangular parallelograms, is compounded (23. 6.) of the ratios of AB to KL, and of AD to KN. Wherefore, the ratio of AG to KQ is compounded of the three ratios of AB to KL, AD to KN, and AE to KO; and the three ratios have already been proved to be equal; therefore, the ratio that is compounded of them, viz. the ratio of the solid AG to the solid KQ, is triplicate of any of them (def. 12. 5.): it is therefore triplicate of the ratio of AB to KL. Therefore, similar solid parallelopipeds, &c. Q. E. D.

Cor. 1. If as AB to KL, so KL to m, and as KL to m, so is m to n, then AB is to n as the solid AG to the solid KQ. For the ratio of AB to n is triplicate of the ratio of AB to KL (def. 12. 5.), and is therefore equal to that of the solid AG to the solid KQ.

Cor. 2. As cubes are similar solids, therefore the cube on AB is to the cube on KL in the triplicate ratio of AB to KL, that is in the same ratio with the solid AG, to the solid KQ. Similar solid parallelopipeds are therefore to one another as the cubes on their homologous sides.

Cor. 3. In the same manner it is proved, that similar prisms are to one another in the triplicate ratio, or in the ratio of the cubes of their homologous sides.

PROP. XII. THEOR.

If two triangular pyramids, which have equal bases and altitudes, be cut by planes that are parallel to the bases, and at equal distances from them, the sections are equal to one another.

Let ABCD and EFGH be two pyramds, having equal bases BDC and FGH, and equal altitudes, viz. the perpendiculars AQ, and ES,

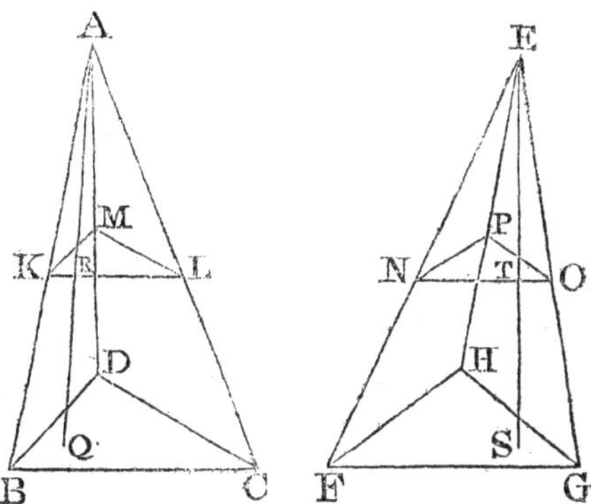

drawn from A and E upon the planes BDC and FGH: and let them be cut by planes parallel to BDC and FGH, and at equal altitudes QR and ST above those planes, and let the sections be the triangles KLM, NOP; KLM and NOP are equal to one another.

Because the plane ABD cuts the parallel planes BDC, KLM, the common sections BD and KM are parallel (14. 2. Sup.). For the same reason, DC and ML are parallel. Since therefore KM and ML are parallel to BD and DC, each to each, though not in the same plane with them, the angle KML is equal to the angle BDC (9. 2. Sup.). In like manner the other angles of these triangles are proved to be equal; therefore, the triangles are equiangular, and consequently similar; and the same is true of the triangles NOP, FGH.

Now, since the straight lines ARQ, AKB meet the parallel planes BDC and KML, they are cut by them proportionally (16. 2. Sup.), or QR : RA :: BK : KA; and AQ : AR :: AB : AK (18. 5.), for the same reason, ES : ET :: EF : EN; therefore AB : AK :: EF : EN, because AQ is equal to ES, and AR to ET. Again, because the triangles ABC, AKL are similar,

AB : AK :: BC : KL; and for the same reason

EF : EN :: FG : NO; therefore,

BC : KL :: FG : NO. And, when four straight lines are proportionals, the similar figures described on them are also proportionals (22. 6.); therefore the triangle BCD is to the triangle KLM as the triangle FGH to the triangle NOP; but the triangles BDC, FGH are equal; therefore, the triangle KLM is also equal to the triangle NOP (1. 5.). Therefore, &c. Q. E. D.

Cor. 1. Because it has been shewn that the triangle KLM is similar to the base BCD; therefore, any section of a triangular pyramid parallel to the base, is a triangle similar to the base. And in the same manner it is shewn, that the sections parallel to the base of a polygonal pyramid are similar to the base.

Cor. 2. Hence also, in polygonal pyramids of equal bases and altitudes, the sections parallel to the bases, and at equal distances from them, are equal to one another.

PROP. XIII. THEOR.

A series of prisms of the same altitude may be circumscribed about any pyramid, such that the sum of the prisms shall exceed the pyramid by a solid less than any given solid.

Let ABCD be a pyramid, and Z* a given solid; a series of prisms having all the same altitude, may be circumscribed about the pyramid ABCD, so that their sum shall exceed ABCD, by a solid less than Z.

Let Z be equal to a prism standing on the same base with the pyramid, viz. the triangle BCD, and having for its altitude the perpendicular drawn from a certain point E in the line AC upon the plane BCD. It is evident, that CE multiplied by a certain number m will be greater than AC; divide CA into as many equal parts as there are units in m, and let these be CF, FG, GH, HA, each of which will be less than CE. Through each of the points F, G, H let planes be made to pass parallel to the plane BCD, making with the sides of the pyramid the sections FPQ, GRS, HTU, which will be all similar to one another, and to the base BCD (1. cor. 12. 3. Sup.). From the point B draw in the plane of the triangle ABC, the straight line BK parallel to CF meeting FP produced in K. In like manner, from D draw DL parallel to CF, meeting FQ in L: Join KL, and it is plain, that the solid KBCDLF is a prism (def. 4. 3. Sup.). By the same construction, let the prisms PM, RO, TV be described. Also, let the straight line IP, which is in the plane of the triangle ABC be produced till it meet BC in h; and let the line MQ, be produced till it meet DC in g: Join hg; then hC gQFP is a prism, and is equal to the prism PM (1. Cor. 8. 3. Sup.). In the same manner is described the prism MS equal to the prism RO, and the prism qU equal to the prism TV. The sum, therefore, of all the inscribed prisms hQ, mS, and qU is equal to the sum of the prisms PM, RO and TV, that is, to the sum of all the circumscribed prisms except the prism BL; wherefore, BL is the excess of the prism circumscribed about the pyramid ABCD above the prisms inscribed with

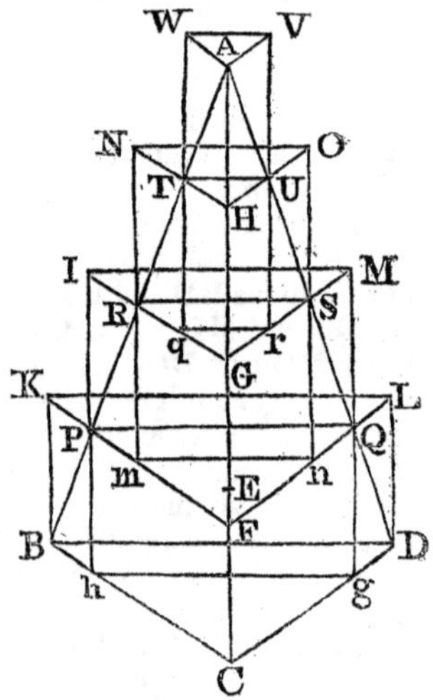

* The Solid Z is not represented in the figure of this, or the following Proposition.

in it. But the prism BL is less than the prism which has the triangle BCD for its base, and for its altitude the perpendicular from E upon the plane BCD; and the prism which has BCD for its base, and the perpendicular from E for its altitude, is by hypothesis equal to the given solid Z; therefore, the excess of the circumscribed, above the inscribed prisms, is less than the given solid Z. But the excess of the circumscribed prisms above the inscribed is greater than their excess above the pyramid ABCD, because ABCD is greater than the sum of the inscribed prisms. Much more, therefore, is the excess of the circumscribed prisms above the pyramid, less than the solid Z. A series of prisms of the same altitude has therefore been circumscribed about the pyramid ABCD, exceeding it by a solid less than the given solid Z. Q. E. D.

PROP. XIV. THEOR.

Pyramids that have equal bases and altitudes are equal to one another.

Let ABCD, EFGH. be two pyramids that have equal bases BCD, FGH, and also equal altitudes, viz. the perpendiculars drawn from the vertices A and E upon the planes, BCD, FGH: The pyramid ABCD is equal to the pyramid EFGH.

If they are not equal, let the pyramid EFGH exceed the pyramid ABCD by the solid Z. Then, a series of prisms of the same altitude

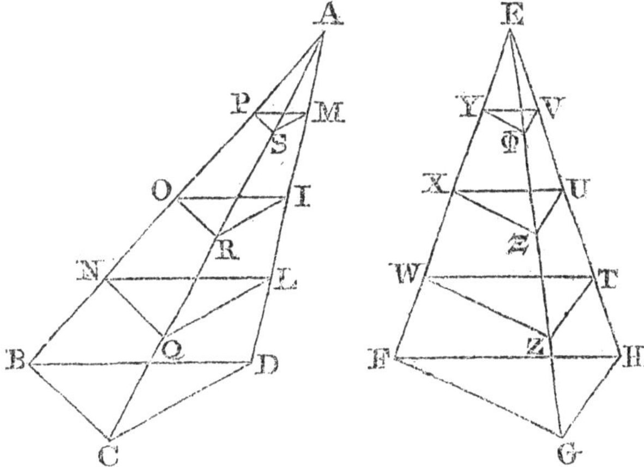

may be described about the pyramid ABCD that shall exceed it, by a solid less than Z (13. 3. Sup.); let these be the prisms that have for their bases the triangles BCD, NQL, ORI, PSM. Divide EH into the same number of equal parts into which AD is divided, viz. HT, TU, UV, VE, and through the points T, U and V, let the sections

TZW, UΞX, VΦY be made parallel to the base FGH. The section NQL is equal to the section WZT (12. 3. Sup.); as also ORI to XΞU, and PSM to YΦV; and therefore also the prisms that stand upon the equal sections are equal (1. Cor. 8. 3. Sup.), that is, the prism which stands on the base BCD, and which is between the planes BCD and NQL is equal to the prism which stands on the base FGH, and which is between the planes FGH and WZT; and so of the rest, because they have the same altitude: wherefore, the sum of all the prisms described about the pyramid ABCD is equal to the sum of all those described about the pyramid EFGH. But the excess of the prisms described about the pyramid ABCD above the pyramid ABCD is less than Z (13. 3. Sup.); and therefore, the excess of the prism described about the pyramid EFGH above the pyramid ABCD is also less than Z. But the excess of the pyramid EFGH above the pyramid ABCD is equal to Z, by hypothesis therefore, the pyramid EFGH exceeds the pyramid ABCD, more than the prisms described about EFGH exceed the same pyramid ABCD. The pyramid EFGH is therefore greater than the sum of the prisms described about it, which is impossible. The pyramids ABCD, EFGH, therefore, are not unequal, that is, they are equal to one another. Therefore, pyramids, &c. Q. E. D.

PROP XV. THEOR.

Every prism having a triangular base may be divided into three pyramids that have triangular bases, and that are equal to another.

Let there be a prism of which the base is the triangle ABC, and let DEF be the triangle opposite the base: The prism ABCDEF may be divided into three equal pyramids having triangular bases.

Join AE, EC, CD; and because ABED is a parallelogram, of which AE is the diameter, the triangle ADE is equal (34. 1.) to the triangle ABE: therefore the pyramid of which the base is the triangle ADE, and vertex the point C, is equal (14. 3. Sup.) to the pyramid, of which the base is the triangle ABE, and vertex the point C. But the pyramid of which the base is the triangle ABE, and vertex the point C, that is, the pyramid ABCE is equal to the pyramid DEFC (14. 3. Sup.), for they have equal bases, viz. the triangles ABC, DEF, and the same altitude, viz. the altitude of the prism ABCDEF. Therefore the three pyramids ADEC, ABEC, DFEC are equal to one another. But the pyramids ADEC, ABEC, DFEC make up the whole prism ABCDEF; therefore, the prism ABCDEF is divided into three equal pyramids. Wherefore, &c. Q. E. D.

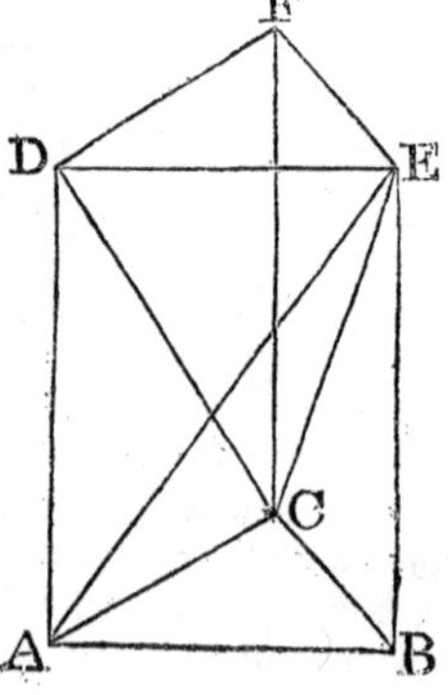

Cor. 1. From this it is manifest, that every pyramid is the third part of a prism which has the same base, and the same altitude with it; for if the base of the prism be any other figure than a triangle, it may be divided into prisms having triangular bases.

Cor. 2. Pyramids of equal altitudes are to one another as their bases; because the prisms upon the same bases, and of the same altitude, are (1. Cor. 38. Sup.) to one another as their bases.

PROP. XVI. THEOR.

If from any point in the circumference of the base of a cylinder, a straight line be drawn perpendicular to the plane of the base, it will be wholly in the cylindric superficies.

Let ABCD be a cylinder of which the base is the circle AEB, DFC the circle opposite to the base, and GH the axis; from E, any point in the circumference AEB, let EF be drawn perpendicular to the plane of the circle AEB: the straight line EF is in the superficies of the cylinder.

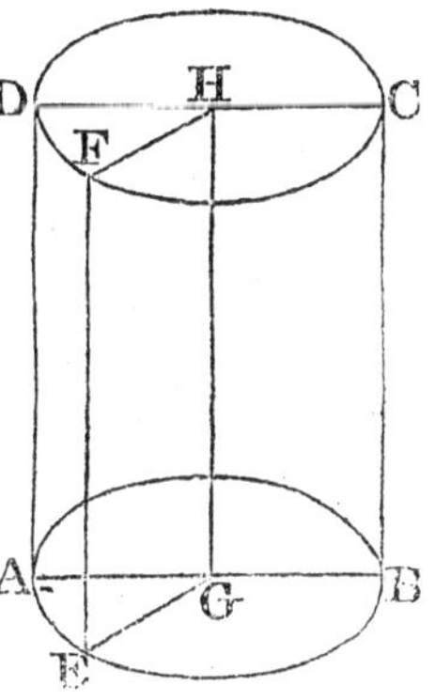

Let F be the point in which EF meets the plane DFC opposite to the base; join EG and FH; and let AGHD be the rectangle (14. def. 3. Sup.) by the revolution of which the cylinder ABCD is described.

Now, because GH is at right angles to GA, the straight line, which by its revolution describes the circle AEB, it is at right angles to all the straight lines in the plane of that circle which meet it in G, and it is therefore at right angles to the plane of the circle AEB. But EF is at right angles to the same plane; therefore, EF and GH are parallel (6. 2. Sup.) and in the same plane. And since the plane through GH and EF cuts the parallel planes AEB, DFC, in the straight lines EG and FH, EG is parallel to FH (14. 2. Sup.). The figure EGHF is therefore a parallelogram, and it has the angle EGH a right angle, therefore it is a rectangle, and is equal to the rectangle AH, because EG is equal to AG. Therefore, when in the revolution of the rectangle AH, the straight line AG coincides with EG the two rectangles AH and EH will coincide, and the straight line AD will coincide with the straight line EF. But AD is always in the superficies of the cylinder, for it describes that superficies; therefore, EF is also in the superficies of the cylinder. Therefore, &c. Q. E. D.

PROP. XVII. THEOR.

A cylinder and a parallelopiped having equal bases and altitudes, are equal to one another.

Let ABCD be a cylinder, and EF a parallelopiped having equal bases, viz. the circle AGB and the parallelogram EH, and having also equal altitudes; the cylinder ABCD is equal to the parallelopiped EF.

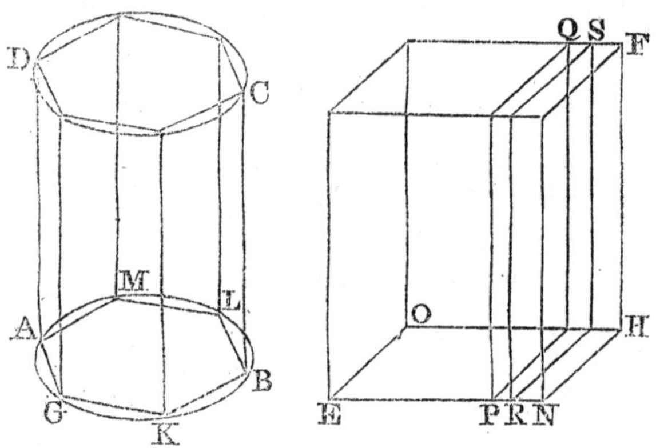

If not, let them be unequal; and first, let the cylinder be less than the parallelopiped EF; and from the parallelopiped EF let there be cut off a part EQ by a plane PQ parallel to NF, equal to the cylinder ABCD. In the circle AGB inscribe the polygon AGKBLM that shall differ from the circle by a space less than the parallelogram PH (Cor. 1. 4. 1. Sup.), and cut off from the parallelogram EH, a part OR equal to the polygon AGKBLM. The point R will fall between P and N. On the polygon AGKBLM let an upright prism AGBCD be constituted of the same altitude with the cylinder, which will therefore be less than the cylinder, because it is within it (16. 3. Sup.); and if through the point R a plane RS parallel to NF be made to pass, it will cut off the parallelopiped ES equal (2. Cor. 8. 3. Sup.) to the prism AGBC, because its base is equal to that of the prism, and its altitude is the same. But the prism AGBC is less than the cylinder ABCD, and the cylinder ABCD is equal to the parallelopiped EQ, by hypothesis; therefore, ES is less than EQ, and it is also greater, which is impossible. The cylinder ABCD, therefore, is not less than the parallelopiped EF; and in the same manner, it may be shewn not to be greater than EF. Therefore they are equal. Q. E. D.

PROP. XVIII. THEOR.

If a cone and cylinder have the same base and the same altitude, the cone is the third part of the cylinder.

Let the cone ABCD, and the cylinder BFKG have the same base, viz. the circle BCD, and the same altitude, viz. the perpendicular from the point A upon the plane BCD, the cone ABCD is the third part of the cylinder BFKG.

If not, let the cone ABCD be the third part of another cylinder LMNO, having the same altitude with the cylinder BFKG, but let the bases BCD and LIM be unequal; and first, let BCD be greater than LIM.

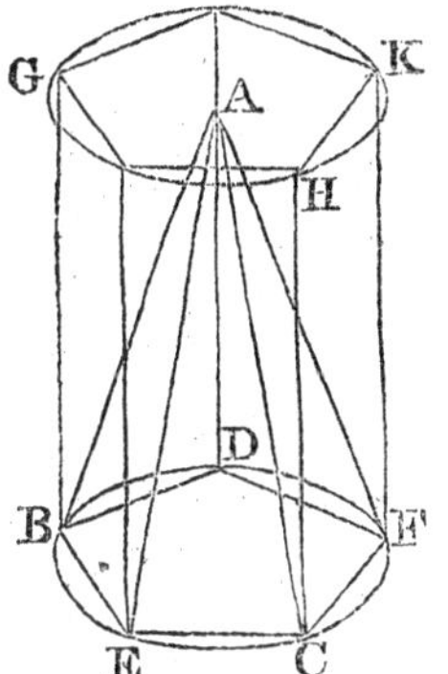

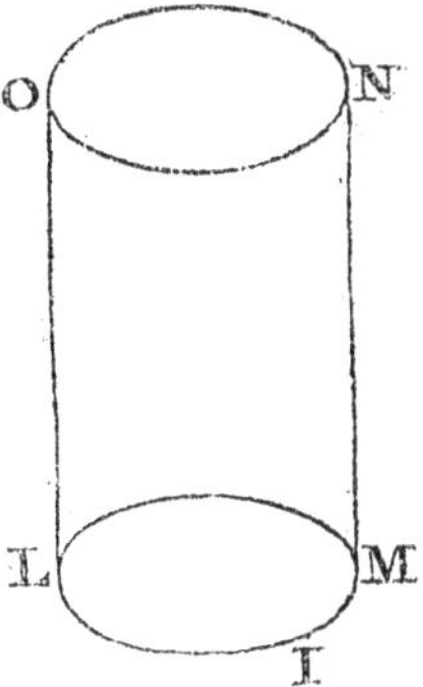

Then, because the circle BCD is greater than the circle LIM, a polygon may be inscribed in BCD, that shall differ from it less than LIM does (4. 1. Sup.), and which, therefore, will be greater than LIM. Let this be the polygon BECFD; and upon BECFD, let there be constituted the pyramid ABECFD, and the prism BCFKHG.

Because the polygon BECFD is greater than the circle LIM, the prism BCFKHG, is greater than the cylinder LMNO, for they have the same altitude, but the prism has the greater base. But the pyramid ABECFD is the third part of the prism (15. 3. Sup.) BCFKHG, therefore it is greater than the third part of the cylinder LMNO. Now, the cone ABECFD is, by hypothesis, the third part of the cylinder LMNO, therefore, the pyramid ABECFD is greater than the cone ABCD, and it is also less, because it is inscribed in the cone, which is impossible. Therefore, the cone ABCD is not less than the third part of the cylinder BFKG: And in the same manner, by circumscribing a polygon about the circle BCD, it may be shewn that the cone ABCD is not greater than the third part of the cylinder BFKG; therefore, it is equal to the third part of that cylinder. Q. E. D.

PROP. XIX. THEOR.

If a hemisphere and a cone have equal bases and altitudes, a series of cylinders may be inscribed in the hemisphere, and another series may be described about the cone, having all the same altitudes with one another, and such that their sum shall differ from the sum of the hemisphere, and the cone, by a solid less than any given solid.

Let ADB be a semicircle, of which the centre is C, and let CD be at right angles to AB; let DB and DA be squares described on DC, draw CE, and let the figure thus constructed revolve about DC: then, the sector BCD, which is the half of the semicircle ADB, will describe a hemisphere having C for its centre (7. def. 3. Sup.), and the triangle CDE will describe a cone, having its vertex to C, and having for its base the circle (11. def. 3. Sup.) described by DE, equal to that described by BC, which is the base of the hemisphere. Let W be any given solid. A series of cylinders may be inscribed in the hemisphere ADB, and another described about the cone ECI, so that their sum shall differ from the sum of the hemisphere and the cone, by a solid less than the solid W.

Upon the base of the hemisphere let a cylinder be constituted equal to W, and let its altitude be CX. Divide CD into such a number of equal parts, that each of them shall be less than CX; let these be CH, HG, GF, and FD. Through the points F, G, H, draw FN, GO, HP parallel to CB, meeting the circle in the points K, L and M; and the straight line CE in the points Q, R and S. From the points K, L, M draw Kf, Lg, Mh perpendicular to GO, HP and CB; and from Q, R, and S, draw Qq, Rr, Ss, perpendicular to the same lines. It is evident, that the figure being thus constructed, if the whole revolve about CD, the rectangles Ff, Gg, Hh will describe cylinders (14. def 3. Sup.) that will be circumscribed by the hemisphere BDA; and the rectangles DN, Fq, Gr, Hs, will also describe cylinders that will circumscribe the cone ICE. Now, it may be demonstrated, as was done of the prisms inscribed in a pyramid (13. 3. Sup.), that the sum of all the cylinders described within the hemisphere, is exceeded by the hemisphere by a solid less than the cylinder

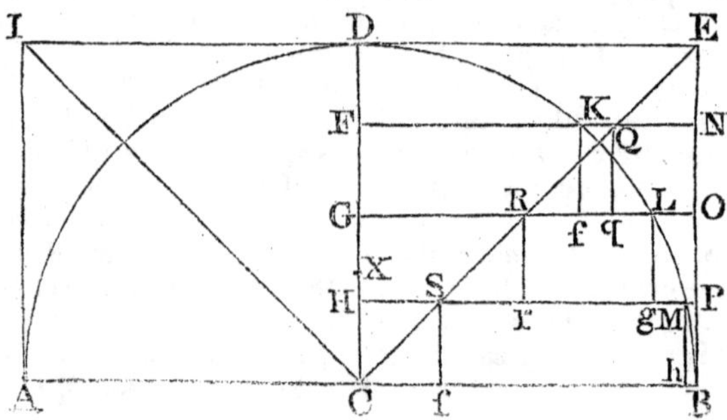

generated by the rectangle HB, that is, by a solid less than W, for the cylinder generated by HB is less than W. In the same manner, it may be demonstrated, that the sum of the cylinders circumscribing the cone ICE is greater than the cone by a solid less than the cylinder generated by the rectangle DN, that is, by a solid less than W. Therefore, since the sum of the cylinders inscribed in the hemisphere, together with a solid less than W, is equal to the hemisphere; and, since the sum of the cylinders described about the cone is equal to the cone together with a solid less than W; adding equals to equals, the sum of all these cylinders, together with a solid less than W, is equal to the sum of the hemisphere and the cone together with a solid less than W. Therefore, the difference between the whole of the cylinders and the sum of the hemisphere and the cone, is equal to the difference of two solids, which are each of them less than W; but this difference must also be less than W, therefore the difference between the two series of cylinders and the sum of the hemisphere and cone is less than the given solid W. Q. E. D.

PROP. XX.

The same things being supposed as in the last proposition, the sum of all the cylinders inscribed in the hemisphere, and described about the cone, is equal to a cylinder, having the same base and altitude with the hemisphere.

Let the figure DCB be constructed as before, and supposed to revolve about CD; the cylinders inscribed in the hemisphere, that is, the cylinders described by the revolution of the rectangles Hh, Gg, Ff, together with those described about the cone, that is, the cylinders described by the revolution of the rectangles Hs, Gr, Fq, and DN are equal to the cylinder described by the revolution of the rectangle DB.

Let L be the point in which GO meets the circle ADB, then, because CGL is a right angle if CL be joined, the circles described with the distances CG and GL are equal to the circle described with the distance CL (2. Cor. 6. 1. Sup.) or GO; now, CG is equal to GR, because CD is equal to DE, and therefore also, the circles described with the distances GR and GL are together equal to the circle described with the distance GO, that is, the circles described by the revolution of GR and GL about the point G, are together equal to the circle described by the revolution of GO about the same point G; therefore also, the cylinders that stand upon the two first of these circles having the common altitudes GH, are equal to the cylinder which stands on the remaining circle, and which has the same altitude GH. The cylinders described by the revolution of the rectangles Gg, and Gr are therefore equal to the cylinder described by the rectangle GP. And as the same may be shown of all the rest, therefore the cylinders described by the rectangles Hh, Gg, Ff, and by the rectangles Hs, Gr,

Fq, DN, are together equal to the cylinder described by DB, that is, to the cylinder having the same base and altitude with the hemisphere. Q. E. D.

PROP. XXI.

Every sphere is two-thirds of the circumscribing cylinder.

Let the figure be constructed as in the two last propositions, and if the hemisphere described by BDC be not equal to two thirds of the cylinder described by BD, let it be greater by the solid W. Then, as the cone described by CDE is one-third of the cylinder (18. 3. Sup.) described by BD, the cone and the hemisphere together will exceed

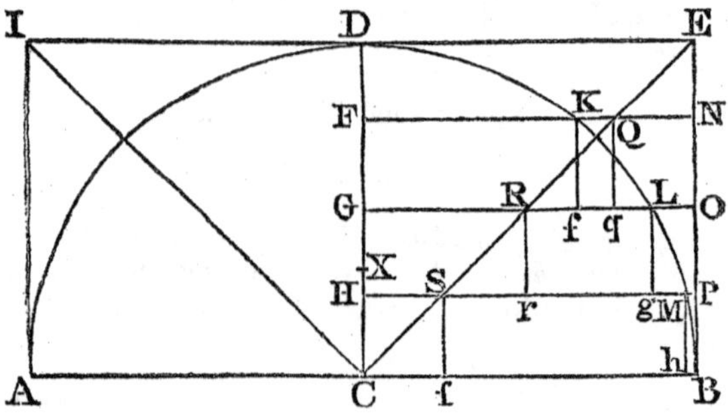

the cylinder by W. But that cylinder is equal to the sum of all the cylinders described by the rectangles Hh, Gg, Ff, Hs, Gr, Fq, DN (20. 3. Sup.); therefore the hemisphere and the cone added together exceed the sum of all these cylinders by the given solid W, which is absurd; for it has been shown (19. 3. Sup.), that the hemisphere and the cone together differ from the sum of the cylinders by a solid less than W. The hemisphere is therefore equal to two-thirds of the cylinder described by the rectangle BD; and therefore the whole sphere is equal to two-thirds of the cylinder described by twice the rectangle BD, that is, to two-thirds of the circumscribing cylinder. Q. E. D.

END OF THE SUPPLEMENT TO THE ELEMENTS.

ELEMENTS

OF

PLANE AND SPHERICAL

TRIGONOMETRY.

ELEMENTS

OF

PLANE TRIGONOMETRY.

TRIGONOMETRY is the application of Arithmetic to Geometry : or, more precisely, it is the application of number to express the relations of the sides and angles of triangles to one another. It therefore necessarily supposes the elementary operations of arithmetic to be understood, and it borrows from that science several of the signs or characters which peculiarly belong to it. Thus, the product of two numbers A and B, is either denoted by A.B or $A\times B$; and the products of two or more into one, or into more than one, as of $A+B$ into C, or of $A+B$ into $C+D$, are expressed thus : $(A+B)$, C, $(A+B)(C+D)$, or sometimes thus, $\overline{A+B}\times C$, and $\overline{A+B}\times\overline{C+D}$,

The quotient of one number A, divided by another B, is written thus, $\frac{A}{B}$.

The sign $\surd$ is used to signify the square root : Thus $\surd M$ is the square root of M, or it is a number which, if multiplied into itself, will produce M. So also, $\surd\overline{M^2+N^2}$ is the square root of M^2+N^2, &c. The elements of Plane Trigonometry, as laid down here, are divided into three sections : the first explains the principles ; the second delivers the rules of calculation ; the third contains the construction of trigonometrical tables, together with the investigation of some theorems, useful for extending trigonometry to the solution of the more difficult problems.

SECTION I.

LEMMA I.

An angle at the centre of a circle is to four right angles as the arch on which it stands is to the whole circumference.

Let ABC be an angle at the centre of the circle ACF, standing on the circumference AC : the angle ABC is to four right angles as the arch AC to the whole circumference ACF.

Produce AB till it meet the circle in E, and draw DBF perpendicular to AE.

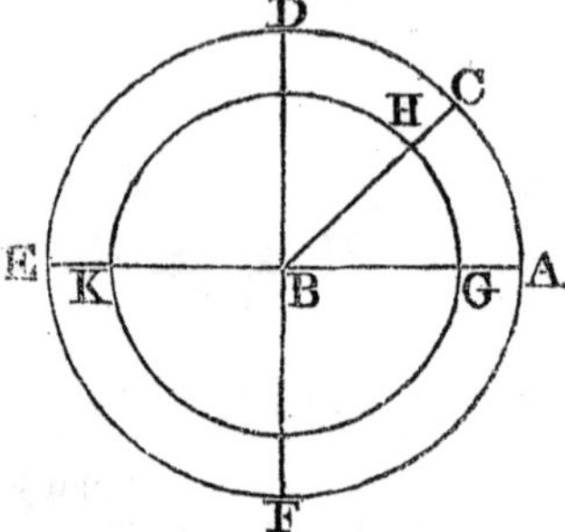

Then, because ABC, ABD are two angles at the centre of the circle ACF, the angle ABC is to the angle ABD as the arch AC to the arch AD, (33. 6.); and therefore also, the angle ABC is to four times the angle ABD as the arch AC to four times the arch AD (4. 5.).

But ABD is a right angle, and therefore four times the arch AD is equal to the whole circumference ACF; therefore, the angle ABC is to four right angles as the arch AC to the whole circumference ACF.

Cor. Equal angles at the centres of different circles stand on arches which have the same ratio to their circumferences. For, if the angle ABC, at the centre of the circles, ACE, GHK, stand on the arches AC, GH, AC is to the whole circumference of the circle ACE, as the angle ABC to four right angles; and the arch HG is to the whole circumference of the circle GHK in the same ratio. Therefore, &c.

DEFINITIONS.

I.

If two straight lines intersect one another in the centre of a circle, the arch of the circumference intercepted between them is called the *Measure* of the angle which they contain. Thus the arch AC is the measure of the angle ABC.

II.

If the circumference of a circle be divided into 360 equal parts, each of these parts is called a *Degree*; and if a degree be divided into 60 equal parts, each of these is called a minute; and if a *Minute* be divided into 60 equal parts, each of them is called a *Second*, and so on. And as many degrees, minutes, seconds, &c. as are in any arch, so many degrees, minutes, seconds, &c. are said to be in the angle measured by that arch.

Cor. 1. Any arch is to the whole circumference of which it is a part, as the number of degrees and parts of a degree contained in it is to the number 360. And any angle is to four right angles as the number of degrees and parts of a degree in the arch, which is the measure of that angle, is to 360.

Cor. 2. Hence also, the arches which measure the same angle, whatever be the radii with which they are described, contain the same number of degrees, and parts of a degree. For the number of degrees and parts of a degree contained in each of these arches has the same ratio to the number 360, that the angle which they measure has to four right angles (Cor. Lem. 1.).

The degrees, minutes, seconds, &c. contained in any arch or angle, are usually written as in this example, 49°. 36′. 24″. 42‴; that is, 49 degrees, 36 minutes, 24 seconds, and 42 thirds.

III.

Two angles, which are together equal to two right angles, or two arches which are together equal to a semicircle, are called the *Supplements* of one another.

IV.

A straight line CD drawn through C, one of the extremities of the arch AC, perpendicular to the diameter passing through the other extremity A, is called the *Sine* of the arch AC, or of the angle ABC, of which AC is the measure.

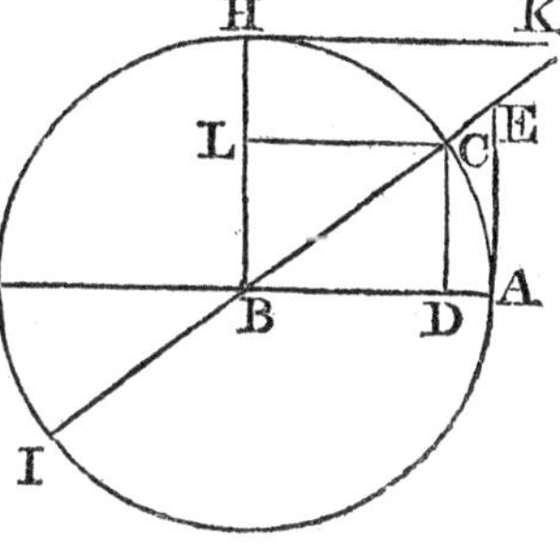

Cor. 1. The sine of a quadrant, or of a right angle, is equal to the radius.

Cor. 2. The sine of an arch is half the chord of twice that arch: this is evident by producing the sine of any arch till it cut the circumference.

V.

The segment DA of the diameter passing through A, one extremity of the arch AC, between the sine CD and the point A, is called the *Versed sine* of the arch AC, or of the angle ABC.

VI.

A straight line AE touching the circle at A, one extremity of the arch AC, and meeting the diameter BC, which passes through C the other extremity, is called the *Tangent* of the arch AC, or of the angle ABC.

Cor. The tangent of half a right angle is equal to the radius.

VII.

The straight line BE, between the centre and the extremity of the tangent AE is called the *Secant* of the arch AC, or of the angle ABC.

Cor. to Def. 4, 6, 7, the sine, tangent and secant of any angle ABC, are likewise the sine, tangent, and secant of its supplements CBF.

It is manifest, from Def. 4. that CD is the sine of the angle CBF. Let CB be produced till it meet the circle again in I; and it is also manifest, that AE is the tangent, and BE the secant, of the angle ABI, or CBF, from Def. 6, 7.

Cor. to Def. 4, 5, 6, 7. The sine, versed sine, tangent, and secant of an arch, which is the measure of any given angle ABC, is to the sine, versed sine, tangent and secant, of any other arch which is the measure of the same angle. as the radius of the first arch is to the radius of the second.

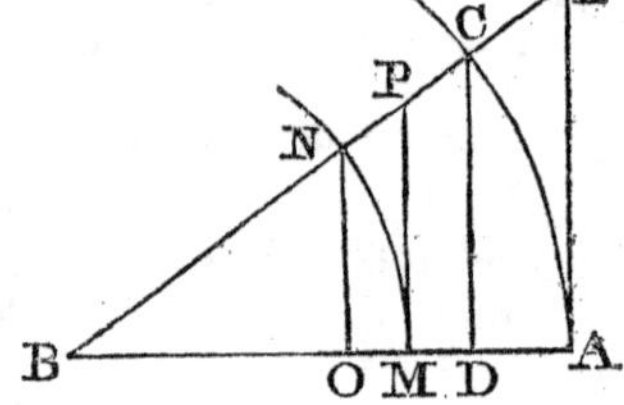

Let AC, MN be measures of the angle ABC, according to Def. 1.; CB the sine, DA the versed sine. AE the tangent, and BE the secant of the arch AC, according to Def. 4, 5, 6, 7; NO the sine, OM the versed sine, MP the tangent, and BP the secant of the arch MN, according to the same definitions. Since CD, NO, AE, MP are parallel, CD : NO :: rad. CB : rad. NB, and AE : MP :: rad. AB : rad. BM, also BE : BP :: AB : BM; likewise because BC : BD :: BN : BO, that is, BA : BD :: BM : BO, by conversion and alternation, AD : MO :: AB : MB. Hence the corollary is manifest. And therefore, if tables be constructed, exhibiting in numbers the sines, tangents, secants, and versed sines of certain angles to a given radius, they will exhibit the ratios of the sines, tagents, &c. of the same angles to any radius whatsoever.

In such tables, which are called Trigonometrical Tables, the radius is either supposed 1, or some number in the series 10, 100, 1000, &c. The use and construction of these tables are about to be explained.

VIII.

The difference between any angle and a right angle, or between any arch and a quadrant, is called the *Complement* of that angle, or of that arch. Thus, if BH be perpendicular to AB, the angle CBH is the complement of the angle ABC, and the arch HC the complement of AC; also the complement of the obtuse angle FBC is the angle HBC, its excess above a right angle; and the complement of the arch FC is HC.

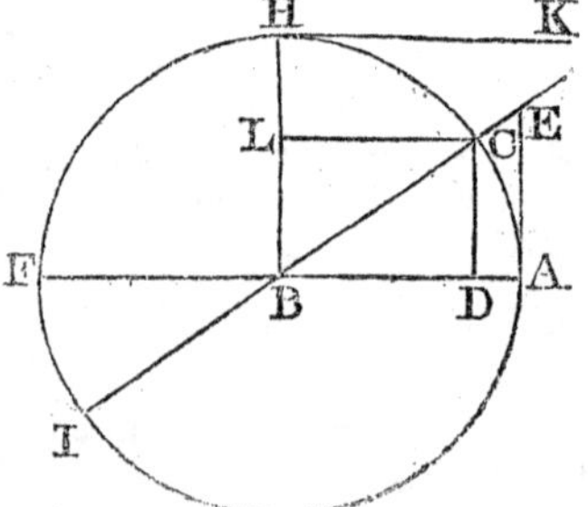

IX.

The sine, tangent, or secant of the complement of any angle is called the *Cosine, Cotangent*, or *Cosecant* of that angle. Thus, let CL or DB, which is equal to CL, be the sine of the angle CBH; HK the tangent, and BK the secant of the same angle: CL or BD is the cosine, HK the cotangent, and BK the cosecant of the angle ABC.

COR. 1. The radius is a mean proportional between the tangent and the cotangent of any angle ABC; that is, tan. ABC×cot. ABC= R^2.

For, since HK, BA are parallel, the angles HKB, ABC are equal, and KHB, BAE are right angles; therefore the triangles BAE, KHB are similar, and therefore AE is to AB, as BH or BA to HK.

COR. 2. The radius is a mean proportional between the cosine and secant of any angle ABC; or
cos. ABC×sec. ABC=R^2.

Since CD, AE are parallel, BD is to BC or BA, as BA to BE.

PROP. I.

In a right angled plane triangle, as the hypotenuse to either of the sides, so the radius to the sine of the angle opposite to that side; and as either of the sides is to the other side, so is the radius to the tangent of the angle opposite to that side.

Let ABC be a right angled plane triangle, of which BC is the hypotenuse. From the centre C, with any radius CD, describe the arch DE; draw DF at right angles to CE, and from E draw EG touching the circle in E, and meeting CB in G; DF is the sine, and EG the tangent of the arch DE, or of the angle C.

The two triangles DFC, BAC are equiangular, because the angles DFC, BAC are right angles, and the angle at C is common. Therefore, CB : BA :: CD : DF; but CD is the radius, and DF the sine of the angle C, (Def. 4.); therefore CB : BA :: R : sin. C.

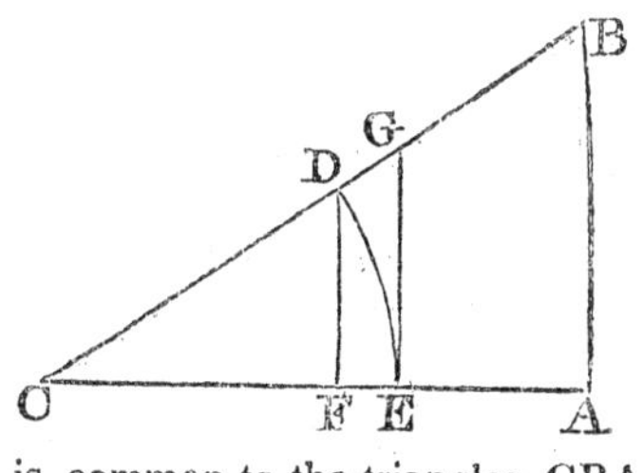

Also, because EG touches the circle in E, CEG is a right angle, and therefore equal to the angle BAC; and since the angle at C is common to the triangles CBA, CGE, these triangles are equiangular, wherefore CA : AB :: CE : EG; but CE is the radius, and EG the tangent of the angle C; therefore, CA : AB :: R : tan C.

COR. 1. As the radius to the secant of the angle C, so the side adjacent to that angle to the hypotenuse. For CG is the secant of the angle G (def. 7.), and the triangles CGE, CBA being equiangular, CA : CB :: CE : CG, that is, CA : CB :: R : sec. C.

COR. 2. If the analogies in this proposition, and in the above corollary be arithmetically expressed, making the radius = 1, they give sin. $C=\frac{AB}{BC}$; tan. $C=\frac{AB}{AC}$, sec. $C=\frac{BC}{AC}$. Also, since sin. C = cos B, because B is the complement of C, cos. $B=\frac{AB}{BC}$, and for the same reason, cos. $C=\frac{AC}{BC}$.

COR. 3. In every triangle, if a perpendicular be drawn from any of the angles on the opposite side, the segments of that side are to one another as the tangents of the parts into which the opposite angle is divided by the perpendicular. For, if in the triangle ABC, AD be drawn perpendicular to the base BC, each of the triangles CAD, ABD being right angled, AD : DC :: R : tan. CAD, and AD : DB : : R : tan. DAB; therefore, ex æquo, DC : DB : : tan. CAD : tan. BAD.

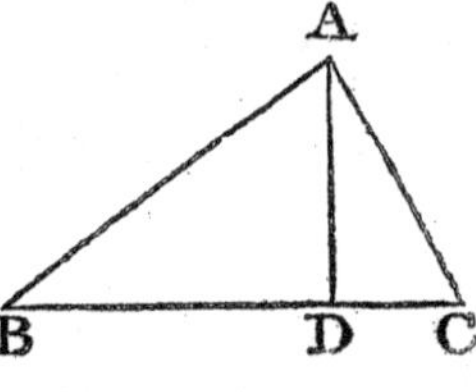

SCHOLIUM.

The proposition, just demonstrated, is most easily remembered, by stating it thus: If in a right angled triangle the hypotenuse be made the radius, the sides become the sines of the opposite angles; and if one of the sides be made the radius, the other side becomes the tangent of the opposite angle, and the hypotenuse the secant of it.

PROP. II.

The sides of a plane triangle are to one another as the sines of the opposite angles.

From A any angle in the triangle ABC, let AD be drawn perpendicular to BC. And because the triangle ABD is right angled at D, AB : AD : : R : sin. B; and for the same reason, AC : AD :: R : sin. C, and inversely, AD : AC :: sin. C : R; therefore, ex æquo inversely, AB : AC : : sin. C : sin. B. In the same manner it may be demonstrated, that AB : BC : : sin. C : sin. A. Therefore, &c. Q. E. D.

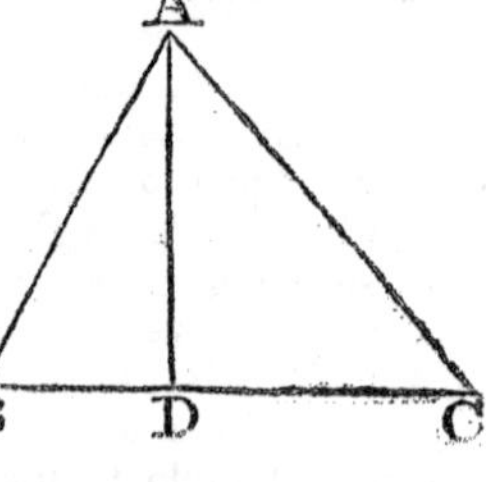

PROP. III.

The sum of the sines of any two arches of a circle, is to the difference of their sines, as the tangent of half the sum of the arches to the tangent of half their difference.

Let AB, AC be two arches of a circle ABCD; let E be the centre, and AEG the diameter which passes through A; sin. AC+sin. AB : sin. AC−sin. AB : : tan, $\frac{1}{2}$ (AC+AB) : tan. $\frac{1}{2}$ (AC−AB).

Draw BF parallel to AG, meeting the circle again in F. Draw BH and CL perpendicular to AE, and they will be the sines of the arches AB and AC; produce CL till it meet the circle again in D; join DF, FC, DE, EB, EC, DB.

Now, since EL from the centre is perpendicular to CD, it bisects the line CD in L and the arch CAD in A: DL is therefore equal to LB, or to the sine of the arch AC; and BH or LK being the sine of AB, DK is the sum of the sines of the arches AC and AB, and CK is the difference of their sines; DAB also is the sum of the arches AC and AB, because AD is equal to AC, and BC is their difference. Now, in the triangle DFC, because FK is perpendicular to DC, (3. cor. 1.) DK : KC : : tan. DFK : tan. CFK; but tan. DFK=tan. $\frac{1}{2}$ arc. BD, because the angle DFK (20. 3.) is the half of DEB, and is therefore measured by half the arch DB. For the same reason, tan. CFK=tan. $\frac{1}{2}$ arc. BC; and consequently, DK : KC : : tan. $\frac{1}{2}$ arc. BD : tan. $\frac{1}{2}$ arc. BC. But DK is the sum of the sines of the arches AB and AC; and KC is the difference of the sines; also BD is the sum of the arches AB and AC, and BC the difference of those arches. Therefore, &c. Q. E. D.

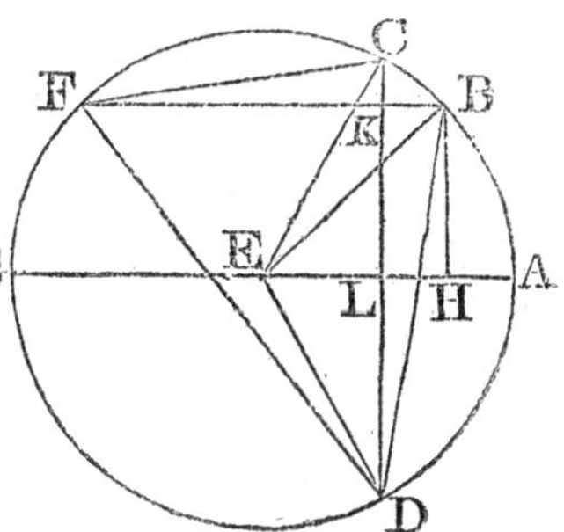

Cor. 1. Because EL is the cosine of AC, and EH of AB, FK is the sum of these cosines, and KB their difference; for FK=$\frac{1}{2}$ FB+EL=EH+EL, and KB=LH=EH−EL. Now, FK : KB : : tan. FDK : tan. BDK : and tan. DFK=cotan. FDK, because DFK is the complement of FDK; therefore, FK : KB : : cotan. DFK : tan. BDK, that is, FK : KB : : cotan. $\frac{1}{2}$ arc. DB : tan. $\frac{1}{2}$ arc. BC. The sum of the cosines of two arches is therefore to the difference of the same cosines as the cotangent of half the sum of the arches to the tangent of half their difference.

Cor. 2. In the right angled triangle FKD, FK : KD : : R : tan. DFK : Now FK=cos. AB+cos. AC, KD=sin. AB+sin. AC, and tan. DFK = tan. $\frac{1}{2}$ (AB + AC), therefore cos. AB+cos. AC : sin. AB + sin. AC :: R : tan $\frac{1}{2}$ (AB + AC).

In the same manner, by help of the triangle FKC, it may be shown that cos. AB+cos. AC : sin. AC−sin. AB :: R : tan. $\frac{1}{2}$ (AC−AB).

Cor. 3. If the two arches AB and AC be together equal to 90°, the tangent of half their sum, that is, of 45°, is equal to the radius. And the arch BC being the excess of DC above DB, or above 90°, the half of the arch BC will be equal to the excess of the half of DC above the half of DB, that is, to the excess of AC above 45°; therefore, when the sum of two arches is 90°, the sum of the sines of those arches is to their difference as the radius to the tangent of the difference between either of them and 45°.

PROP. IV.

The sum of any two sides of a triangle is to their difference, as the tangent of half the sum of the angles opposite to those sides, to the tangent of half their difference.

Let ABC be any plane triangle;
CA+AB : CA−AB :: tan $\frac{1}{2}$ (B+C) : tan. $\frac{1}{2}$ (B−C).
For (2.) CA : AB :: sin. B : sin. C;
and therefore (E. 5.)
CA+AB : CA−AB :: sin. B+sin. C : sin. B−sin. C.
But, by the last, sin. B+sin. C : sin. B−sin. C ::
tan. $\frac{1}{2}$ (B+C) : tan. $\frac{1}{2}$ (B−C); therefore also, (11. 5.)
CA+AB : CA−AB :: tan $\frac{1}{2}$ (B+C) : tan. $\frac{1}{2}$ (B−C).
Q. E. D.

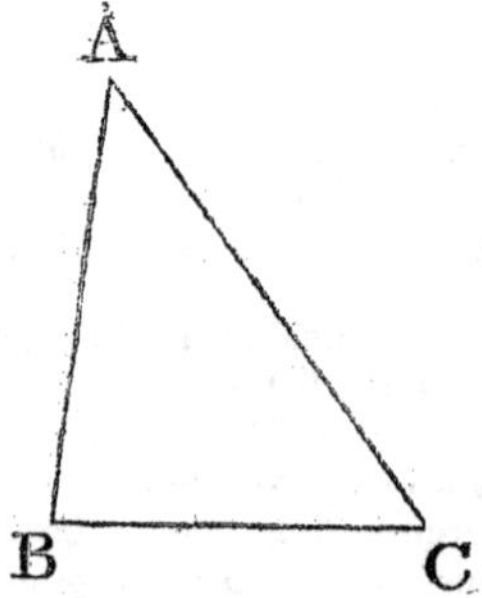

Otherwise, without the 3d.

Let ABC be a triangle; the sum of AB and AC any two sides, is to the difference of AB and AC as the tangent of half the sum of the angles ACB and ABC, to the tangent of half their difference.

About the centre A with the radius AB, the greater of the two sides, describe a circle meeting BC produced in D, and AC produced in E and F. Join DA, EB, FB; and draw FG parallel to CB, meeting EB in G.

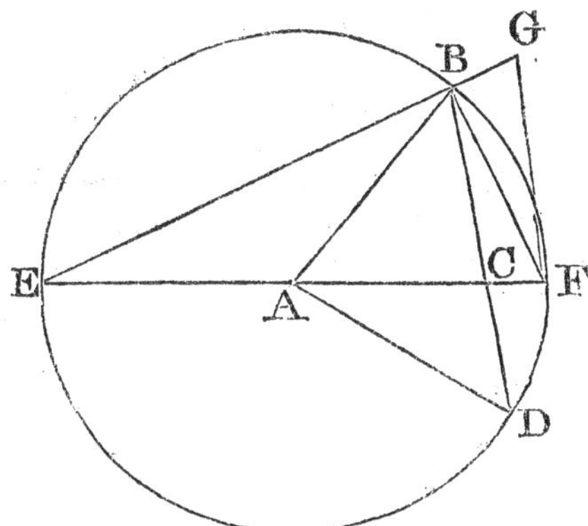

Because the exterior angle EAB is equal to the two interior ABC, ACB, (32. 1.): and the angle EFB, at the circumference is equal to half the angle EAB at the centre (20. 3.): therefore EFB is half the sum of the angles opposite to the sides AB and AC.

Again, the exterior angle ACB is equal to the two interior CAD, ADC, and therefore CAD is the difference of the angles ACB, ADC, that is, of ACB, ABC, for ABC is equal to ADC. Wherefore also DBF, which is the half of CAD, or BFG, which is equal to DBF, is half the difference of the angles opposite to the sides AB, AC.

Now because the angle FBE in a semicircle is a right angle, BE is the tangent of the angle EFB, and BG the tangent of the angle BFG to the radius FB; and BE is therefore to BG as the tangent of half the sum of the angles ACB, ABC to the tangent of half their difference. Also CE is the sum of the sides of the triangle ABC, and CF their difference; and because BC is parallel to FG, CE : CF :: BE : BG, (2. 6.) that is, the sum of the two sides of the triangle ABC is to their difference as the tangent of half the sum of the angles opposite to those sides to the tangent of half their difference. Q. E. D.

PROP. V. THEOR.

If a perpendicular be drawn from any angle of a triangle to the opposite side, or base; the sum of the segments of the base is to the sum of the other two sides of the triangle as the difference of those sides to the difference of the segments of the base.

For (K. 6.), the rectangle under the sum and difference of the segments of the base is equal to the rectangle under the sum and difference of the sides, and therefore (*16. 6.*) the sum of the segments of the base is to the sum of the sides as the difference of the sides to the difference of the segments of the base. Q. E. D.

PROP. VI. THEOR.

In any triangle, twice the rectangle contained by any two sides is to the difference between the sum of the squares of those sides, and the square of the base, as the radius to the cosine of the angle included by the two sides.

Let ABC be any triangle, 2AB.BC is to the difference between AB^2+BC^2 and AC^2 as radius to cos. B.

From A draw AD perpendicular to BC, and (12. and 13. 2.) the difference between the sum of the squares of AB and BC, and the square on AC is equal to 2BC.BD.

But BC.BA : BC.BD : : BA : BD : : R : cos. B, therefore also 2BC.BA : 2BC.BD : : R : cos. B. Now 2BC.BD is the difference between AB^2+BC^2 and AC^2, therefore twice the rectangle AB.BC is to the difference between AB^2+BC^2, and AC^2 as radius to the cosine of B. Wherefore, &c. Q. E. D.

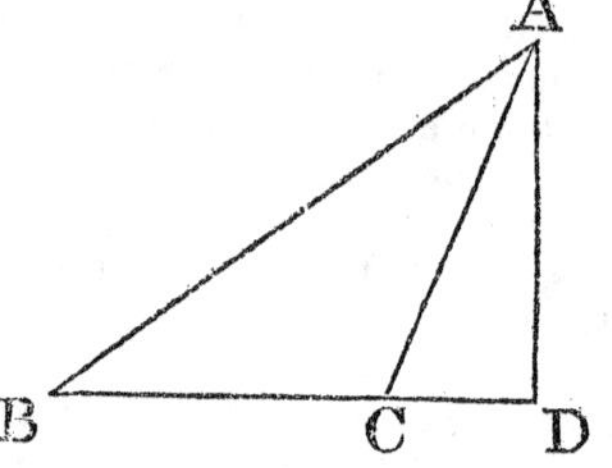

Cor. If the radius=1, BD=BA × cos. B, (1.), and 2BC.BA × cos. B=2BC.BD, and therefore when B is acute, $2BC.BA\times \cos.B = BC^2+BA^2-AC^2$, and adding AC^2 to both; $AC^2+2\cos.B\times BC.BA = BC^2+BA^2$; and taking $2.\cos.B\times BC.BA$ from both, $AC^2=BC^2-2\cos.B\times BC.BA+BA^2$. Wherefore $AC=\sqrt{(BC^2-2\cos.B\times BC.BA+BA^2)}$.

If B is an obtuse angle, it is shown in the same way that $AC=\sqrt{(BC^2+2\cos.B\times BC.BA+BA^2)}$.

PROP. VII.

Four times the rectangle contained by any two sides of a triangle, is to the rectangle contained by two straight lines, of which one is the base or third side of the triangle increased by the difference of the two sides, and the other the base diminished by the difference of the same sides, as the square of the radius to the square of the sine of half the angle included between the two sides of the triangle.

Let ABC be a triangle of which BC is the base, and AB the greater of the two sides; $4AB.AC : (BC+(AB-AC))\times(BC-(AB-AC)) : : R^2 : (\sin.\frac{1}{2}BAC)^2$.

Produce the side AC to D, so that AD=AB; join BD, and draw

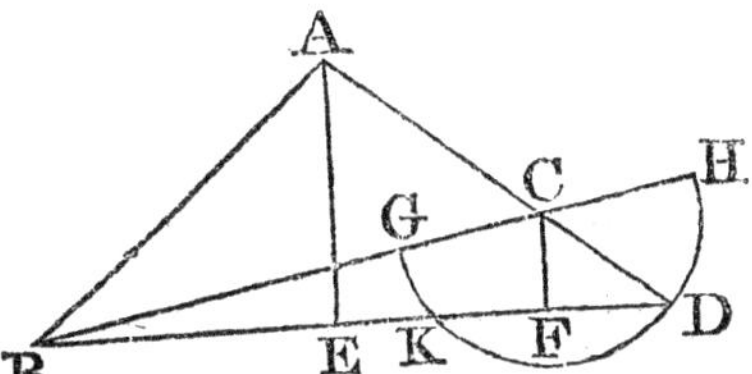

AE, CF at right angles to it; from the centre C with the radius CD describe the semicircle GDH, cutting BD in K, BC in G, and meeting BC produced in H.

It is plain that CD is the difference of the sides, and therefore that BH is the base increased, and BG the base diminished by the difference of the sides; it is also evident, because the triangle BAD is isosceles, that DE is the half of BD, and DF is the half of DK, wherefore DE—DF=the half of BD—DK, (6. 5) that is, EF=$\frac{1}{2}$ BK. And because AE is drawn parallel to CF, a side of the triangle CFD, AC : AD :: EF : ED, (2. 6.); and rectangles of the same altitude being as their bases AC.AD : AD^2 :: EF.ED : ED^2, (1. 6.), and therefore 4AC.AD : AD^2 :: 4EF.ED : ED^2, or alternately, 4AC.AD : 4EF.ED :: AD^2 : ED^2.

But since 4EF=2BK, 4EF.ED=2BK.ED=2ED.BK=DB.BK= HB.BG; therefore, 4AC.AD : DB.BK :: AD^2 : ED^2. Now AD : ED :: R : sin. EAC=sin. $\frac{1}{2}$ BAC (1. Trig.) and AD^2 : ED^2 :: R^2 : (sin. $\frac{1}{2}$ BAC)^2 : therefore, (11. 5.) 4AC.AD : HB.BG :: R^2 : (sin. $\frac{1}{2}$ BAC)^2, or since AB=AD, 4AC.AB : HB.BG :: R^2 : (sin $\frac{1}{2}$ BAC)^2. Now 4AC.AB is four times the rectangle contained by the sides of the triangle; HB.BG is that contained by BC+(AB—AC) and BC—(AB—AC). Therefore, &c. Q. E. D.

Cor. Hence 2 $\sqrt{AC.AD}$: $\sqrt{HB.BG}$ :: R : sin. $\frac{1}{2}$ BAC.

PROP. VIII.

Four times the rectangle contained by any two sides of a triangle, is to the rectangle contained by two straight lines, of which one is the sum of those sides increased by the base of the triangle, and the other the sum of the same sides diminished by the base, as the square of the radius to the square of the cosine of half the angle included between the two sides of the triangle.

Let ABC be a triangle, of which BC is the base, and AB the greater of the other two sides, 4AB.AC : (AB+AC+BC) (AB+AC—BC) :: R^2 : (cos. $\frac{1}{2}$ BAC)^2.

From the centre C, with the radius CB, describe the circle BLM, meeting AC, produced, in L and M. Produce AL to N, so that AN =AB; let AD=AB; draw AE perpendicular to BD; join BN, and let it meet the circle again in P; let CO be perpendicular to BN; and let it meet AE in R.

It is evident that MN=AB+AC+BC; and that LN=AB+AC −BC. Now, because BD is bisected in E, (3. 3.) and DN in A, BN is parallel to AE, and is therefore perpendicular to BD, and the triangles DAE, DNB are equiangular; wherefore, since DN = 2AD, BN=2AE, and BP=2BO=2RE; also PN=2AR.

But because the triangles ARC and AED are equiangular, AC : AD : : AR : AE, and because rectangles of the same altitude are as

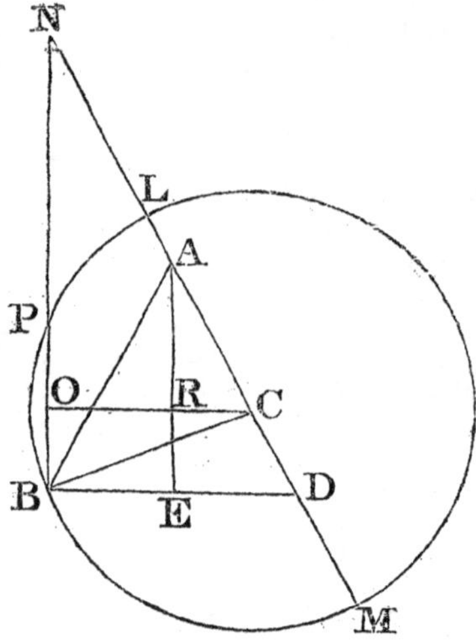

their bases, (1. 6.), AC.AD : AD^2 : : AR.AE : AE^2, and alternately AC.AD : AR.AE : : AD^2 : AE^2, and 4AC.AD : 4AR.AE : : AD^2 : AE^2. But 4AR.AE=2AR × 2AE = NP.NB=MN.NL; therefore 4AC.AD : MN.NL : : AD^2 : AE^2. But AD : AE : : R : cos. DAE (1) = cos. $\frac{1}{2}$ (BAC) : Wherefore 4AC.AD : MN.NL : : R^2 : (cos. $\frac{1}{2}$ $BAC)^2$.

Now 4AC.AD is four times the rectangle under the sides AC and AB, (for AD=AB), and MN.NL is the rectangle under the sum of the sides increased by the base, and the sum of the sides diminished by the base. Therefore, &c. Q. E. D.

Cor. 1. Hence 2 $\sqrt{\text{AC.AB}}$: $\sqrt{\text{MN.NL}}$: : R : cos. $\frac{1}{2}$ BAC.

Cor. 2. Since by Prop. 7. 4AC.AB : (BC + (AB−AC)) (BC−(AB − BC)) : : R^2 : (sin. $\frac{1}{2}$ $BAC)^2$; and as has been now proved 4AC.AB : (AB+AC+BC) (AB+AC−BC) : : R^2 : (cos. $\frac{1}{2}$ $BAC)^2$: therefore ex æquo, (AB+AC+BC) (AB+AC−BC) : (BC + (AB −AC)) (BC−(AB−AC)) : : (cos. $\frac{1}{2}$ $BAC)^2$: (sin. $\frac{1}{2}$ $BAC)^2$. But the cosine of any arch is to the sine, as the radius to the tangent of

the same arch; therefore, $(AB+AC+BC)(AB+AC-BC) : (BC+(AB-AC))(BC-(AB-AC)) :: R^2 : (\tan \frac{1}{2} BAC)^2$; and
$\sqrt{(AB+AC+BC)(AB+AC-BC)} :$
$\sqrt{(BC+AB-AC)(BC-(AB-AC))} :: R : \tan \frac{1}{2} BAC$.

LEMMA II.

If there be two unequal magnitudes, half their difference added to half their sum is equal to the greater; and half their difference taken from half their sum is equal to the less.

Let AB and BC be two unequal magnitudes, of which AB is the greater; suppose AC bisected in D, and AE equal to BC. It is manifest that AC is the sum, and EB the difference of the magnitudes. And because AC is bisected in D, AD is equal to DC: but AE is also equal to BC, therefore DE is equal to DB, and DE or DB is half the difference of the magnitudes. But AB is equal to BD and DA, that is, to half the difference added to half the sum; and BC is equal to the excess of DC, half the sum above DB, half the difference. Therefore, &c. Q. E. D.

A E D B C

Cor. Hence, if the sum and the difference of two magnitudes be given, the magnitudes themselves may be found; for to half the sum add half the difference, and it will give the greater: from half the sum subtract half the difference, and it will give the less.

SECTION II.

OF THE RULES OF TRIGONOMETRICAL CALCULATION.

The General Problem which Trigonometry proposes to resolve is: *In any plane triangle, of the three sides and the three angles, any three being given, and one of these three being a side, to find any of the other three.*

The things here said to be given are understood to be expressed by their numerical values: the angles, in degrees, minutes, &c.; and the sides in feet, or any other known measure.

The reason of the restriction in this problem to those cases in which at least one side is given, is evident from this, that by the angles alone being given, the magnitudes of the sides are not determined. Innumerable triangles, equiangular to one another, may exist, without the sides of any one of them being equal to those of any other; though the ratios of their sides to one another will be the same in them all, (4. 6.). If therefore, only the three angles are given, nothing can be determined of the triangle but the ratios of the sides, which may be found by trigonometry, as being the same with the ratios of the sines of the opposite angles.

For the conveniency of calculaton, it is usual to divide the general problem into two; according as the triangle has, or has not, one of its angles a right angle.

PROB. I.

In a right angled triangle, of the three sides, and three angles, any two being given, besides the right angle, and one of those two being a side, it is required to find the other three.

It is evident, that when one of the acute angles of a right angled triangle is given, the other is given, being the complement of the former to a right angle; it is also evident that the sine of any of the acute angles is the cosine of the other.

This problem admits of several cases, and the solutions, or rules for calculation, which all depend on the first Proposition, may be conveniently exhibited in the form of a table; where the first column contains the things given; the second, the things required; and the third, the rules or propositions by which they are found.

GIVEN.	SOUGHT.	SOLUTION.	
CB and B, the hypotenuse and angle.	AC. AB.	R : sin B :: CB : AC. R : cos B :: CB : AB	1 2
AC and C, a side and one of the acute angles.	BC. AB.	Cos C : R :: AC : BC. R : tan C :: AC : AB.	3 4
CB and BA, the hypotenuse and a side.	C. AC.	CB : BA :: R : sin C. R : cos C :: CB : AC.	5 6
AC and AB, the two sides.	C. CB.	AC : AB :: R : tan C. Cos C : R :: AC : CB.	7 8

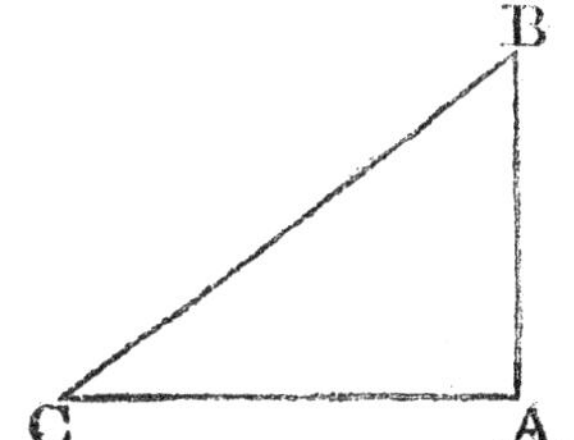

Remarks on the Solutions in the table.

In the second case, when AC and C are given to find the hypotenuse BC, a solution may also be obtained by help of the secant, for CA : CB :: R : sec. C.; if, therefore, this proportion be made R : sec. C :: AC : CB, CB will be found.

In the third case, when the hypotenuse BC and the side AB are given to find AC, this may be done either as directed in the Table, or by the 47th of the first; for since $AC^2 = BC^2 - BA^2$. $AC = \sqrt{BC^2 - BA^2}$. This value of AC will be easy to calculate by logarithms, if the quantity $BC - BA^2$ be separated into two multipliers, which may be done; because (Cor. 5. 2.), $BC^2 - BA^2 = (BC+BA)(BC-BA.)$. Therefore $AC = \sqrt{(BC+BA)(BC-BA)}$.

When AC and AB are given, BC may be found from the 47th, as in the preceding instance, for $BC = \sqrt{BA^2+AC^2}$. But BA^2+AC^2 cannot be separated into two multipliers; and therefore, when BA and AC are large numbers, this rule is inconvenient for computation by logarithms. It is best in such cases to seek first for the tangent of C, by the analogy in the Table, AC : AB :: R : tan. C; but if C itself is not required, it is sufficient, having found tan. C by this proportion, to take from the Trigonometric Tables the cosine that corresponds to tan. C, and then to compute CB from the proportion cos. C : R :: AC : CB.

PROBLEM II.

In an oblique angled triangle, of the three sides and three angles, any three being given, and one of these three being a side, it is required to find the other three.

This problem has four cases, in each of which the solution depends on some of the foregoing propositions.

CASE I.

Two angles A and B, and one side AB, of a triangle ABC, being given, to find the other sides.

SOLUTION.

Because the angles A and B are given, C is also given, being the supplement of A+B; and, (2.)
Sin. C : sin. A : : AB : BC; also,
Sin. C : sin. B : : AB : AC.

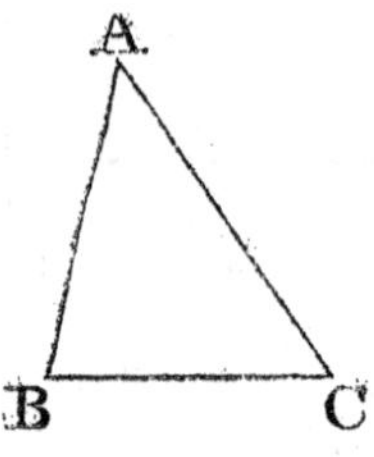

CASE II.

Two sides AB and AC, and the angle B opposite to one of them, being given, to find the other angles A and C, and also the other side BC.

SOLUTION.

The angle C is found from this proportion, AC : AB : : sin B : sin C. Also, A=180°−B−C; and then, sin. B : sin A : : AC : CB, by Case 1.

In this case, the angle C may have two values; for its sine being found by the proportion above, the angle belonging to that sine may either be that which is found in the tables, or it may be the supplement of it, (Cor. def. 4.). This ambiguity, however, does not arise from any defect in the solution, but from a circumstance essential to the problem, viz. that whenever AC is less than AB, there are two triangles which have the sides AB, AC, and the angle at B of the same magnitude in each, but which are nevertheless unequal, the angle opposite to AB in the one, being the supplement of that which is opposite to it in the other. The truth of this appears by describing from the centre A with the radius AC, an arch intersecting BC in C

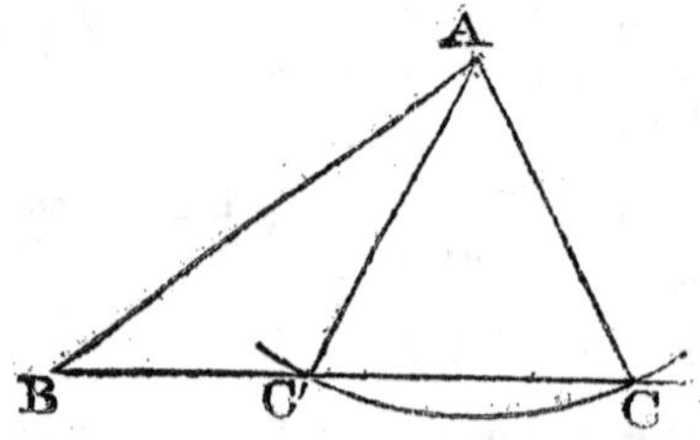

and C′; then, if AC and AC′ be drawn, it is evident that the triangles ABC, ABC′ have the side AB and the angle at B common, and the sides AC and AC′ equal, but have not the remaining side of the one equal to the remaining side of the other, that is, BC to BC′, nor their other angles equal, viz. BC′A to BCA, nor BAC′ to BAC. But in these triangles the angles ACB, AC′B are the supplements of one another. For the triangle CAC′ is isosceles, and the angle ACC′= the AC′C, and therefore, AC′B, which is the supplement of AC′C, is also the supplement of ACC′ or ACB; and these two angles, ACB, AC′B are the angles found by the computation above.

From these two angles, the two angles BAC, BAC′ will be found: the angle BAC is the supplement of the two angles ACB, ABC, (32. 1.), and therefore its sine is the same with the sine of the sum of ABC and ACB. But BAC′ is the difference of the angles ACB, ABC: for it is the difference of the angles AC′C and ABC, because AC′C, that is, ACC′ is equal to the sum of the angles ABC, BAC′, (32. 1.). Therefore, to find BC, having found C, make sin C : sin (C+B) :: AB : BC; and again, sin C : sin (C—B) :: AB : BC.

Thus, when AB is greater than AC, and C consequently greater than B, there are two triangles which satisfy the conditions of the question. But when AC is greater than AB, the intersections C and C′ fall on opposite sides of B, so that the two triangles have not the same angle at B common to them, and the solution ceases to be ambiguous, the angle required being necessarily less than B, and therefore an acute angle.

CASE III.

Two sides AB and AC, and the angle A, between them, being given to find the other angles B and C, and also the side BC.

SOLUTION.

First, make $AB+AC : AB-AC :: \tan\frac{1}{2}(C+B) : \tan\frac{1}{2}(C-B.)$. Then, since $\frac{1}{2}(C+B)$ and $\frac{1}{2}(C-B)$ are both given, B and C may be found. For $B=\frac{1}{2}(C+B)+\frac{1}{2}(C-B,)$ and $C=\frac{1}{2}(C+B)-\frac{1}{2}(C-B)$. (Lem. 2.).

To find BC.

Having found B, make sin B : sin A :: AC : BC.

But BC may also be found without seeking for the angle B and C; for $BC=\sqrt{AB^2-2\cos A\times AB.AC+AC^2}$, Prop. 6.

This method of finding BC is extremely useful in many geometrical investigations, but it is not very well adapted for computation by logarithms, because the quantity under the radical sign cannot be separated into simple multipliers. Therefore, when AB and AC are expressed by large numbers, the other solution, by finding the angles, and then computing BC, is preferable.

CASE IV.

The three sides AB, BC, AC, being given, to find the angles A, B, C.

SOLUTION I.

Take F such that BC : BA+AC :: BA−AC : F, then F is either the sum or the difference of BD, DC, the segments of the base, (5.). If F be greater than BC, F is the sum, and BC the difference of BD, DC; but, if F be less than BC, BC is the sum, and F the difference of BD and DC. In either case, the sum of BD and DC, and their difference being given, BD and DC are found. (Lem. 2.)

Then, (1.) CA : CD :: R : cos. C; and BA : BD :: R : cos. B; wherefore C and B are given, and consequently A.

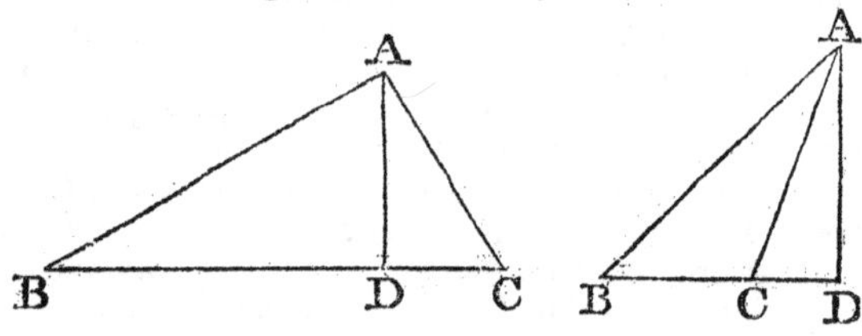

SOLUTION II.

Let D be the difference of the sides AB, AC. Then (Cor. 7.) $2\sqrt{AB.AC} : \sqrt{(BC+D)(BC-D)} :: R : \sin\frac{1}{2} BAC$.

SOLUTION III.

Let S be the sum of the sides BA and AC. Then (1 Cor. 8.) $2\sqrt{AB.AC} : \sqrt{(S+BC)(S-BC)} :: R : \cos\frac{1}{2} BAC$.

SOLUTION IV.

S and D retaining the significations above, (2 Cor. 8.) $\sqrt{(S+BC)(S-BC)} : \sqrt{(BC+D)(BC-D)} :: R : \tan\frac{1}{2} BAC$.

It may be observed of these four solutions, that the first has the advantage of being easily remembered, but that the others are rather more expeditious in calculation. The second solution is preferable to the third, when the angle sought is less than a right angle; on the other hand, the third is preferable to the second, when the angle sought is greater than a right angle; and in extreme cases, that is, when the angle sought is very acute or very obtuse, this distinction is very material to be considered. The reason is, that the sines of angles, which are nearly = 90°, or the cosines of angles, which are nearly = 0, vary very little for a considerable variation in the corresponding angles, as may be seen from looking into the tables of sines and cosines. The consequence of this is, that when the sine or cosine of such an angle is given, (that is, a sine or cosine nearly equal to the radius,) the angle itself cannot be very accurately found. If, for instance, the natural sine .9998500 is given, it will be immediately perceived from the tables, that the arch corresponding is between 89°, and 89°, 1′; but it cannot be found true to seconds, because the sines of 89° and of 89°, 1′, differ only by 50 (in the two last places,) whereas the arches themselves differ by 60 seconds. Two arches, therefore, that differ by 1″, or even by more than 1″, have the same sine in the tables, if they fall in the last degree of the quadrant.

The fourth solution, which finds the angle from its tangent, is not liable to this objection; nevertheless, when an arch approaches very near to 90, the variations of the tangents become excessive, and are too irregular to allow the proportional parts to be found with exactness, so that when the angle sought is extremely obtuse, and its half of consequence very near to 90, the third solution is the best.

It may always be known, whether the angle sought is greater or less than a right angle by the square of the side opposite to it being greater or less than the squares of the other two sides.

SECTION III.

CONSTRUCTION OF TRIGONOMETRICAL TABLES.

In all the calculations performed by the preceding rules, tables of sines and tangents are necessarily employed, the construction of which remains to be explained.

The tables usually contain the sines, &c. to every minute of the quadrant from 1′ to 90°, and the first thing required to be done, is to compute the sine of 1′, or of the least arch in the tables.

1. If ADB be a circle, of which the centre is C, DB any arch of that circle, and the arch DBE double of DB; and if the chords DE, DB be drawn, and also the perpendiculars to them from C, viz. CF, CG, it has been demonstrated, (8. 1. Sup.) that CG is a mean propor-

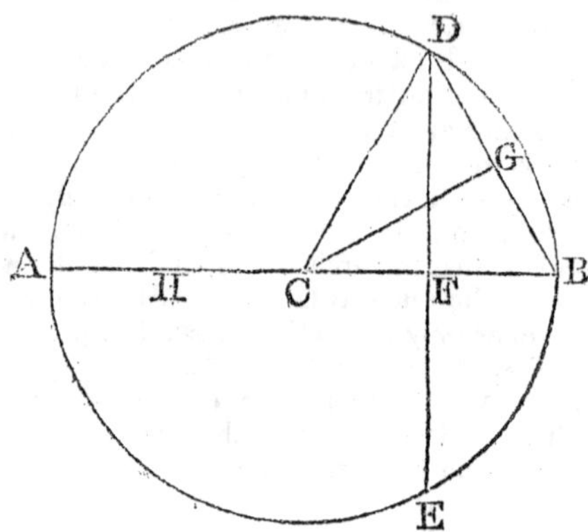

tional between AH, half the radius, and AF, the line made up of the radius and the perpendicular CF. Now CF is the cosine of the arch BD, and CG the cosine of the half of BD; whence the cosine of the half of any arch BD, of a circle of which the radius $=1$, is a mean proportional between $\frac{1}{2}$ and $1+\cos$ BD. Or, for the greater generality, supposing $A=$ any arch, $\cos\frac{1}{2}A$ is a mean proportional between $\frac{1}{2}$ and $1+\cos A$, and therefore $(\cos\frac{1}{2}A)^2=\frac{1}{2}(1+\cos A)$ or $\cos\frac{1}{2}A=\sqrt{\frac{1}{2}(1+\cos A)}$.

2. From this theorem, (which is the same that is demonstrated (9. 1. Sup.), only that it is here expressed trigonometrically), it is evident, that if the cosine of any arch be given, the cosine of half that arch may be found. Let BD, therefore, be equal to 60°, so that the chord BD $=$ radius, then the cosine or perpendicular CF was shown (9. 1. Sup.) to be $=\frac{1}{2}$, and therefore cos. $\frac{1}{2}$ BD, or $\cos 30°=\sqrt{\frac{1}{2}(1+\frac{1}{2})}=\sqrt{\frac{3}{4}}=\frac{\sqrt{3}}{2}$. In the same manner, $\cos 15°=\sqrt{\frac{1}{2}(1+\cos 30°)}$, and $\cos 7°, 30'=\sqrt{\frac{1}{2}(1+\cos 15°)}$, &c. In this way the cosine of 3°, 45′, of 1°, 52′, 30″, and so on, will be computed, till after twelve bisections of the arch of 60°, the cosine of 52″. 44‴. 93⁗. 45_{v}. is found. But from the cosine of an arch its sine may be found, for if from the square of the radius, that is, from 1, the square of the cosine be taken away, the remainder is the square of the sine, and its square root is the sine itself. Thus the sine of 52″. 44‴. 03⁗. 45_{v}. is found.

3. But it is manifest, that the sines of very small arches are to one another nearly as the arches themselves. For it has been shown that the number of the sides of an equilateral polygon inscribed in a circle may be so great, that the perimeter of the polygon and the cir-

cumference of the circle may differ by a line less than any given line, or, which is the same, may be nearly to one another in the ratio of equality. Therefore their like parts will also be nearly in the ratio of equality, so that the side of the polygon will be to the arch which it subtends nearly in the ratio of equality; and therefore, half the side of the polygon to half the arch subtended by it, that is to say, the sine of any very small arch will be to the arch itself, nearly in the ratio of equality. Therefore, if two arches are both very small, the first will be to the second as the sine of the first to the sine of the second. Hence, from the sine of 52″. 44‴. 03⁗. 45_v. being found, the sine of 1′ becomes known; for, as 52″. 44‴. 03⁗. 45_v. to 1, so is the sine of the former arch to the sine of the latter. Thus the sine of 1′ is found=0.0002908882.

4. The sine 1′ being thus found, the sines of 2′, of 3′, or of any number of minutes, may be found by the following proposition.

THEOREM.

Let AB, AC, AD be three such arches, that BC the difference of the first and second is equal to CD the difference of the second and third; the radius is to the cosine of the common difference BC as the sine of AC, the middle arch, to half the sum of the sines of AB and AD, the extreme arches.

Draw CE to the centre: let BF, CG, and DH perpendicular to AE, be the sines of the arches AB, AC, AD. Join BD, and let it meet CE in I; draw IK perpendicular to AE, also BL and IM perpendicular to DH. Then, because the arch BD is bisected in C, EC is at right angles to BD, and bisects it in I; also BI is the sine, and EI the cosine of BC or CD. And, since BD is bisected in I, and IM is parallel to BL, (2. 6.), LD is also bisected in M. Now BF is equal to HL, therefore, BF+DH=DH+HL=DL+2LH=2LM+2LH=2MH or 2KI; and therefore IK is half the sum of BF and DH. But because the triangles CGE, IKE are equiangular, CE : EI : : CG : IK, and it has been shown that EI=cos BC, and IK=½ (BF+DH); therefore R : cos BC : : sin AC : ½ (sin AB + sin AD). Q. E. D.

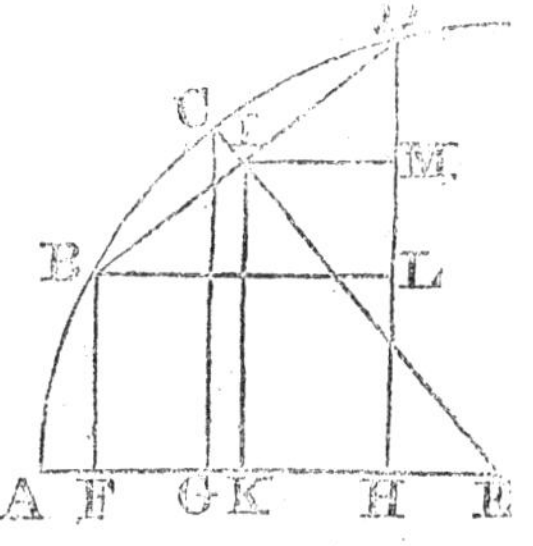

COR. Hence, if the point B coincide with A,
R : cos. BC : : sin. BC : ½ sin. BD, that is, the radius is to the cosine of any arch, as the sine of the arch is to half the sine of twice the arch; or if any arch=A, ½ sin. 2A=sin. A×cos. A, or sin. 2A=2 sin. A×cos. A.

Therefore also, sin 2′=2′ sin 1′×cos 1′; so that from the sine and cosine of one minute the sine of 2′ is found.

Again, 1′, 2′, 3′, being three such arches that the difference between the first and second is the same as between the second and third, R : cos 1′ :: sin 2′ : $\frac{1}{2}$ (sin 1′+sin 3′), or sin 1′+sin 3′=2 cos 1′+sin 2′, and taking sin 1′ from both, sin 3′=2 cos 1′×sin. 2′ – sin 1.

In like manner, sin 4′=2′ cos 1′×sin 3′–sin 2′,

sin 5′=2′ cos 1′×sin 4′–sin 3′,

sin 6′=2′ cos 1′×sin 5′–sin 4′, &c.

Thus a table containing the sines for every minute of the quadrant may be computed; and as the multiplier, cos 1′ remains always the same, the calculation is easy.

For computing the sines of arches that differ by more than 1′, the method is the same. Let A, A+B, A+2B be three such arches, then, by this theorem, R : cos B :: sin (A+B) : $\frac{1}{2}$ (sin A+sin (A+2B)); and therefore making the radius 1,

sin A+sin (A+2B)=2 cos B ×sin (A+B),

or sin (A+2B)=2 cos B×sin (A+B)–sin A.

By means of these theorems, a table of the sines, and consequently also of the cosines, of arches of any number of degrees and minutes, from 0 to 90, may be constructed. Then, because $\tan A=\frac{\sin A}{\cos A}$, the table of tangents is computed by dividing the sine of any arch by the cosine of the same arch. When the tangents have been found in this manner as far as 45°, the tangents for the other half of the quadrant may be found more easily by another rule. For the tangent of an arch above 45° being the co-tangent of an arch as much under 45°; and the radius being a mean proportional between the tangent and co-tangent of any arch, (1 Cor. def. 9.), it follows, if the difference between any arch and 45° be called D, that tan (45°–D) : 1 :: 1 : tan (45°+D), so that $\tan(45°+D)=\frac{1}{\tan(45°-D)}$.

Lastly, the secants are calculated from (Cor. 2. def. 9.) where it is shown that the radius is a mean proportional between the cosine and the secant of any arch, so that if A be any arch, $\sec A=\frac{1}{\cos A}$.

The versed sines are found by subtracting the cosines from the radius.

5. The preceding Theorem is one of four, which, when arithmetically expressed, are frequently used in the application of trigonometry to the solution of problems.

1mo, If in the last Theorem, the arch AC=A, the arch BC=B, and the radius EC=1, then AD=A+B, and AB = A–B; and by what has just been demonstrated,

1 : cos B :: sin A : $\frac{1}{2}$ sin (A+B)+$\frac{1}{2}$ sin (A–B),

and therefore

sin A×cos B = $\frac{1}{2}$ sin (A+B) + $\frac{1}{2}$ (A–B).

2do, Because BF, IK, DH are parallel, the straight lines BD and FH are cut proportionally, and therefore FH, the difference of the straight lines FE and HE, is bisected in K; and therefore, as was shown in the last Theorem, KE is half the sum of FE and HE, that is, of the cosines of the arches AB and AD. But because of the similar triangles EGC, EKI, EC : EI : : GE : EK; now, GE is the cosine of AC, therefore,

$$R : \cos BC : : \cos AC : \tfrac{1}{2}\cos AD + \tfrac{1}{2}\cos AB,$$

$$\text{or } 1 : \cos B : : \cos A : \tfrac{1}{2}\cos(A+B) + \tfrac{1}{2}\cos(A-B);$$

and therefore,

$$\cos A \times \cos B = \tfrac{1}{2}\cos(A+B) + \tfrac{1}{2}\cos(A-B);$$

3tio, Again, the triangles IDM, CEG are equiangular, for the angles KIM, EID are equal, being each of them right angles, and therefore, taking away the angle EIM, the angle DIM is equal to the angle EIK, that is, to the angle ECG; and the angles DMI, CGE are also equal, being both right angles, and therefore the triangles IDM, CGE have the sides about their equal angles proportionals, and consequently, EC : CG : : DI : IM; now, IM is half the difference of the cosines FE and EH, therefore,

$$R : \sin AC : : \sin BC : \tfrac{1}{2}\cos AB - \tfrac{1}{2}\cos AD,$$

$$\text{or } 1 : \sin A : : \sin B : \tfrac{1}{2}\cos(A-B) - \tfrac{1}{2}\cos(A+B);$$

and also,

$$\sin A \times \sin B = \tfrac{1}{2}\cos(A-B) - \tfrac{1}{2}\cos(A+B).$$

4to, Lastly, in the same triangles ECG, DIM, EC : EG : : ID : DM; now, DM is half the difference of the sines DH and BE, therefore,

$$R : \cos AC : : \sin BC : \tfrac{1}{2}\sin AD - \tfrac{1}{2}\sin AB,$$

$$\text{or } 1 : \cos A : : \sin B : \tfrac{1}{2}\sin(A+B) - \tfrac{1}{2}\sin(A+B);$$

and therefore,

$$\cos A \times \sin B = \tfrac{1}{2}\sin(A+B) - \tfrac{1}{2}\sin(A-B).$$

6. If therefore A and B be any two arches whatsoever, the radius being supposed 1;

I. $\sin A \times \cos B = \tfrac{1}{2}\sin(A+B) + \tfrac{1}{2}\sin(A-B)$.
II. $\cos A \times \cos B = \tfrac{1}{2}\cos(A-B) + \tfrac{1}{2}\cos(A+B)$.
III. $\sin A \times \sin B = \tfrac{1}{2}\cos(A-B) - \tfrac{1}{2}\cos(A+B)$.
IV. $\cos A \times \sin B = \tfrac{1}{2}\sin(A+B) - \tfrac{1}{2}\sin(A-B)$.

From these four Theorems are also deduced other four.

For adding the first and fourth together,

$$\sin A \times \cos B + \cos A \times \sin B = \sin(A+B).$$

Also, by taking the fourth from the first

$$\sin A \times \cos B - \cos A \times \sin B = \sin(A-B).$$

Again, adding the second and third,

$$\cos A \times \cos B + \sin A \times \sin B = \cos(A-B);$$

And, lastly, subtracting the third from the second,

$$\cos A \times \cos B - \sin A \times \sin B = \cos(A+B).$$

7. Again, since by the first of the above theorems, $\sin A\times\cos B=\frac{1}{2}\sin(A+B)+\frac{1}{2}\sin(A-B)$, if $A+B=S$, and $A-B=D$, then (Lem. 2.) $A=\frac{S+D}{2}$, and $B=\frac{S-D}{2}$; wherefore $\sin\frac{S+D}{2}\times\cos\frac{S-D}{2}=\frac{1}{2}\sin S+\frac{1}{2}\sin D$. But as S and D may be any arches whatever, to preserve the former notation, they may be called A and B, which also express any arches whatever: thus,

$$\sin\frac{A+B}{2}\times\cos\frac{A-B}{2}=\frac{1}{2}\sin A+\frac{1}{2}\sin B\text{, or}$$

$$2\sin\frac{A+B}{2}\times\cos\frac{A-B}{2}=\sin A+\sin B.$$

In the same manner, from Theor. 2 is derived,

$2\cos\frac{A+B}{2}\times\cos\frac{A-B}{2}=\cos B+\cos A$. From the 3d,

$2\sin\frac{A+B}{2}\times\sin\frac{A-B}{2}=\cos B-\cos A$; and from the 4th,

$2\cos\frac{A+B}{2}\times\sin\frac{A-B}{2}=\sin A-\sin B$.

In all these Theorems, the arch B is supposed less than A.

8. Theorems of the same kind with respect to the tangents of arches may be deduced from the preceding. Because the tangent of any arch is equal to the sine of the arch divided by its cosine, $\tan(A+B)=\frac{\sin(A+B)}{\cos(A+B)}$. But it has just been shown, that $\sin(A+B)=\sin A\times\cos B+\cos A\times\sin B$, and that $\cos(A+B)=\cos A\times\cos B-\sin A\times\sin B$; therefore $\tan(A+B)=\frac{\sin A\times\cos B+\cos A\times\sin B}{\cos A\times\cos B-\sin A\times\sin B}$, and dividing both the numerator and denominator of this fraction by $\cos A\times\cos B$, $\tan(A+B)=\frac{\tan A+\tan B}{1\ \tan A\times\tan B}$. In like manner, $\tan(A-B)=\frac{\tan A\ \tan B}{1+\tan A\times\tan B}$.

9. If the theorem demonstrated in Prop. 3. be expressed in the same manner with those above, it gives

$$\frac{\sin A+\sin B}{\sin A-\sin B}=\frac{\tan\frac{1}{2}(A+B)}{\tan\frac{1}{2}(A-B)}.$$

Also by Cor. 1, to the 3d,

$$\frac{\cos A+\cos B}{\cos A-\cos B}=\frac{\cot\frac{1}{2}(A+B)}{\tan\frac{1}{2}(A-B)}.$$

And by Cor. 2, to the same proposition,

$$\frac{\sin A+\sin B}{\cos A+\cos B}=\frac{\tan\frac{1}{2}(A+B)}{R},$$ or since R is here supposed = 1, $$\frac{\sin A+\sin B}{\cos A+\cos B}=\tan\tfrac{1}{2}(A+B).$$

10. In all the preceding theorems, R, the radius is supposed = 1, because in this way the propositions are most concisely expressed, and are also most readily applied to trigonometrical circulation. But if it be required to enunciate any of them geometrically, the multiplier R, which has disappeared, by being made = 1, must be restored, and it will always be evident from inspection in what terms this multiplier is wanting. Thus, Theor. 1, $2\sin A\times\cos B=\sin(A+B)+\sin(A-B)$, is a true proposition, taken arithmetically; but taken geometrically, is absurd, unless we supply the radius as a multiplier of the terms on the right hand of the sine of equality. It then becomes $2\sin A\times\cos B=R(\sin(A+B)+\sin(A-B))$; or twice the rectangle under the sine of A, and the cosine of B equal to the rectangle under the radius, and the sum of the sines of A+B and A−B.

In general, the number of *linear multipliers*, that is, of lines whose numerical values are multiplied together, must be the same in every term, otherwise we will compare unlike magnitudes with one another.

The propositions in this section are useful in many of the higher branches of the Mathematics, and are the foundation of what is called the *Arithmetic of sines*.

ELEMENTS

OF

SPHERICAL

TRIGONOMETRY.

PROP. I.

If a sphere be cut by a plane through the centre, the section is a circle, having the same centre with the sphere, and equal to the circle by the revolution of which the sphere was described.

For all the straight lines drawn from the centre to the superficies of the sphere are equal to the radius of the generating semicircle, (Def. 7. 3. Sup.). Therefore the common section of the spherical superficies, and of a plane passing through its centre, is a line, lying in one plane, and having all its points equally distant from the centre of the sphere; therefore it is the circumference of a circle, (Def. 11. 1.), having for its centre the centre of the sphere, and for its radius the radius of the sphere, that is, of the semicircle by which the sphere has been described. It is equal, therefore, to the circle, of which that semicircle was a part. Q. E. D.

DEFINITIONS.

I.

Any circle, which is a section of a sphere by a plane through its centre, is called a great circle of the sphere.

Cor. All great circles of a sphere are equal; and any two of them bisect one another.

They are all equal, having all the same radii, as has just been shown; and any two of them bisect one another, for as they have the same centre, their common section is a diameter of both, and therefore bisects both.

II.

The pole of a great circle of a sphere is a point in the superficies of the sphere, from which all straight lines drawn to the circumference of the circle are equal.

III.

A spherical angle is an angle on the superficies of a sphere, contained by the arches of two great circles which intersect one another; and is the same with the inclination of the planes of these great circles.

IV.

A spherical triangle is a figure, upon the superficies of a sphere, comprehended by three arches of three great circles, each of which is less than a semicircle.

PROP. II.

The arch of a great circle, between the pole and the circumference of another great circle, is a quadrant.

Let ABC be a great circle, and D its pole; if DC, an arch of a great circle, pass through D, and meet ABC in C, the arch DC is a quadrant.

Let the circle, of which CD is an arch, meet ABC again in A, and let AC be the common section of the planes of these great circles, which will pass through E, the centre of the sphere: Join DA, DC. Because AD=DC, (Def. 2.), and equal straight lines, in the same circle, cut off equal arches, (28. 3.) the arch AD = the arch DC; but ADC is a semicircle, therefore the arches AD, DC are each of them quadrants. Q. E. D.

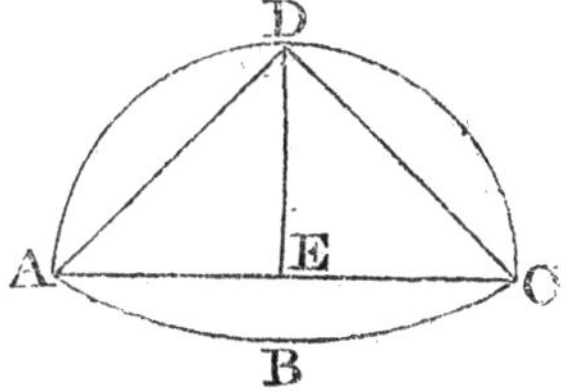

Cor. 1. If DE be drawn, the angle AED is a right angle; and DE being therefore at right angles to every line it meets with in the plane of the circle ABC, is at right angles to that plane, (4. 2. Sup.). Therefore the straight line drawn from the pole of any great circle to the centre of the sphere is at right angles to the plane of that circle; and, conversely, a straight line drawn from the centre of the sphere perpendicular to the plane of any greater circle, meets the superficies of the sphere in the pole of that circle.

Cor. 2. The circle ABC has two poles, one on each side of its plane, which are the extremities of a diameter of the sphere perpendicular to the plane ABC; and no other points but these two can be poles of the circle ABC.

PROP. III.

If the pole of a great circle be the same with the intersection of other two great circles: the arch of the first mentioned circle intercepted between the other two, is the measure of the spherical angle which the same two circles make with one another.

Let the great circles BA, CA on the superficies of a sphere, of which the centre is D, intersect one another in A, and let BC be an arch of another great circle, of which the pole is A; BC is the measure of the spherical angle BAC.

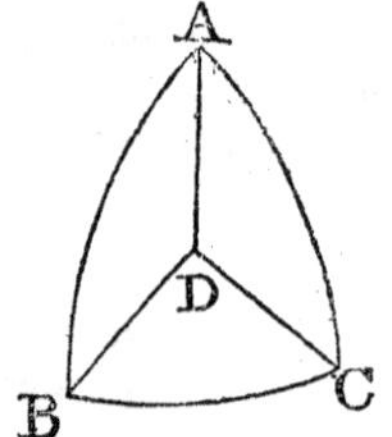

Join AD, DB, DC; since A is the pole of BC, AB, AC are quadrants, (2.), and the angles ADB, ADC are right angles: therefore (4. def. 2 Sup.), the angle CDB is the inclination of the planes of the circles AB, AC, and is (def. 3.) equal to the spherical angle BAC; but the arch BC measures the angle BDC, therefore it also measures the spherical angle BAC.* Q. E. D.

Cor. If two arches of great circles, AB and AC, which intersect one another in A, be each of them quadrants, A will be the pole of the great circle which passes through E and C the extremities of those arches. For since the arches AB and AC are quadrants, the angles ADB, ADC are right angles, and AD is therefore perpendicular to the plane BDC, that is, to the plane of the great circle which passes through B and C. The point A is therefore (Cor. 1. 2.) the pole of the great circle which passes through B and C.

PROP. VI.

If the planes of two great circles of a sphere be at right angles to one another, the circumference of each of the circles passes through the poles of the other; and if the circumference of one great circle pass through the poles of another, the planes of these circles are at right angles.

Let ACBD, AEBF be two great circles, the planes of which are right angles to one another, the poles of the circle AEBF are in the circumference ACBD, and the poles of the circle ACBD in the circumference AEBF.

From G the centre of the sphere, draw GC in the plane ACBD perpendicular to AB. Then because GC in the plane ACBD, at

* When in any reference no mention is made of a Book, or of the Plane Trigonometry, the Spherical Trigonometry is meant.

right angles to the plane AEBF, is at right angles to the common section of the two planes, it is (Def. 2. 2. Sup.) also at right angles to the plane AEBF, and therefore (Cor. 1. 2.) C is the pole of the circle AEBF; and if CG be produced to D, D is the other pole of the circle AEBF.

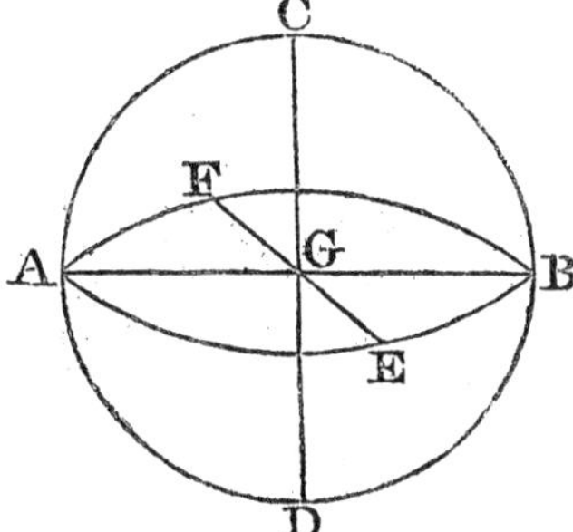

In the same manner, by drawing GE in the plane AEBF, perpendicular to AB, and producing it to F, it has shown that E and F are the poles of the circle ACBD. Therefore, the poles of each of these circles are in the circumference of the other.

Again, If C be one of the poles of the circle AEBF, the great circle ACBD which passes through C, is at right angles to the circle AEBF. For, CG being drawn from the pole to the centre of the circle AEBF, is at right angles (Cor. 1. 2.) to the plane of that circle; and therefore, every plane passing through CG (17. 2. Sup.) is at right angles to the plane AEBF; now, the plane ACBD passes through CG. Therefore, &c. Q. E. D.

Cor. 1. If of two great circles, the first passes through the poles of the second, the second also passes through the poles of the first. For, if the first passes through the poles of the second, the plane of the first must be at right angles to the plane of the second, by the second part of this proposition; and therefore, by the first part of it, the circumference of each passes through the poles of the other.

Cor. 2. All greater circles that have a common diameter have their poles in the circumference of a circle, the plane of which is perpendicular to that diameter.

PROP. V.

In isosceles spherical triangles the angles at the base are equal.

Let ABC be a spherical triangle, having the side AB equal to the side AC; the spherical angles ABC and ACB are equal.

Let C be the centre of the sphere; join DB, DC, DA, and from A on the straight lines DB, DC, draw the perpendiculars AE, AF; and from the points E and F draw in the plane DBC the straight lines EG, FG perpendicular to DB and DC, meeting one another in G: Join AG.

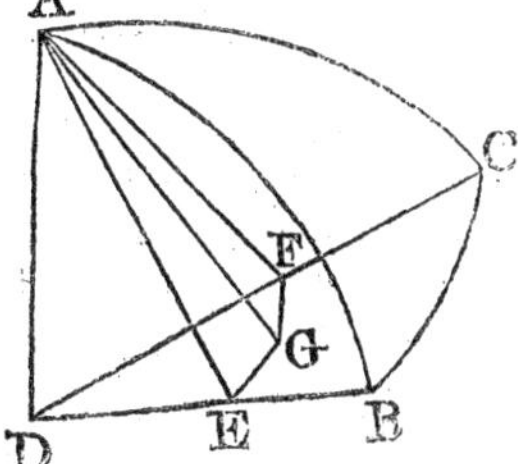

Because DE is at right angles to each of the straight lines AE, EG, it is at right angles to the plane AEG, which passes through AE, EG (4. 2. Sup.); and therefore, every plane that passes through DE is at right angles to the plane AEG (17. 2. Sup.); wherefore, the plane DBC is at right angles to the plane AEG. For the same reason, the plane DBC is at right angles to the plane AFG, and therefore AG, the common section of the planes AFG, AEG is at right angles (18. 2. Sup.) to the plane DBC, and the angles AGE, AGF are consequently right angles.

But since the arch AB is equal to the arch AC, the angle ADB is equal to the angle ADC. Therefore the triangles ADE, ADF, have the angles EDA, FDA, equal, as also the angles AED, AFD, which are right angles; and they have the side AD common, therefore the other sides are equal, viz. AE to AF, (26. 1.), and DE to DF. Again, because the angles AGE, AGF are right angles, the squares on AG and GE are equal to the square of AE; and the squares of AG and GF to the square of AF. But the squares of AE and AF are equal, therefore the squares of AG and GE are equal to the squares of AG and GF, and taking away the common square of AG, the remaining squares of GE and GF are equal, and GE is therefore equal to GF. Wherefore, in the triangles AFG, AEG, the side GF is equal to the side GE, and AF has been proved to be equal to AE, and the base AG is common; therefore, the angle AFG is equal to the angle AEG (8. 1.). But the angle AFG is the angle which the plane ADC makes with the plane DBC (4. def. 2. Sup.) because FA and FG, which are drawn in these planes, are at right angles to DF, the common section of the planes. The angle AFG (3. def.) is therefore equal to the spherical angle ACB; and, for the same reason, the angle AEG is equal to the spherical angle ABC. But the angles AFG, AEG are equal. Therefore the spherical angles ACB, ABC are also equal. Q. E. D.

PROP. VI.

If the angles at the base of a spherical triangle be equal, the triangle is isosceles.

Let ABC be a spherical triangle having the angles ABC, ACB equal to one another; the sides AC and AB are also equal.

Let D be the centre of the sphere; join DB, DC, DA, and from A on the straight lines DB, DC, draw the perpendiculars AE, AF; and from the points E and F, draw in the plane DBC the straight lines EG, FG perpendicular to DB and DC, meeting one another in G; join AG.

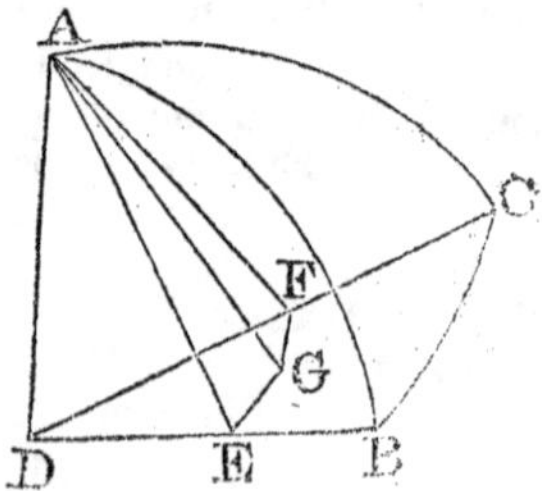

Then, it may be proved, as was done in the last proposition, that AG is at right angles to the plane BCD, and that therefore the angles AGF, AGE are right angles, and also that the angles AFG, AEG are equal to the angles which the planes DAC, DAB make with the plane DBC. But because the spherical angles ACB, ABC are equal, the angles which the planes DAC, DAB make with the plane DBC are equal, (3. def.) and therefore the angles AFG, AEG are also equal. The triangles AGE, AGF have therefore two angles of the one equal to two angles of the other, and they have also the side AG common, wherefore they are equal, and the side AF is equal to the side AE.

Again, because the triangles ADF, ADE are right angled at F and E, the squares of DF and FA are equal to the square of DA, that is, to the squares of DE and DA; now, the square of AF is equal to the square of AE, therefore the square of DF is equal to the square of DE, and the side DF to the side DE. Therefore, in the triangles DAF, DAE, because DF is equal to DE and DA common, and also AF equal to AE, the angle ADF is equal to the angle ADE; therefore also the arches AC and AB, which are the measures of the angles ADF and ADE, are equal to one another; and the triangle ABC is isosceles. Q. E. D.

PROP. VII.

Any two sides of a spherical triangle are greater than the third.

Let ABC be a spherical triangle, any two sides AB, BC are greater than the third side AC.

Let D be the centre of the sphere; join DA, DB, DC.

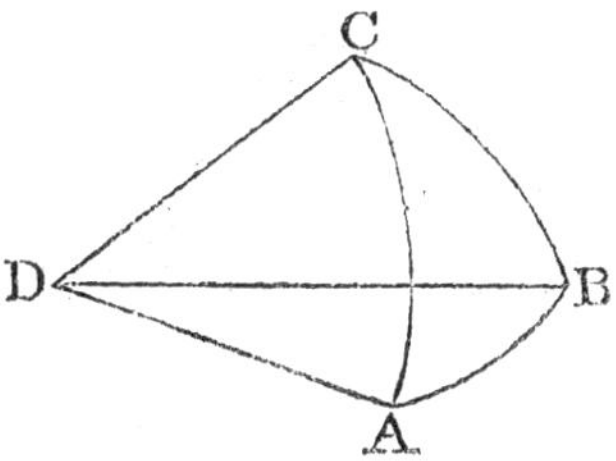

The solid angle at D is contained by three plane angles ADB, ADC, BDC; any two of which, ADB, BDC are greater (20. 2. Sup.) than the third ADC; and therefore any two of the arches AB, AC, BC, which measure these angles, as AB and BC, must also be greater than the third AC. Q. E. D.

PROP. VIII.

The three sides of a spherical triangle are less than the circumference of a great circle.

Let ABC be a spherical triangle as before, the three sides AB, BC, AC are less than the circumference of a great circle.

Let D be the centre of the sphere: The solid angle at D is contained by three plane angles BDA, BDC, ADC, which together are less than four right angles (21. 2. Sup.) therefore the sides AB, BC, AC, which are the measures of these angles, are together less than four quadrants described with the radius AD, that is, than the circumference of a great circle. Q. E. D.

PROP. IX.

In a spherical triangle the greater angle is opposite to the greater side; and conversely.

Let ABC be a spherical triangle, the greater angle A is opposed to the greater side BC.

Let the angle BAD be made equal to the angle B, and then BD, DA will be equal, (6.), and therefore AD, DC are equal to BC; but AD, DC are greater than AC (7.), therefore BC is greater than AC, that is, the greater angle A is opposite to the greater side BC. The converse is demonstrated as Prop. 19. 1. Elem. Q. E. D.

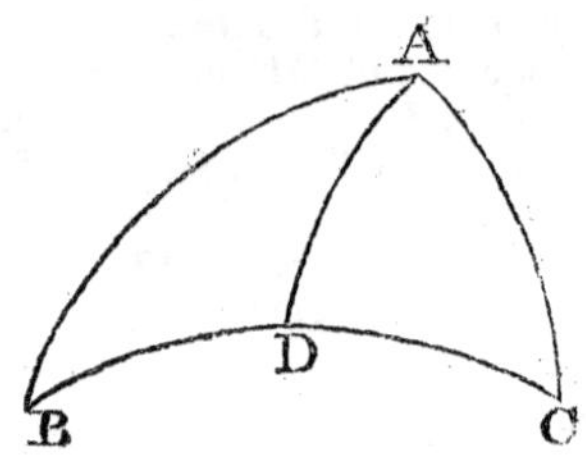

PROP. X.

According as the sum of two of the sides of a spherical triangle, is greater than a semicircle, equal to it, or less, each of the interior angles at the base is greater than the exterior and opposite angle at the base, equal to it, or less; and also the sum of the two interior angles at the base greater than two right angles, equal to two right angles, or less than two right angles.

Let ABC be a spherical triangle, of which the sides are AB and BC; produce any of the two sides as AB, and the base AC, till they meet again in D; then, the arch ABD is a semicircle, and the spherical angles at A and D are equal, because each of them is the inclination of the circle ABD to the circle ACD.

1. If AB, BC be equal to a semicircle, that is, to AD, BC will be equal to BD, and therefore (5.) the angle D, or the angle A will be equal to the angle BCD, that is, the interior angle at the base equal to the exterior and opposite.

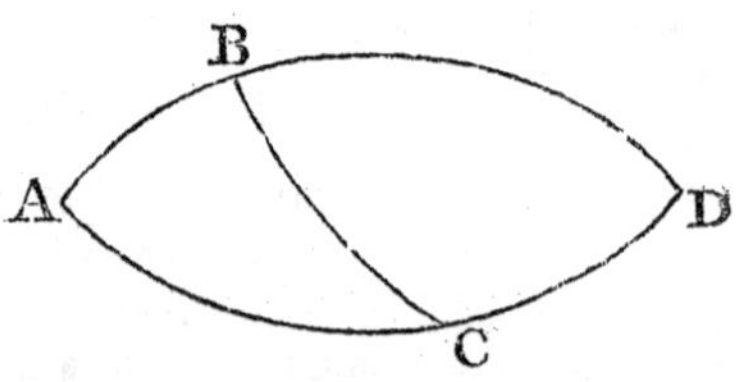

2. If AB, BC together be greater than a semicircle, that is, greater than ABD, BC will be greater than BD; and therefore (9.), the angle D, that is, the angle A, is greater than the angle BCD.

3. In the same manner it is shown, if AB, BC together be less than a semicircle, that the angle A is less than the angle BCD.

Now, since the angles BCD, BCA are equal to two right angles, if the angle A be greater than BCD, A and ACB together will be greater than two right angles. If A be equal to BCD, A and ACB together, will be equal to two right angles; and if A be less than BCD, A and ACB will be less than two right angles. Q. E. D.

PROP. XI.

If the angular points of a spherical triangle be made the poles of three great circles, these three circles by their intersections will form a triangle, which is said to be supplemental to the former; and the two triangles are such, that the sides of the one are the supplements of the arches which measure the angles of the other.

Let ABC be a spherical triangle; and from the points A, B, and C as poles, let the great circles FE, ED, DF be described, intersecting one another in F, D and E; the sides of the triangle FED are the supplements of the measures of the angles A, B, C, viz. FE of the angle BAC, DE of the angle ABC, and DF of the angle ACB: And again, AC is the supplement of the angle DFE, AB of the angle FED, and BC of the angle EDF.

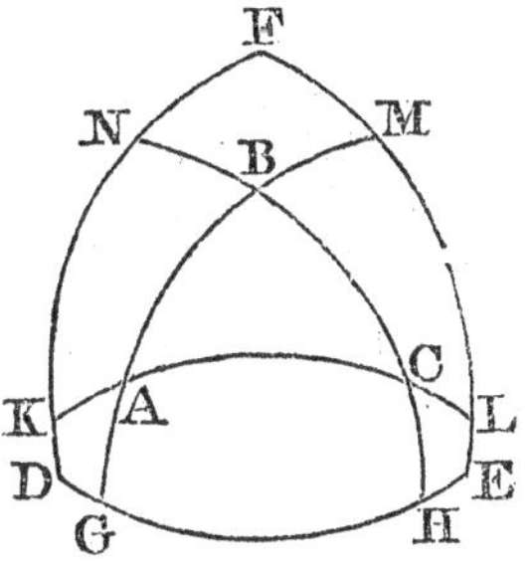

Let AB produced meet DE, EF in G, M; let AC meet FD, FE in K, L; and let BC meet FD, DE in N, H.

Since A is the pole of FE, and the circle AC passes through A, EF will pass through the pole of AC (1. Cor. 4.) and since AC passes through C, the pole of FD, FD will pass through the pole of AC; therefore the pole of AC is in the point F, in which the arches DF, EF intersect each other. In the same manner, D is the pole of BC, and E the pole of AB.

And since F, E are the poles of AL, AM, the arches FL and EM (2.) are quadrants, and FL, EM together, that is, FE and ML together, are equal to a semicircle. But since A is the pole of ML, ML is the measure of the angle BAC (3.), consequently FE is the supplement of the measure of the angle BAC. In the same manner, ED, DF are the supplements of the measures of the angles, ABC, BCA.

Since likewise CN, BH are quadrants, CN and BH together, that is, NH and BC together, are equal to a semicircle; and since D is the pole of NH, NH is the measure of the angle FDE, therefore the measure of the angle FDE is the supplement of the side BC. In the same manner, it is shown that the measures of the angles DEF, EFD are the supplements of the sides AB, AC in the triangle ABC. Q. E. D.

PROP. XII.

The three angles of a spherical triangle are greater than two, and less than six, right angles.

The measures of the angles A, B, C, in the triangle ABC, together with the three sides of the supplemental triangle DEF, are (11.) equal to three semicircles; but the three sides of the triangle FDE, are (8.) less than two semicircles; therefore the measures of the angles A, B, C are greater than a semicircle; and hence the angles A, B, C are greater than two right angles.

And because the interior angles of any triangle, together with the exterior, are equal to six right angles, the interior alone are less than six right angles. Q. E. D.

PROP. XIII.

If to the circumference of a great circle, from a point, in the surface of the sphere, which is not the pole of that circle, arches of great circles be drawn; the greatest of these arches is that which passes through the pole of the first-mentioned circle, and the supplement of it is the least; and of the other arches, that which is nearer to the greatest is greater than that which is more remote.

Let ADB be the circumference of a great circle, of which the pole is H, and let C be any other point; through C and H let the semicircle ACB be drawn meeting the circle ADB in A and B; and let the arches CD, CE, CF also be described. From C draw CG perpendicular to AB, and then, because the circle AHCB which passes through H, the pole of the circle ADB, is at right angles to ADB, CG is perpendicular to the plane ADB. Join GD, GE, GF, CA, CD, CE, CF, CB.

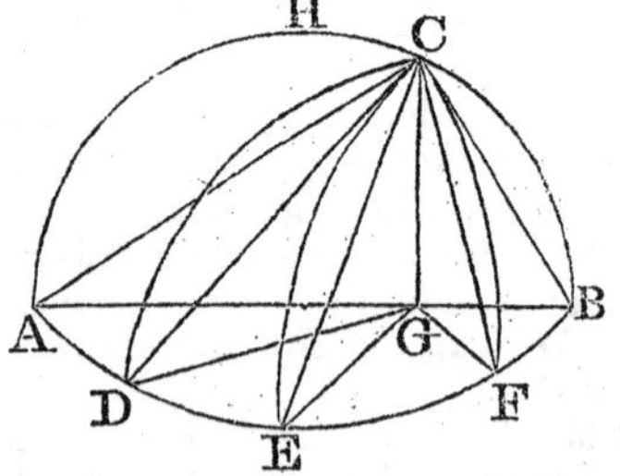

Because AB is the diameter of the circle ADB, and G a point in it, which is not the centre, (for the centre is in the point where the perpendicular from H meets AB), therefore AG, the part of the diameter in which the centre is, is the greatest, (7. 3.), and GB the least of all the straight lines that can be drawn from G to the circumference; and GD, which is nearer to AB, is greater than GE, which is more remote. But the triangles CGA, CGD are right angled at G, and therefore $AC^2 = AG^2 + GC^2$, and $DC^2 = DG^2 + GC^2$; but $AG^2 + GC^2 > DG^2 + GC^2$; because $AG > DG$; therefore $AC^2 > DC^2$, and $AC > DC$. And because the chord AC is greater than the chord DC, the arch AC is greater than the arch DC. In the same manner, since GD is greater than GE, and GE than GF, it is shown that CD is greater than CE, and CE than CF. Wherefore also the arch CD is greater than the arch CE, and the arch GE greater than the arch CF, and CF than CB, that is, of all the arches of greater circles drawn from C to the circumference of the circle ADB, AC which passes through the pole H, is the greatest, and CB its supplement is the least; and of the others, that which is nearer to AC the greatest, is greater than that which is more remote. Q. E. D.

PROP. XIV.

In a right angled spherical triangle, the sides containing the right angle are of the same affection with the angles opposite to them, that is, if the sides be greater or less than quadrants, the opposite angles will be greater or less than right angles, and conversely.

Let ABC be a spherical triangle, right angled at A, any side AB will be of the same affection with the opposite angle ACB.

Produce the arches AC, AB, till they meet again in D, and bisect AD in E. Then ACD, ABD are semicircles, and AE an arch of 90°. Also, because CAB is by hypothesis a right angle, the plane of the circle ABD is perpendicular to the plane of the circle ACD, so that the pole of ACD is in ABD, (cor. 1. 4.), and is therefore the point E. Let EC be an arch of a great circle passing through E and C.

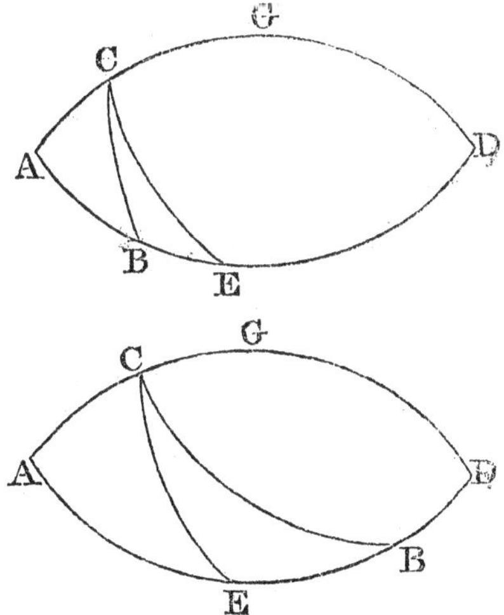

Then because E is the pole of the circle ACD, EC is a (2.) quadrant, and the plane of the circle EC (4.) is at right angles to the plane of the circle ACD, that is, the spherical angle ACE is a right angle; and therefore, when AB is less than AE, the angle ACB, being less than ACE, is less than a right angle. But when AB is greater than AE, the angle ACB is greater than ACE, or than a right angle. In the same way may the converse be demonstrated. Therefore, &c. Q. E. D.

PROP. XV.

If the two sides of a right angled spherical triangle about the right angle be of the same affection, the hypotenuse will be less than a quadrant; and if they be of different affection, the hypotenuse will be greater than a quadrant.

Let ABC be a right angled spherical triangle; according as the two sides AB, AC are of the same or of different affection, the hypotenuse BC will be less, or greater than a quadrant.

The construction of the last proposition remaining, bisect the semicircle ACD in G, then AG will be an arch of 90°, and G will be the pole of the circle ABD.

1. Let AB, AC be each less than 90°. Then, because C is a point on the surface of the sphere, which is not the pole of the circle ABD, the arch CGD, which passes through G the pole of ABD is greater than CE, (13.), and CE greater than CB. But CE is a quadrant, as was before shown, therefore CB is less than a quadrant. Thus also it is proved of the right angled triangle CDB, (right angled at D), in which each of the sides CD, DB is greater than a quadrant, that the hypotenuse BC is less than a quadrant.

2. Let AC be less, and AB greater than 90°. Then because CB falls between CGD and CE, it is greater (13.) than CE, that is, than a quadrant. Q. E. D.

Cor. 1. Hence conversely, if the hypotenuse of a right angled triangle be greater or less than a quadrant, the sides will be of different or the same affection.

Cor. 2. Since (14.) the oblique angles of a right angled spherical triangle have the same affection with the opposite sides, therefore, according as the hypotenuse is greater or less than a quadrant, the oblique angles will be of different, or of the same affection.

Cor. 3. Because the sides are of the same affection with the opposite angles, therefore when an angle and the side adjacent are of the same affection, the hypotenuse is less than a quadrant: and conversely.

PROP. XVI.

In any spherical triangle, if the perpendicular upon the base from the opposite angle fall within the triangle, the angles at the base are of the same affection; and if the perpendicular fall without the triangle, the angles at the base are of different affection.

Let ABC be a spherical triangle, and let the arch CD be drawn from C perpendicular to the base AB.

1. Let CD fall within the triangle; then, since ADC, BDC are right angled spherical triangles, the angles A, B must each be of the same affection with CD, (14).

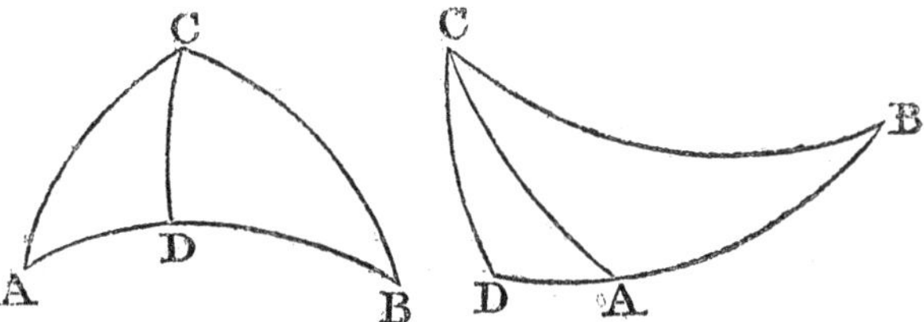

2. Let CD fall without the triangle; then (14.) the angle B is of the same affection with CD; and the angle CAD is of the same affection with CD; therefore the angle CAD and B are of the same affection, and the angle CAB and B are therefore of different affections. Q. E. D.

Cor. Hence, if the angles A and B be of the same affection, the perpendicular will fall within the base; for if it did not, A and B would be of different affection. And if the angles A and B be of different affection, the perpendicular will fall without the triangle; for, if it did not, the angles A and B would be of the same affection, contrary to the supposition.

PROP. XVII.

If to the base of a spherical triangle a perpendicular be drawn from the opposite angle, which either falls within the triangle, or is the nearest of the two that fall without; the least of the segments of the base is adjacent to the least of the sides of the triangle, or to the greatest, according as the sum of the sides is less or greater than a semicircle.

Let ABEF be a great circle of a sphere, H its pole, and GHD any circle passing through H, which therefore is perpendicular to the circle ABEF. Let A and B be two points in the circle ABEF, on opposite sides of the point D, and let D be nearer to A than to B, and let C be any point in the circle GHD between H and D. Through the points A and C, B and C, let the arches AC and BC be drawn, and let them be produced till they meet the circle ABEF in the points E and F, then the arches ACE, BCF are semicircles. Also ACB, ACF, CFE, ECB, are four spherical triangles contained by arches of the same circles, and having the same perpendiculars CD and CG.

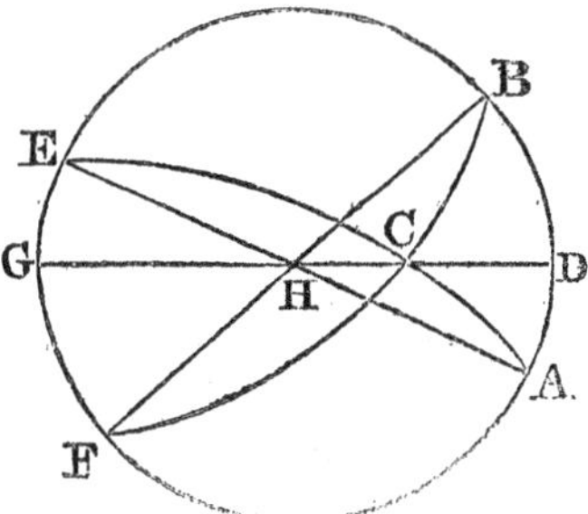

1. Now because CE is nearer to the arch CHG than CB is, CE is greater than CB, and therefore CE and CA are greater than CB and CA, wherefore CB and CA are less than a semicircle; but because AD, is by supposition less than DB, AC is also less than CB, (13.), and therefore in this case, viz. when the perpendicular falls within the triangle, and when the sum of the sides is less than a semicircle, the last segment is adjacent to the least side.

2. Again, in the triangle FCA the two sides FC and CA are less than a semicircle; for since AC is less than CB, AC and CF are less than BC and CF. Also, AC is less than CF, because it is more remote from CHG than CF is; therefore in this case also, viz. when the perpendicular falls without the triangle, and when the sum of the sides is less than a semicircle, the least segment of the base AD is adjacent to the least side.

3. But in the triangle FCE the two sides FC and CE are greater than a semicircle; for, since FC is greater than CA, FC and CE are greater than AC and CE. And because AC is less than CB, EC is greater than CF, and EC is therefore nearer to the perpendicular CHG than CF is, wherefore EG is the least segment of the base, and is adjacent to the greater side.

4. In the triangle ECB the two sides EC, CB are greater than a semicircle; for, since by supposition CB is greater than CA, EC and CB are greater than EC and CA. Also, EC is greater than CB, wherefore in this case, also, the least segment of the base EG is adjacent to the greatest side of the triangle. Therefore, when the sum of the sides is greater than a semicircle, the least segment of the base is adjacent to the greatest side, whether the perpendicular fall within or without the triangle: and it has been shown, that when the sum of the sides is less than a semicircle, the least segment of the base is adjacent to the least of the sides, whether the perpendicular fall within or without the triangle. Wherefore, &c. Q. E. D.

PROP. XVIII.

In right angled spherical triangles, the sine of either of the sides about the right angle, is to the radius of the sphere, as the tangent of the remaining side is to the tangent of the angle opposite to that side.

Let ABC be a triangle, having the right angle at A; and let AB be either of the sides, the sine of the side AB will be to the radius, as the tangent of the other side AC to the tangent of the angle ABC, opposite to AC. Let D be the centre of the sphere; join AD, BD, CD, and let AF be drawn perpendicular to BD, which therefore will be the sine of the arch AB, and from the point F, let there be drawn in the plane BDC the straight line FE at right angles to BD, meeting DC in E, and let AE be joined. Since therefore the straight

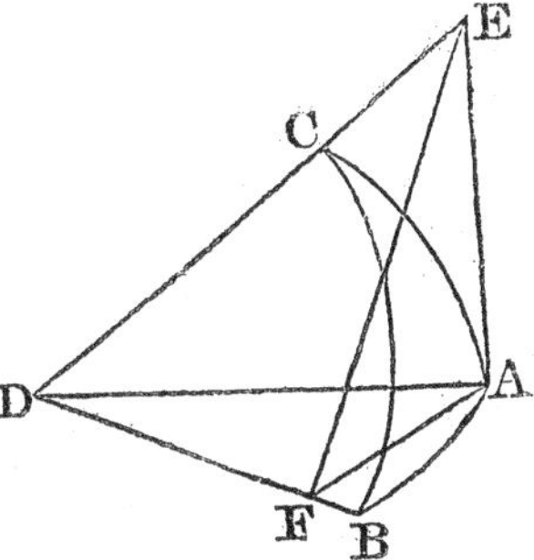

line DE is at right angles to both FA and FE, it will also be at right angles to the plane AEF (4. 2. Sup.); wherefore the plane ABD, which passes through DF, is perpendicular to the plane AEF (17. 2. Sup.), and the plane AEF perpendicular to ABD: But the plane ACD or AED, is also perpendicular to the same ABD, because the spherical angle BAC is a right angle: Therefore AE, the common section of the planes AED, AEF, is at right angles to the plane ABD, (18. 2. Sup.), and EAF, EAD are right angles. Therefore AE is the tangent of the arch AC; and in the rectilineal triangle AEF, having a right angle at A, AF is to the radius as AE to the tangent of the angle AFE, (1. Pl. Tr.); but AF is the sine of the arch AB, and AE the tangent of the arch AC; and the angle AFE is the inclination of the planes CBD, ABD, (4. def. 2. Sup.), or is equal to the spherical angle ABC: Therefore the sine of the arch AB is to the radius as the tangent of the arch AC to the tangent of the opposite angle ABC. Q. E. D.

Cor. Since by this proposition, sin AB : R : : tan AC : tan ABC; and because R : cot ABC : : tan ABC : R (1. Cor. def. 9, Pl. Tr.) by equality, sin AB : cot ABC : : tan AC : R.

PROP. XIX.

In right angled spherical triangles the sine of the hypotenuse is to the radius as the sine of either side is to the sine of the angle opposite to that side.

Let the triangle ABC be right angled at A, and let AC be either of the sides; the sine of the hypotenuse BC will be to the radius as the sine of the arch AC is to the sine of the angle ABC.

Let D be the centre of the sphere, and let CE be drawn perpendicular to DB, which will therefore be the sine of the hypotenuse BC, and from the point E let there be drawn in the plane ABD the straight line EF perpendicular to DB, and let CF be joined: then CF will be at right angles to the plane ABD, because as was shown of EA in the preceding proposition, it is the common section of two planes DCF, ECF, each perpendicular to the plane ADB. Wherefore CFD, CFE are right angles, and CF is the sine of the arch AC; and in the triangle CFE having the right angle CFE

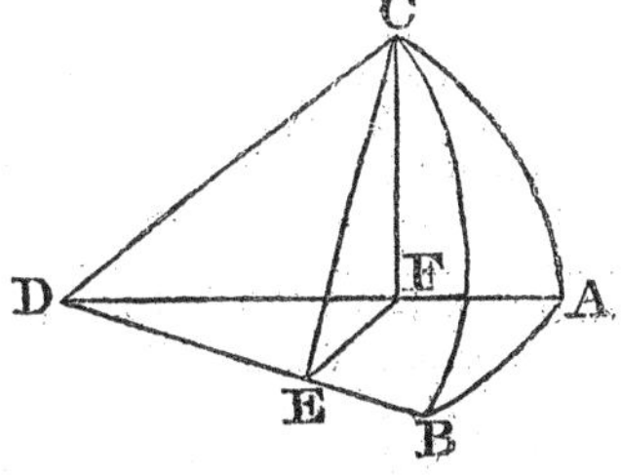

CE is to the radius, as CF to the sine of the angle CEF (1. Pl. Tr.). But, since CE, FE are at right angles to DEB, which is the common section of the planes CBD, ABD, the angle CEF is equal to the inclination of these planes, (4. def. 2. Sup.), that is, to the spherical angle ABC. Therefore the sine of the hypotenuse CB, is to the radius, as the sine of the side AC to the sine of the opposite angle ABC. Q. E. D.

PROP. XX.

In right angled sphercial triangles, the cosine of the hypotenuse is to the radius as the cotangent of either of the angles is to the tangent of the remaining angle.

Let ABC be a spherical triangle, having a right angle at A, the cosine of the hypotenuse BC is to the radius as the cotangent of the angle ABC to the tangent of the angle ACB.

Describe the circle DE, of which B is the pole, and let it meet AC in F, and the circle BC in E; and since the circle BD passes through

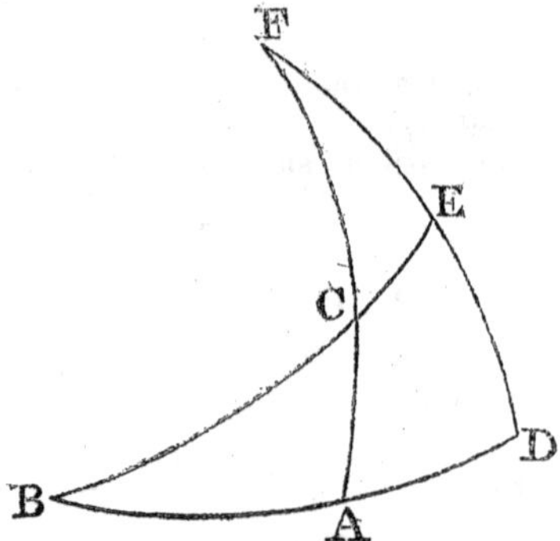

the pole B, of the circle DF, DF must pass through the pole of BD, (4.). And since AC is perpendicular to BD, the plane of the circle AC is perpendicular to the plane of the circle BAD, and therefore AC must also (4.) pass through the pole of BAD; wherefore, the pole of the circle BAD is in the point F, where the circles AC, DE, intersect. The arches FA, FD are therefore quadrants, and likewise the arches BD, BE. Therefore, in the triangle CEF, right angled at the point E, CE is the complement of BC, the hypotenuse of the triangle ABC; EF is the complement of the arch ED, the measure of the angle ABC, and FC, the hypotenuse of the triangle CEF, is the complement of AC, and the arch AD, which is the measure of the angle CFE, is the complement of AB.

But (18.) in the triangle CEF, sin CE : R : : tan EF : tan ECF, that is, in the triangle ACB, cos BC : R : : cot ABC : tan ACB. Q. E. D.

Cor. Because cos BC : R : : cot ABC : tan ACB, and (Cor. 1. def. 9. Pl. Tr.) cot ACB : R : : R : tan ACB, ex æquo, cot ACB : cos BC : : R : cot ABC.

PROP. XXI.

In right angled spherical triangles, the cosine of an angle is to the radius as the tangent of the side adjacent to that angle is to the tangent of the hypotenuse.

The same construction remaining; In the triangle CEF, sin FE : R : : tan CE : tan CFE (18.); but sin EF=cos ABC; tan CE=cot BC, and tan CFE = cot AB, therefore cos ABC : R : : cot BC : cot AB. Now, because (Cor. 1. def. 9, Pl. Tr.) cot BC : R : : R : tan BC, and cot AB : R : : R : tan AB, by equality inversely, cot BC : cot AB : : tan AB : tan BC; therefore (11. 5.) cos ABC : R : : tan AB : tan BC. Therefore, &c. Q. E. D.

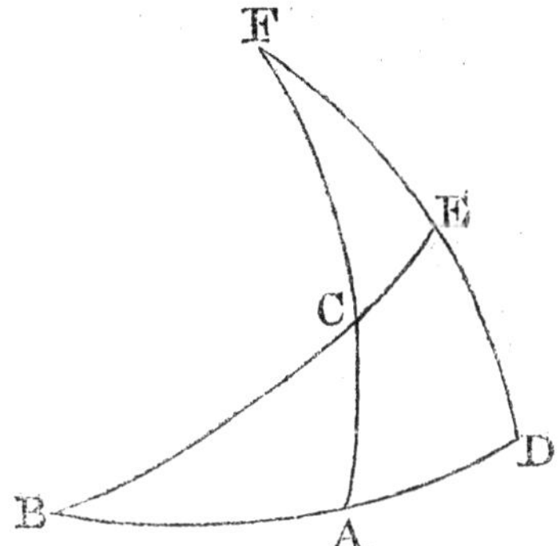

Cor. 1. From the demonstration it is manifest, that the tangents of any two arches AB, BC are reciprocally proportional to their cotangents.

Cor. 2. Because cos ABC : R : : tan AB : tan BC, and R : cos BC : : tan BC : R, by equality, cos ABC : cot BC : : tan AB : R. That is, the cosine of any of the oblique angles is to the cotangent of the hypotenuse, as the tangent of the side adjacent to the angle is to the radius.

PROP. XXII.

In right angled spherical triangles, the cosine of either of the sides is to the radius, as the cosine of the hypotenuse is to the cosine of the other side.

The same construction remaining: In the triangle CEF, sin CF : R : : sin CE : sin CFE, (19.); but sin CF = cos CA, sin CE=cos BC, and sin CFE=cos AB; therefore cos CA : R : : cos BC : cos AB. Q. E. D.

PROP. XXIII.

In right angled spherical triangles, the cosine of either of the sides is to the radius, as the cosine of the angle opposite to that side is to the sine of the other angle.

The same construction remaining: In the triangle CEF, sin CF : R : : sin EF : sin ECF, (19.); but sin CF = cos CA, sin EF=cos ABC, and sin ECF=sin BCA: therefore, cos CA : R : : cos ABC : sin BCA. Q. E. D.

PROP. XXIV.

In spherical triangles, whether right angled or oblique angled, the sines of the sides are proportional to the sines of the angles opposite to them.

First, Let ABC be a right angled triangle, having a right angle at A; therefore, (19.) the sine of the hypotenuse BC is to the radius, (or the sine of the right angle at A), as the sine of the side AC to the sine of the angle B. And, in like manner, the sine of BC is to the sine of the angle A, as the sine of AB to the sine of the angle C; wherefore (11. 5.) the sine of the side AC is to the sine of the angle B, as the sine of AB to the sine of the angle C.

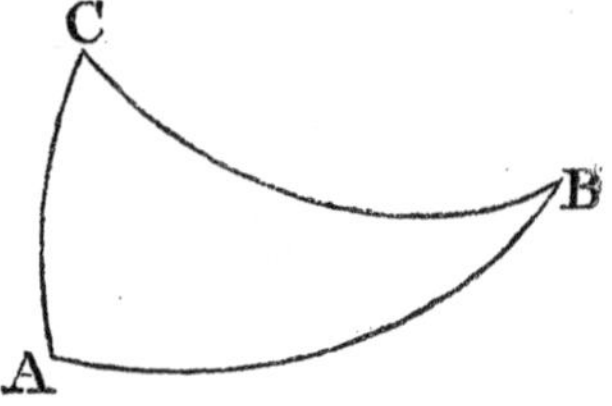

Secondly, Let ABC be an oblique angled triangle, the sine of any of the sides BC will be to the sine of any of the other two AC, as the side of the angle A opposite to BC, is to the sine of the angle B opposite to AC. Through the point C, let there be drawn an arch of a great circle CD perpendicular to AB; and in the right angled trian-

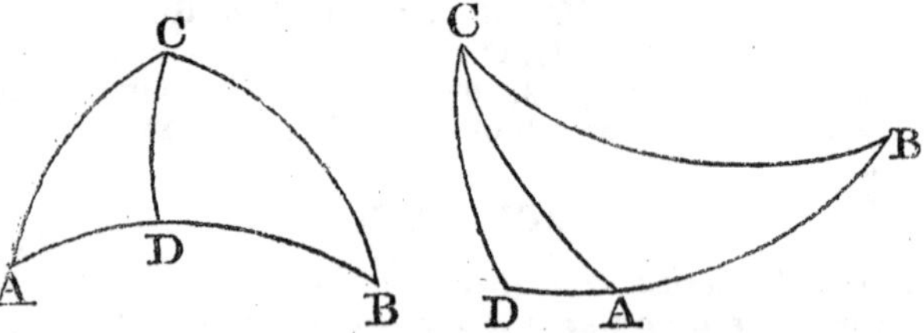

gle BCD, sin BC : R :: sin CD : sin B, (19.); and in the triangle ADC, sin AC : R :: sin CD : sin A; wherefore, by equality inversely, sin BC : sin AC :: sin A : sin B. In the same manner, it may be proved that sin BC : sin AB : : sin A : sin C, &c. Therefore, &c. Q. E. D.

PROP. XXV.

In oblique angled spherical triangles, a perpendicular arch being drawn from any of the angles upon the opposite side, the cosines of the angles at the base are proportional to the sines of the segments of the vertical angle.

Let ABC be a triangle, and the arch CD perpendicular to the base BA; the cosine of the angle B will be to the cosine of the angle A, as the sine of the angle BCD to the sine of the angle ACD.

For having drawn CD perpendicular to AB, in the right angled triangle BCD, (23.) cos CD : R :: cos B : sin DCB; and in the right angled triangle ACD, cos CD : R :: cos A : sin ACD; therefore (11. 5.) cos B : sin DCB :: cos A : sin ACD, and alternately, cos B : cos A :: sin BCD : sin ACD. Q. E. D.

PROP. XXVI.

The same things remaining, the cosines of the sides BC, CA, are proportional to the cosines of BD, DA, the segments of the base.

For in the triangle BCD, (22.), cos BC : cos BD :: cos DC : R, and in the triangle ACD, cos AC : cos AD :: cos DC : R; therefore (11. 5.) cos BC : cos BD :: cos AC : cos AD, and alternately, cos BC : cos AC :: cos BD : cos AD. Q. E. D.

PROP. XXVII.

The same construction remaining, the sines of BD, DA the segments of the base are reciprocally proportional to the tangents of B and A, the angles at the base.

In the triangle BCD, (18.), sin BD : R :: tan DC : tan B; and in the triangle ACD, sin AD : R :: tan DC : tan A; therefore, by equality inversely, sin BD : sin AD :: tan A : tan B. Q. E. D.

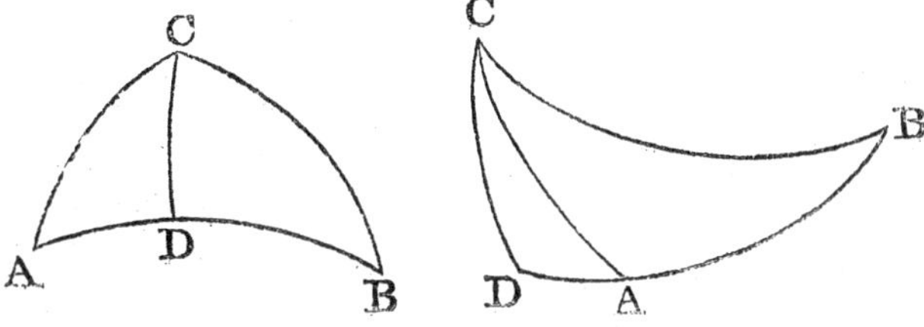

PROP. XXVIII.

The same construction remaining, the cosines of the segments of the vertical angle are reciprocally proportional to the tangents of the sides.

Because (21.), cos BCD : R :: tan CD : tan BC, and also cos ACD : R :: tan CD : tan AC, by equality inversely, cos BCD : cos ACD :: tan AC : tan BC. Q. E. D.

PROP. XXIX.

If from an angle of a spherical triangle there be drawn a perpendicular to the opposite side, or base, the rectangle contained by the tangents of half the sum, and of half the difference of the segments of the base is equal to the rectangle contained by the tangents of half the sum, and of half the difference of the two sides of the triangle.

Let ABC be a spherical triangle, and let the arch CD be drawn from the angle C at right angles to the base AB, tan $\frac{1}{2}(m+n)\times$tan $\frac{1}{2}(m-n)=\frac{1}{2}$ tan $(a+b)\times\frac{1}{2}$ tan $(a-b)$.

Let BC$=a$, AC$=b$; DD$=m$, AD$=n$. Because (26.) cos a : cos b :: cos m : cos n, (E. 5.) cos $a+b$: cos $a-$cos b :: cos $m+$cos n : cos $m-$cos n. But (1. Cor. 3. Pl. Trig.), cos $a+$cos b : cos $a-$cos b :: cot $\frac{1}{2}(a+b)$: tan $\frac{1}{2}(a-b)$, and also, cos $m+$cos n : cos $m-$cos n :: cot $\frac{1}{2}(m+n)$: tan $\frac{1}{2}(m-n)$. Therefore, (11. 5.) cot $\frac{1}{2}(a+b)$: tan $\frac{1}{2}(a-b)$ :: cot $\frac{1}{2}(m+n)$: tan $\frac{1}{2}(m-n)$. And because rectangles of the same altitude are as their bases, tan $\frac{1}{2}(a+b)\times$ cot $\frac{1}{2}(a+b)$: tan $\frac{1}{2}(a+b)\times$ tan $\frac{1}{2}(a-b)$ :: tan $\frac{1}{2}(m+n)\times$ cot $\frac{1}{2}(m+n)$: tan $\frac{1}{2}(m\times n)+$tan $\frac{1}{2}(m-n)$. Now the first and third terms of this proportion are equal, being each equal to the square of the radius, (1. Cor. Pl. Trig), therefore the remaining two are equal, (9. 5.) or tan $\frac{1}{2}(m+n)\times$ tan $\frac{1}{2}(m-n)=$ tan $\frac{1}{2}(a+b)\times$ tan $\frac{1}{2}(a-b)$; that is, tan $\frac{1}{2}$ (BD+AD) $\times$ tan $\frac{1}{2}$ (BD−AD) $=$ tan $\frac{1}{2}$ (BC+AC) $\times$ tan $\frac{1}{2}$ (BC−AC). Q. E. D.

Cor. 1. Because the sides of equal rectangles are reciprocally proportional, tan $\frac{1}{2}$ (BD+AD) : tan $\frac{1}{2}$ (BC+AC) :: tan $\frac{1}{2}$ (BC—AC) : tan $\frac{1}{2}$ (BD−AD).

Cor. 2. Since, when the perpendicular CD falls within the triangle, BD+AD $=$ AB, the base; and when CD falls without the triangle BD−AD$=$AB, therefore in the first case, the proportion in the last corollary becomes tan $\frac{1}{2}$ (AB) : tan $\frac{1}{2}$ (BC + AC) :: tan $\frac{1}{2}$ (BC—AC) : tan $\frac{1}{2}$ (BD−AD); and in the second case, it becomes by inversion and alternation, tan $\frac{1}{2}$ (AB) : tan $\frac{1}{2}$ (BC + AC) :: tan $\frac{1}{2}$ (BC—AC) : tan $\frac{1}{2}$ (BD+AD).

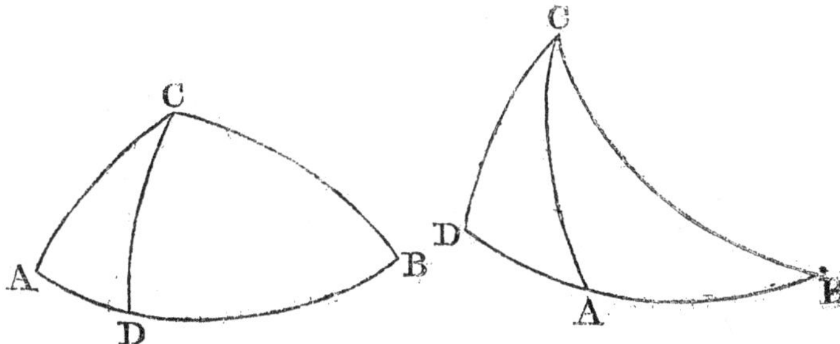

SCHOLIUM.

The preceding proposition, which is very useful in spherical trigonometry, may be easily remembered from its analogy to the proposition in plane trigonometry, that the rectangle under half the sum, and half the difference of the sides of a plane triangle, is equal to the rectangle under half the sum, and half the difference of the segments of the base. See (K. 6.), also 4th Case Pl. Tr. We are indebted to Napier for this and the two following theorems, which are so well adapted to calculation by Logarithms, that they must be considered as three of the most valuable propositions in Trigonometry.

PROP. XXX.

If a perpendicular be drawn from an angle of a spherical triangle to the opposite side or base, the sine of the sum of the angles at the base is to the sine of their difference as the tangent of half the base to the tangent of half the difference of its segments, when the perpendicular falls within; but as the co-tangent of half the base to the co-tangent of half the sum of the segments, when the perpendicular falls without the triangle: And the sine of the sum of the two sides is to the sine of their difference as the co-tangent of half the angle contained by the sides, to the tangent of half the difference of the angles which the perpendicular makes with the same sides, when it falls within, or to the tangent of half the sum of these angles, when it falls without the triangle.

If ABC be a spherical triangle, and AD a perpendicular to the base BC, sin (C+B) : sin (C−B) : : tan $\frac{1}{2}$ BC : tan $\frac{1}{2}$ (BD−DC), when AD falls within the triangle; but sin (C+B) : sin (C−B) : : cot $\frac{1}{2}$ BC : cot $\frac{1}{2}$ (BD+DC), when AC falls without. And again,

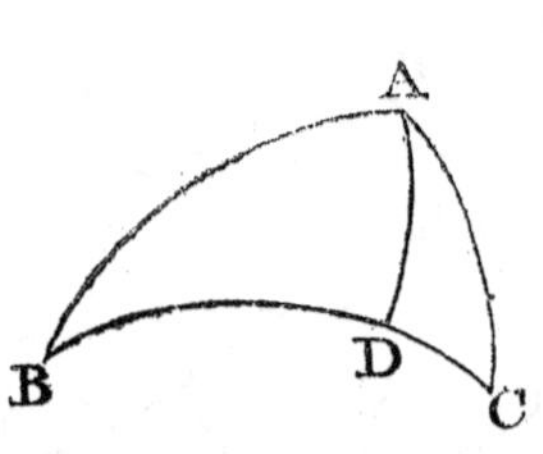

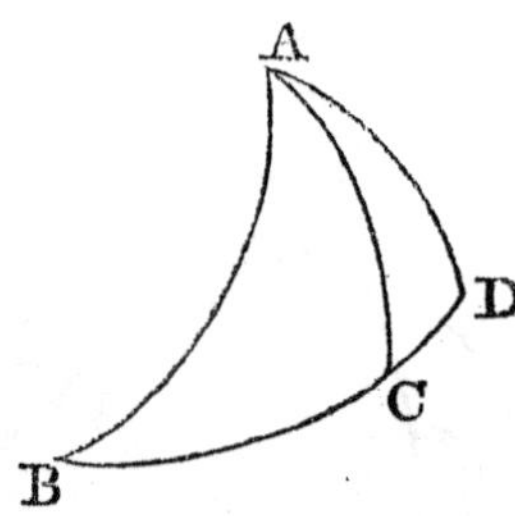

sin (AB+AC) : sin (AB−AC) : : cot $\frac{1}{2}$ BAC : tan $\frac{1}{2}$ (BAD−CAD), when AD falls within : but when AD falls without the triangle,
sin (AB+AC) : sin (AB−AC) : : cot $\frac{1}{2}$ BAC : tan $\frac{1}{2}$ (BAD+CAD).

For in the triangle BAC (27.), tan B : tan C : : sin CD : sin BD, and therefore (E. 5.), tan C+tan B : tan C−tan B : : sin BD+sin CD : sin BD−sin CD. Now, (by the annexed Lemma) tan C+tan B : tan C−tan B : : sin (C+B) : sin (C—B), and sin BD+sin CD : sin BD−sin CD : : tan $\frac{1}{2}$ (BD+CD) : tan $\frac{1}{2}$ (BD−CD), (3. Pl. Trig.), therefore, because ratios which are equal to the same ratio are equal to one another (11. 5.), sin (C+B) : sin (C−B) : : tan $\frac{1}{2}$ (BD+CD) : tan $\frac{1}{2}$ (BD−CD).

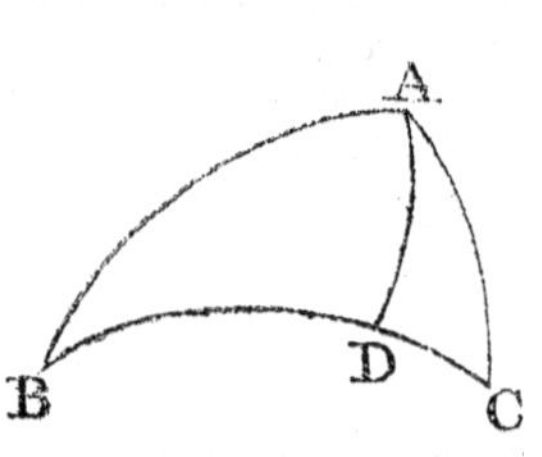

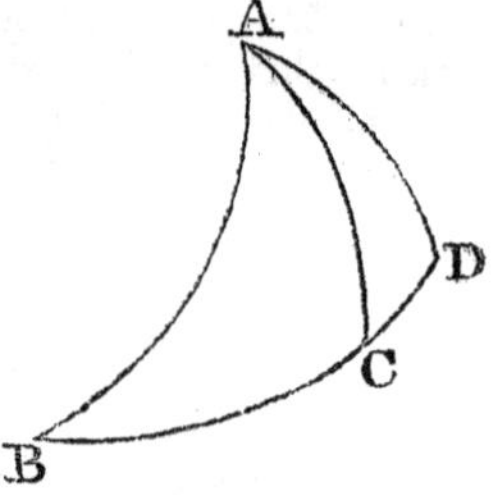

Now when AD is within the triangle, BD+CD=BC, and therefore sin (C+B) : sin (C−B) : : tan $\frac{1}{2}$ BC : tan $\frac{1}{2}$ (BD—CD). And again when AD is without the triangle, BD−CD=BC, and therefore sin (C+B) : sin (C−B) : : tan $\frac{1}{2}$ (BD+CD) : tan $\frac{1}{2}$ BC, or because the tangents of any two arches are reciprocally as their co-tangents, in (C+B) : sin (C−B) : : cot $\frac{1}{2}$ BC : cot $\frac{1}{2}$ (BD+CD).

The second part of the proposition is next to be demonstrated. Because (28.) tan AB : tan AC : : cos CAD : cos BAD, tan AB+tan AC : tan AB tan AC : : cos CAD+cos BAD : cos CAD−cos BAD. But (Lemma) tan AB+tan AC : tan AB−tan AC : : sin (AB+AC) : sin (AB−AC), and (1. cor. 3. Pl. Trig.) cos CAD+cos BAD : cos CAD−cos BAD : : cot $\frac{1}{2}$ (BAD+CAD) : tan $\frac{1}{2}$ (BAD−CAD). Therefore (11. 5.) sin (AB+AC) : sin (AB−AC) : : cot $\frac{1}{2}$ (BAD+CAD) : tan $\frac{1}{2}$ (BAD−CAD). Now, when AD is within the triangle, BAD+CAD=BAC, and therefore sin (AB+AC) : sin (AB—AC) : : cot $\frac{1}{2}$ BAC : tan $\frac{1}{2}$ (BAD−CAD).

But if AD be without the triangle, BAD$-$CAD$=$BAC, and therefore sin (AB$+$AC) : sin (AB$-$AC) : :
cot $\frac{1}{2}$ (BAD$+$CAD) : tan $\frac{1}{2}$ BAC ; or because
cot $\frac{1}{2}$ (BAD$+$CAD) : tan $\frac{1}{2}$ BAC : : cot $\frac{1}{2}$ BAC :
tan $\frac{1}{2}$ (BAD$+$CAD), sin (AB$+$AC) : sin (AB$-$AC) : : cot $\frac{1}{2}$ BAC : tan $\frac{1}{2}$ (BAD$+$CAD). Wherefore, &c. Q. E. D.

LEMMA.

The sum of the tangents of any two arches, is to the difference of their tangents, as the sine of the sum of the arches, to the sine of their difference.

Let A and B be two arches, tan A $+$ tan B : tan A $-$ tan B : : sin (A$+$B) : (A$-$B).

For, by § 6. page 243, sin A$\times$cos B$+$cos A$\times$sin B$=$sin (A$+$B), and therefore dividing all by cos A cos B, $\frac{\sin A}{\cos A}+\frac{\sin B}{\cos B}$ $=\frac{\sin (A+B)}{\cos A\times\cos B}$, that is, because $\frac{\sin A}{\cos A}=$ tan A, tan A$+$tan B $=\frac{\sin (A+B)}{\cos A\times\cos B}$. In the same manner it is proved that tan A$-$tan B $=\frac{\sin (A-B)}{\cos A\times\cos B}$. Therefore tan A $+$ tan B : tan A$-$tan B : : sin (A$+$B) : sin (A$-$B). Q. E. D.

PROP. XXXI.

The sine of half the sum of any two angles of a spherical triangle is to the sine of half their difference, as the tangent of half the side adjacent to these angles is to the tangent of half the difference of the sides opposite to them; and the cosine of half the sum of the same angles is to the cosine of half their difference, as the tangent of half the side adjacent to them, to the tangent of half the sum of the sides opposite.

Let C$+$B$=$2S, C$-$B$=$2D, the base BC$=$2B, and the difference of the segments of the base, or BD$-$CD$=$2X. Then, because (30.) sin (C$+$B) : sin (C$-$B) : : tan $\frac{1}{2}$ BC : tan $\frac{1}{2}$ (BD$-$CD), sin 2S : sin 2D : : tan B : tan X. Now, sin 2S$=$sin (S$+$S)$=$2 sin S $\times$ cos S, (Sect. III. cor. Pl. Tr.). In the same manner, sin 2D$=$2 sin D $\times$ cos D. Therefore sin S $\times$ cos S : sin D $\times$ cos D : : tan B : tan X.

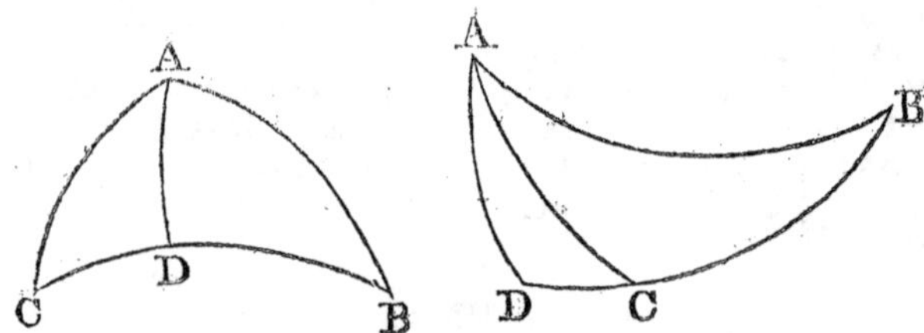

Again, in the spherical triangle ABC it has been proved, that $\sin C+\sin B : \sin C-\sin B :: \sin AB+\sin AC : \sin AB-\sin AC$, and since $\sin C+\sin B=2 \sin \frac{1}{2}(C+B)+\cos \frac{1}{2}(C-B)$, (Sect. III. 7. Pl. Tr.)$=2 \sin S\times\cos D$; and $\sin C-\sin B=2 \cos \frac{1}{2}(C+B)\times\sin \frac{1}{2}(C-B) = 2 \cos S\times\sin D$. Therefore $2 \sin S\times\cos D : 2 \cos S\times\sin D :: \sin AB+\sin AC : \sin AB-\sin AC$. But (3. Pl. Tr.) $\sin AB+\sin AC : \sin AB-\sin AC :: \tan \frac{1}{2}(AB+AC) : \tan \frac{1}{2}(AB-AC) :: \tan \Sigma : \tan \Delta$, Σ being equal to $\frac{1}{2}(AB+AC)$ and Δ to $\frac{1}{2}(AB-AC)$. Therefore $\sin S\times\cos D : \cos S\times\sin D :: \tan \Sigma : \tan \Delta$. Since then $\frac{\tan X}{\tan B}=\frac{\sin D\times\cos D}{\sin S\times\cos S}$; and $\frac{\tan \Delta}{\tan \Sigma}=\frac{\cos S\times\sin D}{\sin S\times\cos D}$, by multiplying equals by equals, $\frac{\tan X}{\tan B}\times\frac{\tan \Delta}{\tan \Sigma}=\frac{(\sin D)^2\times\cos S\times\cos D}{(\sin S)^2\times\cos S\times\cos D}=\frac{(\sin D)^2}{(\sin S)^2}$.

But (29.) $\frac{\tan \frac{1}{2}(BD-DC)}{\tan \frac{1}{2}(AB-AC)}=\frac{\tan \frac{1}{2}(AB+AC)}{\tan \frac{1}{2} BC}$, that is, $\frac{\tan X}{\tan \Delta}=\frac{\tan \Sigma}{\tan B}$, and therefore, $\frac{\tan X}{\tan B}=\frac{\tan \Sigma\times\tan \Delta}{(\tan B)^2}$, as also $\frac{\tan X}{\tan B}=\frac{\tan \Delta}{\tan \Sigma}=\frac{\tan \Delta^2}{(\tan B)^2}$.

But $\frac{\tan X}{\tan B}\times\frac{\tan \Delta}{\tan \Sigma}=\frac{(\sin D)^2}{(\sin S)^2}$; whence $\frac{(\tan \Delta)^2}{(\tan B)^2}=\frac{(\sin D)^2}{(\sin S)^2}$; and $\frac{\tan \Delta}{\tan B}=\frac{\sin D}{\sin S}$, or $\sin S : \sin D :: \tan B : \tan \Delta$, that is, $\sin (C+B) : \sin (C-B) :: \tan \frac{1}{2} BC : \tan \frac{1}{2}(AB-AC)$; which is the first part of the proposition.

Again, since $\frac{\tan \Delta}{\tan \Sigma}=\frac{\cos S\times\sin D}{\sin S\times\cos D}$, or inversely $\frac{\tan \Sigma}{\tan \Delta}=\frac{\sin S\times\cos D}{\cos S\times\sin D}$; and since $\frac{\tan X}{\tan B}=\frac{\sin D\times\cos D}{\sin S\times\cos S}$; therefore by multiplication, $\frac{\tan X}{\tan B}\times\frac{\tan \Sigma}{\tan \Delta}=\frac{(\cos D)^2}{(\cos S)^2}$.

But it was already shown that $\frac{\tan X}{\tan B}=\frac{\tan \Sigma\times\tan \Delta}{(\tan B)^2}$, wherefore also $\frac{\tan X}{\tan B}\times\frac{\tan \Sigma}{\tan \Delta}=\frac{(\tan \Sigma)^2}{(\tan B)^2}$.

Now, $\frac{\tan X}{\tan B}\times\frac{\tan \Sigma}{\tan \Delta}=\frac{(\cos D)^2}{(\cos S)^2}$, as has just been shown.

Therefore $\frac{(\cos D)^2}{(\cos S)^2}=\frac{(\tan \Sigma)^2}{(\tan B)^2}$ and consequently $\frac{\cos D}{\cos S}=\frac{\tan \Sigma}{\tan B}$, or cos

S : cos D : : tan B : tan Σ, that is, cos (C+B) : cos (C−B) : : tan $\frac{1}{2}$ BC : tan $\frac{1}{2}$ (C + B); which is the second part of the proposition. Therefore, &c. Q. E. D.

Cor. 1. By applying this proposition to the triangle supplemental to ABC (11. 3.), and by considering, that the sine of half the sum or half the difference of the supplements of two arches, is the same with the sine of half the sum or half the difference of the arches themselves: and that the same is true of the cosines, and of the tangents of half the sum or half the difference of the supplements of two arches: but that the tangent of half the supplement of an arch is the same with the cotangent of half the arch itself; it will follow, that the sine of half the sum of any two sides of a spherical triangle, is to the sine of half their difference as the cotangent of half the angle contained between them, to the tangent of half the difference of the angles opposite to them: and also that the cosine of half the sum of these sides, is to the cosine of half their difference, as the cotangent of half the angle contained between them, to the tangent of half the sum of the angles opposite to them.

Cor. 2. If therefore A, B, C be the three angles of a spherical triangle, a, b, c the sides opposite to them,

I. $\sin \frac{1}{2}(A+B) : \sin \frac{1}{2}(A-B) :: \tan \frac{1}{2}c : \tan \frac{1}{2}(a-b)$.
II. $\cos \frac{1}{2}(A+B) : \cos \frac{1}{2}(A-B) :: \tan \frac{1}{2}c : \tan \frac{1}{2}(a+b)$.
III. $\sin \frac{1}{2}(a+b) : \sin \frac{1}{2}(a-b) :: \tan \frac{1}{2}C : \tan \frac{1}{2}(A-B)$.
IV. $\cos \frac{1}{2}(a+b) : \cos \frac{1}{2}(a-b) :: \tan \frac{1}{2}C : \tan \frac{1}{2}(A+B)$.

PROBLEM I.

In a right angled spherical triangle, of the three sides and three angles, any two being given, besides the right angle, to find the other three.

This problem has sixteen cases, the solutions of which are contained in the following table, where ABC is any spherical triangle right angled at A.

GIVEN.	SOUGHT.	SOLUTION.	
BC and B.	AC.	R : sin BC : : sin B : sin AC, (19).	1
	AB.	R : cos B : : tan BC : tan AB, (21).	2
	C.	R : cos BC : : tan B : cot C, (20).	3
AC and C.	AB.	R : sin AC : : tan C : tan AB, (18).	4
	BC.	cos C : R : : tan AC : tan BC, (21).	5
	B.	R : cos AC : : sin C : cos B, (23).	6
AC and B.	AB.	tan B : tan AC : : R : sin AB, (18).	7
	BC.	sin B : sin AC : : R : sin BC, (19).	8
	C.	cos AC : cos B : : R : sin C, (23).	9
AC and BC.	AB.	cos AC : cos BC : : R : cos AB, (22).	10
	B.	sin BC : sin AC : : R : sin B, (19).	11
	C.	tan BC : tan AC : : R : cos C, (21).	12
AB and AC.	BC.	R : cos AB : : cos AC : cos BC, (22).	13
	B.	sin AB : R : : tan AC : tan B, (18).	14
	C.	sin AC : R : : tan AB : tan C, (18).	14
B and C.	AB.	sin B : cos C : : R : cos AB, (23).	15
	AC.	sin C : cos B : : R : cos AC, (23).	15
	BC.	tan B : cot C : : R : cos BC, (20).	16

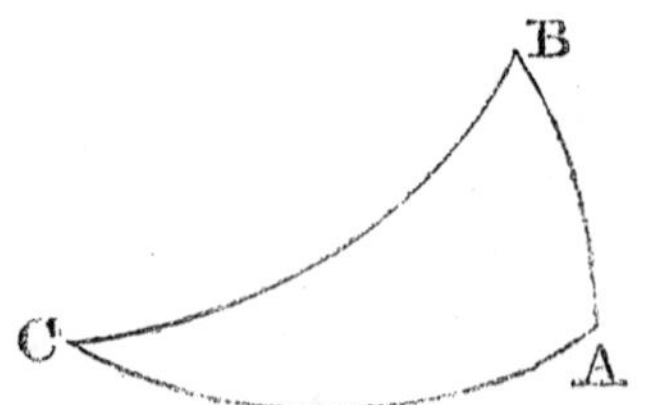

TABLE for determining the affections of the Sides and Angles found by the preceding rules.

AC and B of the same affection, (14).	1
If BC∠90°, AB and B of the same affection, otherwise different, (Cor. 15).	2
If BC∠90°, C and B of the same affection, otherwise different, (15.)	3
AB and C are of the same affection, (14.)	4
If AC and C are of the same affection, BC∠90°; otherwise BC∠90°, (Cor. 15.)	5
B and AC are of the same affection, (14.)	6
Ambiguous.	7
Ambiguous.	8
Ambiguous.	9
When BC∠90°, AB and AC of the same; otherwise of different affection, (15.)	10
AC and B of the same affection, (14.)	11
When BC∠90°, AC and C of the same; otherwise of different affection, (Cor. 15.)	12
BC∠90°, when AB and AC are of the same affection, (1. Cor. 15.)	13
B and AC of the same affection, (14.)	14
C and AB of the same affection, (14.)	14
AB and C of the same affection, (14.)	15
AC and B of the same affection, (14.)	15
When B and C are of the same affection, BC∠90°, otherwise, BC>90°, (15.)	16

The cases marked ambiguous are those in which the thing sought has two values, and may either be equal to a certain angle, or to the supplement of that angle. Of these there are three, in all of which the things given are a side, and the angle opposite to it; and accordingly, it is easy to show, that two right angled spherical triangles may always be found, that have a side and the angle opposite to it the same in both, but of which the remaining sides, and the remaining angle of the one, are the supplements of the remaining sides and the remaining angle of the other, each of each.

Through the affection of the arch or angle found may in all the other cases be determined by the rules in the second of the preceding tables, it is of use to remark, that all these rules except two, may be reduced to one, viz. that *when the thing found by the rules in the first table is either a tangent or a cosine; and when, of the tangents or cosines employed in the computation of it, one only belongs to an obtuse angle, the angle required is also obtuse.*

Thus, in the 15th case, when cos AB is found, if C be an obtuse angle, because of cos C, AB must be obtuse; and in case 16, if either B or C be obtuse, BC is greater than 90°, but if B and C are either both acute, or both obtuse, BC is less than 90°.

It is evident, that this rule does not apply when that which is found is the sine of an arch; and this, besides the three ambiguous cases, happens also in other two, viz. the 1st and 11th. The ambiguity is obviated, in these two cases, by this rule, that the sides of a spherical right angled triangle are of the same affection with the opposite angles.

Two rules are therefore sufficient to remove the ambiguity in all the cases of the right angled triangle, in which it can possibly be removed.

It may be useful to express the same solutions as in the annexed table. Let A be at the right angle as in the figure, and let the side opposite to it be a; let b be the side opposite to B, and c the side opposite to C.

GIVEN.	SOUGHT.	SOLUTION.	
a and B.	b.	$\sin b = \sin a \times \sin B$.	1
	c.	$\tan c = \tan a \times \cos B$.	2
	C.	$\cot C = \cos a \times \tan B$.	3
b and C.	c.	$\tan c = \sin b \times \tan C$.	4
	a.	$\tan a = \frac{\tan b}{\cos C}$.	5
	B.	$\cos B = \cos b \times \sin C$.	6
b and B.	c.	$\sin c = \frac{\tan b}{\tan B}$.	7
	a.	$\sin a = \frac{\sin b}{\sin B}$.	8
	C.	$\sin C = \frac{\cos B}{\cos b}$.	9
a and b.	c.	$\sin c = \frac{\cos a}{\cos b}$.	10
	B.	$\sin B = \frac{\sin b}{\sin a}$.	11
	C.	$\cos C = \frac{\tan b}{\tan a}$.	12
b and c.	a.	$\cos a = \cos b \times \cos c$.	13
	B.	$\tan B = \frac{\tan b}{\sin c}$.	14
	C.	$\tan C = \frac{\tan c}{\sin b}$.	14
B and C.	c.	$\cos c = \frac{\cos C}{\sin B}$.	15
	b.	$\cos b = \frac{\cos B}{\sin C}$.	15
	a.	$\cos a = \frac{\cot C}{\tan B}$.	16

PROBLEM II.

In any oblique angled spherical triangle, of the three sides and three angles, any three being given, it is required to find the other three.

In this Table, the references (c. 4.), (c. 5.), &c. are to the cases in the preceding Table, (16.), (27.), &c. to the propositions in Spherical Trigonometry.

<table>
<tr><th></th><th>GIVEN.</th><th>SOUGHT.</th><th>SOLUTION.</th></tr>
<tr><td>1</td><td rowspan="2">Two sides AB, AC, and the included angle A.</td><td>One of the other angles, B.</td><td>Let fall the perpendicular CD from the unknown angle, not required, on AB.
R : cos A :: tan AC : tan AD, (c. 2.); therefore BD is known, and sin BD : sin AD :: tan A : tan B, (27.); B and A are of the same or different affection, according as AB is greater or less than BD, (16.).</td></tr>
<tr><td>2</td><td>The third side BC.</td><td>Let fall the perpendicular CD from one of the unknown angles on the side AB.
R : cos A :: tan AC : tan AD, (c. 2.); therefore BD is known, and cos AD : cos BD :: cos AC : cos BC, (26.); according as the segments AD and DB are of the same or different affection, AC and CB will be of the same or different affection.</td></tr>
</table>

TABLE continued.

	GIVEN.	SOUGHT.	SOLUTION.
3	Two angles, A and ACB, and AC, the side between them.	The side BC.	From C the extremity of AC near the side sought, let fall the perpendicular CD on AB. R : cos AC :: tan A : cot ACD, (c. 3.); therefore BCD is known, and cos BCD : cos ACD :: tan AC : tan BC, (28.). BC is less or greater than 90°, according as the angles A and BCD are of the same, or different affection.
4		The third angle B.	Let fall the perpendicular CD from one of the given angles on the opposite side AB. R : cos AC :: tan A : cot ACD, (c. 3.); therefore the angle BCD is given, and sin ACD : sin BCD :: cos A : cos B, (25.); B and A are of the same or different affection, according as CD falls within or without the triangle, that is, according as ACB is greater or less than BCD, (18.).

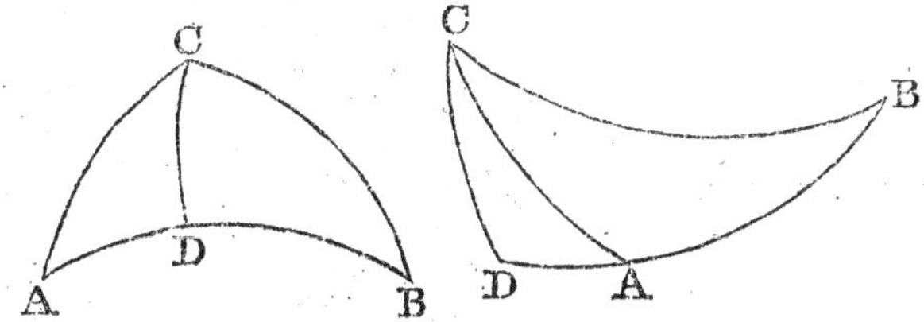

TABLE continued.

	GIVEN	SOUGHT.	SOLUTION.
5	Two sides AC and BC, and an angle, A opposite to one of them, BC.	The angle B opposite to the other given side AC.	Sin BC : sin AC :: sin A : sin B, (24.) The affection of B is ambiguous, unless it can be determined by this rule, that according as AC + BC is greater or less than 180°, A + B is greater or less than 180°, (10.)
6		The angle ACB contained by the given sides AC and BC.	From ACB the angle sought draw CD perpendicular to AB; then R : cos AC :: tan A : cot ACD, (c. [illegible].); and tan B[illegible] : tan AC :: cos ACD : cos BCD, (28.) ACD ± BCD = ACB, and ACB is ambiguous, because of the ambiguous sign + or −.
7		The third side AB.	Let fall the perpendicular CD from the angle C, contained by the given sides, upon the side AB. R : cos A :: tan AC : tan AD, (c. 2.); cos AC : cos BC :: cos AD : cos BD, (26.) AB=AD ± BD, wherefore AB is ambiguous.

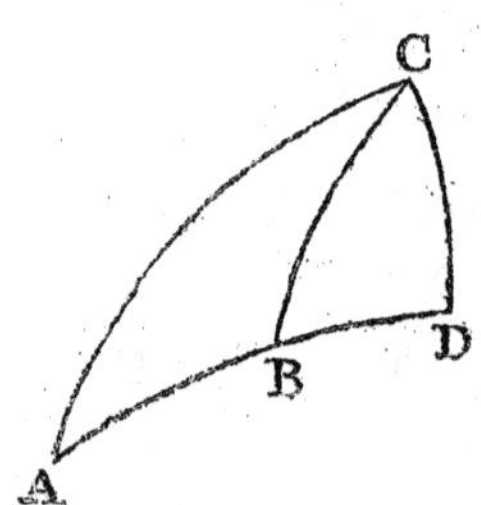

TABLE continued.

	GIVEN.	SOUGHT.	SOLUTION.
8	Two angles A, B, and a side AC opposite to one of them, B.	The side BC opposite to the other given angle A.	Sin B : sin A : : sin AC : sin BC, (24.) ; the affection of BC is uncertain, except when it can be determined by this rule, that according as $A+B$ is greater or less than 180°, $AC + BC$ is also greater or less than 180°, (10.).
9		The side AB adjacent to the given angles A, B.	From the unknown angle C, draw CD perpendicular to AB ; then R : cos A : : tan AC : tan AD, (c. 2.) ; tan B : tan A : : sin AD : sin BD. BD is ambiguous ; and therefore $AB = AD \pm BD$ may have four values, some of which will be excluded by this condition, that AB must be less than 180°.
10		The third angle ACB.	From the angle required, C, draw CD perpendicular to AB. R : cos AC : : tan A : cot ACD, (c. 3.), cos A : cos B : : sin ACD : sin BCD, (25.). The affection of BCD is uncertain, and therefore $ACB = ACD \pm BCD$, has four values, some of which may be excluded by the condition, that ACB is less than 180°.
11	The three sides, AB, AC, and BC.	One of the angles A.	From C one of the angles not required, draw CD perpendicular to AB. Find an arch E such that tan $\frac{1}{2}$ AB : tan $\frac{1}{2}$ $(AC+BC)$: : tan $\frac{1}{2}$ $(AC-BC)$: tan $\frac{1}{2}$ E ; then, if AB be greater than E, AB is the sum, and E the difference of AD and DB ; but if AB be less than E, E is the sum and AB the difference of AD, DB, (29.). In either case, AD and DB are known, and tan AC : tan AD : : R : cos A.

TABLE continued.

	GIVEN.	SOUGHT.	SOLUTION.
21	The three angles A, B, C.	One of the sides BC.	Suppose the supplements of the three given angles, A, B, C, to be, a, b, c, and to be the sides of a spherical triangle. Find, by the last case, the angle of this triangle, opposite to the side a, and it will be the supplement of the side of the given triangle opposite to the angle A, that is, of BC, (11.); and therefore BC is found.

In the foregoing table, the rules are given for ascertaining the affection of the arch or angle found, whenever it can be done: Most of these rules are contained in this one rule, which is of general application, viz. *that when the thing found is either a tangent or a cosine, and of the tangents or cosines employed in the computation of it, either one or three belong to obtuse angles, the angle found is also obtuse.* This rule is particularly to be attended to in cases 5, and 7, where it removes part of the ambiguity.

It may be necessary to remark with respect to the 11th case, that the segments of the base computed there are those cut off by the nearest perpendicular; and also, that when the sum of the sides is less than 180°, the least segment is adjacent to the least side of the triangle: otherwise to the greatest, (17.).

The last table may also be conveniently expressed in the following manner, denoting the side opposite to the angle A, by a, to B by b, and to C by c; and also the segments of the base, or of opposite angle, by x and y.

	GIVEN.	SOUGHT.	SOLUTION.
1	Two sides b and c, and the angle between them A.	B.	Find x, so that $\tan x = \tan b \times \cos A$; then $\tan B = \frac{\sin x \times \tan A}{\sin (c-x)}$.
2		a	Find x, as above, then $\cos a = \frac{\cos b \times \cos (c-x)}{\cos x.}$.
3	Angles A and C and side b.	a	Find x, so that $\cot x = \cos b \times \tan A$: then $\tan a = \frac{\tan b \times \cos x}{\cos (c-x)}$.
4		B	Find x, as above, then $\cos B = \frac{\cos A \times \sin (c-x)}{\sin x}$.
5	Sides a and b and angle A.	B	$\sin B = \frac{\sin b \times \sin A}{\sin a}$.
6		C	Find x, so that $\cot x = \cos b \times \tan A$: then $\cos C = \frac{\cos x \times \tan b}{\tan a}$.
7		c	Find x, so that $\tan x = \tan b \times \cos A$; and find y, so that $\cos y = \frac{\cos a \times \cos x}{\cos b}$. $c = x \pm y$.

TABLE continued.

	GIVEN.	SOUGHT.	SOLUTION.
8	The angles A and B and the side b.	a	$\sin a = \frac{\sin b \times \sin A}{\sin B}$.
9		c	Find x, so that $\tan x = \tan b \times \cos A$; and y, so that $\sin y = \frac{\sin x \times \tan A}{\tan B}$. $c = x \pm y$.
10		C	Find x, so that $\cot x = \cos b \times \tan A$; and also y, so that $\sin y = \frac{\sin x \times \cos B}{\cos A}$. $c = x \pm y$.
11	a, b, c.	A	Let $a+b+c=s$. $\sin \frac{1}{2} A = \frac{\sqrt{\sin(\frac{1}{2}s-b) \times \sin(\frac{1}{2}s-c.)}}{\sqrt{\sin b \times \sin c}}$ or $\cos \frac{1}{2} A = \frac{\sqrt{\sin \frac{1}{2} s \times \sin(\frac{1}{2}s-a.)}}{\sqrt{\sin b \times \sin c}}$
12	A, B, C.	a	Let $A+B+C=S$. $\sin \frac{1}{2} a = \frac{\sqrt{\cos \frac{1}{2} S \times \cos(\frac{1}{2} S - A)}}{\sqrt{\sin B \times \sin C}}$ or $\cos \frac{1}{2} a = \frac{\sqrt{\cos(\frac{1}{2}S - B) \frac{1}{2} \cos(S - C)}}{\sqrt{\sin B \times \sin C}}$

APPENDIX

TO

SPHERICAL

TRIGONOMETRY,

CONTAINING

NAPIER'S RULES OF THE CIRCULAR PARTS.

THE rule of the *Circular Parts*, invented by NAPIER, is of great use in Spherical Trigonometry, by reducing all the theorems employed in the solution of right angled triangles to two. These two are not new propositions, but are merely enunciations, which, by help of a particular arrangement and classification of the parts of a triangle, include all the six propositions, with their corollaries, which have been demonstrated above from the 18th to the 23d inclusive. They are perhaps the happiest example of artificial memory that is known.

DEFINITIONS.

I.

If in a spherical triangle, we set aside the right angle, and consider only the five remaining parts of the triangle, viz. the three sides and the two oblique angles, then the two sides which contain the right angle, and the complements of the other three, namely, of the two angles and the hypotenuse, are called the *Circular Parts.*

Thus, in the triangle ABC right angled at A, the circular parts are AC, AB with the complements of B, BC, and C. These parts are called circular; because, when they are named in the natural order of their succession, they go round the triangle.

II.

When of the five circular parts any one is taken, for the middle part, then of the remaining four, the two which are immediately adjacent to it, on the right and left, are called the adjacent parts; and the other two, each of which is separated from the middle by an adjacent part, are called opposite parts.

Thus in the right angled triangle ABC, A, being the right angle, AC AB, 90°—B, 90°—BC, 90°—C, are the circular parts, by Def. 1.; and if any one as AC be reckoned the middle part, then AB and 90°—C, which are contiguous to it on different sides, are called adjacent parts; and 90°—B, 90°—BC are the opposite parts. In like manner

if AB is taken for the middle part, AC and 90°—B are the adjacent parts: 90°—BC, and 90°—C are the opposite. Or if 90°—BC be the middle part, 90°—B, 90°—C are adjacent; AC and AB opposite, &c.

This arrangement being made, the rule of the circular part is contained in the following

PROPOSITION.

In a right angled spherical triangle, the rectangle under the radius and the sine of the middle part, is equal to the rectangle under the tangents of the adjacent parts; or to the rectangle under the cosines of the opposite parts.

The truth of the two theorems included in this enunciation may be easily proved, by taking each of the five circular parts in succession for the middle part, when the general proposition will be found to coincide with some one of the analogies in the table already given for the resolution of the cases of right angled spherical triangles. Thus, in the triangle ABC, if the complement of the hypotenuse BC be taken as the middle part, 90°—B, and 90°—C, are the adjacent parts, AB and AC the opposite. Then the general rule gives these theorems, $R\times\cos BC=\cot B\times\cot C$; and $R\times\cos BC=\cos AB\times\cos AC$. The former of these coincides with the cor. to the 20th; and the latter with the 22d.

To apply the foregoing general proposition, to resolve any case of a right angled spherical triangle, consider which of the three quantities named (the two things given and the one required) must be made the middle term, in order that the other two may be equidistant from it, that is, may be both adjacent, or both opposite; then one or other of the two theorems contained in the above enunciation will give the value of the thing required.

Suppose, for example, that AB and BC are given, to find C; it is evident that if AB be made the middle part, BC and C are the opposite parts, and therefore R × sin AB = sin C × sin BC, for sin C = cos (90°—C), and cos (90°—BC) = sin BC, and consequently

$\sin C = \frac{\sin AB}{\sin BC}$.

Again, suppose that BC and C are given to find AC; it is obvious that C is in the middle between the adjacent parts AC and (90°—BC), therefore R × cos C = tan AC × cot BC, or tan AC = $\frac{\cos C}{\cot BC}$ = cos C + tan BC; because, as has been shown above, $\frac{1}{\cot BC}$ = tan BC.

In the same way may all the other cases be resolved. One or two trials will always lead to the knowledge of the part which in any given case is to be assumed as the middle part; and a little practice will make it easy, even without such trials, to judge at once which of them is to be so assumed. It may be useful for the learner to range the names of the five circular parts of the triangle round the circumference of a circle, at equal distances from one another, by which means the middle part will be immediately determined

Besides the rule of the *circular parts*, Napier derived from the last of the three theorems ascribed to him above, (schol. 29.) the solutions of all the cases of oblique angled triangles These solutions are as follows: A, B, C, denoting the three triangles of a spherical triangle, and a, b, c, the sides opposite to them.

I.

Given two sides b, c, and the angle A between them.

To find the angles B and C.

$$\tan \tfrac{1}{2} (B-C) = \cot \tfrac{1}{2} A \times \frac{\sin \frac{1}{2} (b-c)}{\sin \frac{1}{2} (b+c)}. \quad (31.) \text{ cor. } 1.$$

$$\tan \tfrac{1}{2} (B+C) = \cot \tfrac{1}{2} A \times \frac{\cos \frac{1}{2} (b-c)}{\cos \frac{1}{2} (b+c)}. \quad (31.) \text{ cor. } 1.$$

To find the third side a,

sin B : sin A :: sin b : sin a.

II.

Given the two sides b, c, and the angle B opposite to one of them.

To find C, and the angle opposite to the other side.

sin b : sin c :: sin B : sin C.

To find the contained angle A.

$$\cot \tfrac{1}{2} A = \tan \tfrac{1}{2} (B-C) \times \frac{\sin \frac{1}{2} (b+c)}{\sin \frac{1}{2} (b-c)}. \quad (31.) \text{ cor. } 1.$$

To find the third side a.

Sin B : sin A :: sin b : sin a.

III.

Given two angles A and B, and the side c between them.

To find the other two sides a, b.

$$\tan\tfrac{1}{2}(b-a) = \tan\tfrac{1}{2}c \times \frac{\sin\frac{1}{2}(A-B)}{\sin\frac{1}{2}(A+B)}. \quad (31.)$$

$$\tan\tfrac{1}{2}(b+a) = \tan\tfrac{1}{2}c \times \frac{\cos\frac{1}{2}(A-B)}{\cos\frac{1}{2}(A+B)}. \quad (31.)$$

To find the third angle C.

sin a : sin c :: sin A : sin C.

IV.

Given two angles A and B, and the side a, opposite to one of them.

To find b, the side opposite to the other.

sin A : sin B :: sin a : sin b.

To find c, the side between the given angles.

$$\tan\tfrac{1}{2}c = \tfrac{1}{2}(a-b) \times \frac{\sin\frac{1}{2}(A+B)}{\sin\frac{1}{2}(A-B)}. \quad (31.)$$

To find the third angle C.

sin a : sin c :: sin A : sin C.

The other two cases, when the three sides are given to find the angles, or when the three angles are given to find the sides, are resolved by the 29th, (the first of NAPIER's Propositions,) in the same way as in the table already given for the case of the oblique angled triangle.

There is a solution of the case of the three sides being given, which it is often very convenient to use, and which is set down here, though the proposition on which it depends has not been demonstrated.

Let a, b, c, be the three given sides, to find the angle A, contained between b, and c.

If Rad. $= 1$, and $a + b + c = s$,

$$\sin \tfrac{1}{2} \text{A} = \frac{\sqrt{\sin (\frac{1}{2} s - b) \times \sin \frac{1}{2} (s - c)}}{\sqrt{\sin b \times \sin c}} \text{; or,}$$

$$\cos \tfrac{1}{2} \text{A} = \frac{\sqrt{\sin (\frac{1}{2} s \times \sin \frac{1}{2} (s - a)}}{\sqrt{\sin b \times \sin c}}.$$

In like manner, if the three angles, A, B, C are given to find c, the side between A and B.

Let $\text{A} + \text{B} + \text{C} = \text{S}$,

$$\sin \tfrac{1}{2} c = \frac{\sqrt{\cos \frac{1}{2} \text{S} \times \cos (\frac{1}{2} \text{S} - \text{A})}}{\sqrt{\sin \text{B} \times \sin \text{C}}} \text{; or,}$$

$$\cos \tfrac{1}{2} c = \frac{\sqrt{\cos (\frac{1}{2} \text{S} - \text{B}) \times \cos (\frac{1}{2} \text{S} - \text{C})}}{\sqrt{\sin \text{B} \times \sin \text{C}}}.$$

These theorems, on account of the facility with which Logarithms are applied to them, are the most convenient of any for resolving the two cases to which they refer. When A is a very obtuse angle, the second theorem, which gives the value of the cosine of its half, is to be used; otherwise the first theorem, giving the value of the sine of its half is preferable. The same is to be observed with respect to the side c, the reason of which was explained, Plane Trig. Schol.

END OF SPHERICAL TRIGONOMETRY.

NOTES

ON THE

ELEMENTS.

NOTES

ON THE

FIRST BOOK OF THE ELEMENTS.

DEFINITIONS.

I.

In the definitions a few changes have been made, of which it is necessary to give some account. One of these changes respects the first definition, that of a point, which Euclid has said to be, 'That which has no parts, or which has no magnitude.' Now, it has been objected to this definition, that it contains only a negative, and that it is not convertible, as every good definition ought certainly to be. That it is not convertible is evident, for though every point is unextended, or without magnitude, yet every thing unextended or without magnitude, is not a point. To this it is impossible to reply, and therefore it becomes necessary to change the definition altogether, which is accordingly done here, a point being defined to be, *that which has position but not magnitude.* Here the affirmative part includes all that is essential to a point, and the negative part excludes every thing that is not essential to it. I am indebted for this definition to a friend, by whose judicious and learned remarks I have often profited.

II.

After the second definition Euclid has introduced the following, " the extremities of a line are points."

Now, this is certainly not a definition, but an inference from the definitions of a point and of a line. That which terminates a line can have no breadth, as the line in which it is has none; and it can have no length, as it would not then be a termination, but a part of that which it is supposed to terminate. The termination of a line can therefore have no magnitude, and having necessarily position, it is a point. But as it is plain, that in all this we are drawing a consequence from two definitions already laid down, and not giving a new definition, I have taken the liberty of putting it down as a corollary to the second definition, and have added, *that the intersections of one line with another are points*, as this affords a good illustration of the nature of a point, and is an inference exactly of the same kind with the preceding. The same thing nearly has been done with the fourth definition where that which Euclid gave as a separate definition, is made a corollary to the fourth, because it is in fact an inference deduced from comparing the definitions of a superficies and a line.

As it is impossible to explain the relation of a superficies, a line and a point to one another, and to the solid in which they all originate, better than Dr. Simson has done, I shall here add, with very little change, the illustration given by that excellent Geometer.

"It is necessary to consider a solid, that is, a magnitude which has length, breadth, and thickness, in order to understand aright the definitions of a point, line and superficies; for these all arise from a solid, and exist in it; The boundary, or boundaries which contain a solid, are called superficies, or the boundary which is common to two solids which are contiguous, or which divides one solid into two contiguous parts, is called a superficies; Thus, if BCGF be one of the boundaries which contain the solid ABCDEFGH, or which is the common boundary of this solid, and the solid BKLCFNMG, and is therefore in the one as well as the other solid, it is called a superficies, and has no thickness; For if it have any, this thickness must either be a part of the thickness of the solid AG, or the solid BM, or a part of the thickness of each of them. It cannot be a part of the thickness of the solid BM; because, if this solid be removed from the solid AG, the superficies BCGF, the boundary of the solid AG, remains still the same as it was. Nor can it be a part of the thickness of the solid AG: because if this be removed from the solid BM, the superficies BCGF, the boundary of the solid BM, does nevertheless remain; therefore the superficies BCGF has no thickness, but only length and breadth.

"The boundary of a superficies is called a line; or a line is the common boundary of two superficies that are contiguous, or it is that which divides one superficies into two contiguous parts: Thus, if BC be one of the boundaries which contain the superficies ABCD, or which is the common boundary of this superficies, and of the superficies, KBCL, which is contiguous to it, this boundary BC is called a line, and has no breadth; For, if it have any, this must be part either of the breadth of the superficies ABCD or of the superficies KBCL, or part of each of them. It is not part of the breadth of the superficies KBCL; for if this superficies be removed from the superficies ABCD, the line BC which is the boundary of the superficies ABCD remains the same as it was. Nor can the breadth that BC is supposed to have, be a part of the breadth of the superficies ABCD; because, if this be removed from

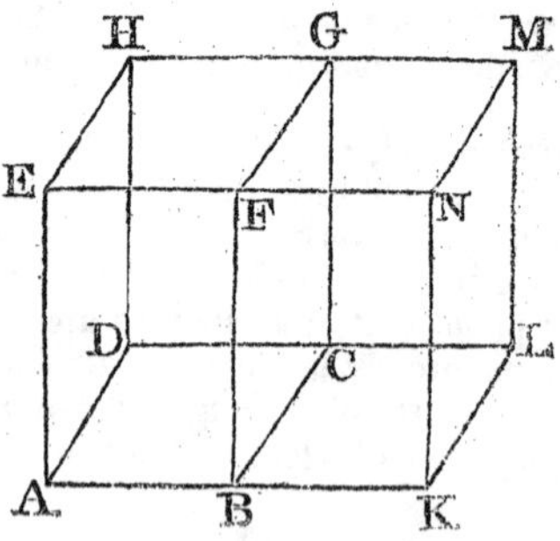

the superficies KBCL, the line BC, which is the boundary of the superficies KBCL, does nevertheless remain: Therefore the line BC has no breadth. And because the line BC is in a superficies, and that a superficies has no thickness, as was shown; therefore a line has neither breadth nor thickness, but only length.

"The boundary of a line is called a point, or a point is a common boundary or extremity of two lines that are contiguous: Thus, if B be the extremity of the line AB, or the common extremity of the two lines AB, KB, this extremity is called a point, and has no length: For if it have any, this length must either be part of the length of the line AB, or of the line KB. It is not part of the length of KB; for if the line KB be removed from AB, the point B, which is the extremity of the line AB, remains the same as it was; Nor is it part of the length of the line AB; for if AB be removed from the line KB, the point B, which is the extremity of the line KB, does nevertheless remain: Therefore the point B has no length: And because a point is in a line, and a line has neither breadth nor thickness, therefore a point has no length, breadth, nor thickness. And in this manner the definitions of a point, line, and superficies are to be understood."

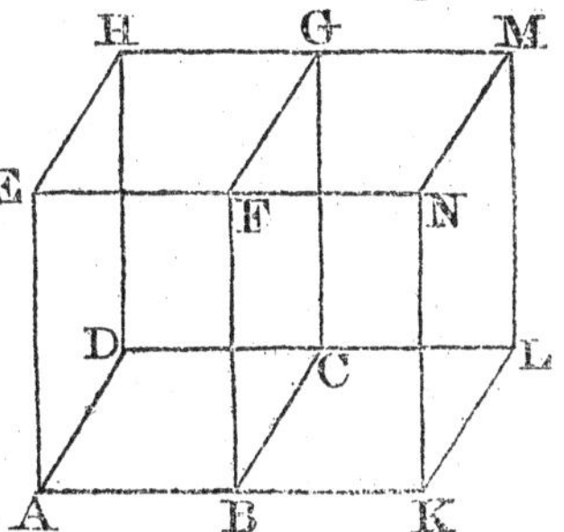

III.

Euclid has defined a straight line to be a line which (as we translate it) "lies evenly between its extreme points." This definition is obviously faulty, the word evenly standing as much in need of an explanation as the word straight, which it is intended to define. In the original, however, it must be confessed, that this inaccuracy is at least less striking than in our translation; for the word which we render *evenly* is ἐξίσȣ, *equally*, and is accordingly translated *ex æquo*, and *equaliter* by Commandine and Gregory. The definition, therefore, is, that a straight line is one which lies equally between its extreme points: and if by this we understand a line that lies between its extreme points so as to be related exactly alike to the space on the one side of it, and to the space on the other, we have a definition that is perhaps a little too metaphysical, but which certainly contains in it the essential character of a straight line. That Euclid took the definition in this sense however, is not certain, because he has not attempted to deduce from it any property whatsoever of a straight line; and indeed, it should seem not easy to do so, without employing some reasonings of a more metaphysical kind than he has any where admitted into his Elements.

To supply the defects of his definition, he has therefore introduced the Axiom, that *two straight lines cannot inclose a space;* on which Axiom it is, and not on his definition of a straight line, that his demonstrations are founded. As this manner of proceeding is certainly not so regular and scientific as that of laying down a definition, from which the properties of the thing defined may be logically deduced, I have substituted another definition of a straight line in the room of Euclid's. This definition of a straight line was suggested by a remark of Boscovich, who, in his Notes on the philosophical Poem of Professor Stay, says, "Rectam lineam rectæ congruere totam toti "in infinitum productum si bina puncta unius binis alterius congruant, "patet ex ipsa admodum clara rectitudinis idea quam habemus." (Supplementum in lib. 3. § 550.) Now, that which Mr. Boscovich would consider as an inference from our idea of straightness, seems itself to be the essence of that idea, and to afford the best criterion for judging whether any given line be straight or not. On this principle we have given the definition above, *If there be two lines which cannot coincide in two points, without coinciding altogether, each of them is called a straight line.*

This definition was otherwise expressed in the two former editions: it was said, that lines are straight lines which cannot coincide in part, without coinciding altogether. This was liable to an objection, viz. that it defined straight *lines*, but not a straight *line;* and though this in truth is but a mere cavil, it is better to leave no room for it. The definition in the form now given is also more simple.

From the same definition, the proposition which Euclid gives as an Axiom, that two straight lines cannot inclose a space, follows as a necessary consequence. For, if two lines inclose a space, they must intersect one another in two points, and yet, in the intermediate part, must not coincide; and therefore by the definition they are not straight lines. It follows in the same way, that two straight lines cannot have a common segment, or cannot coincide in part, without coinciding altogether.

After laying down the definition of a straight line, as in the first Edition, I was favoured by Dr. Reid of Glasgow with the perusal of a MS. containg many excellent observations on the first Book of Euclid, such as might be expected from a philosopher distinguished for the accuracy as well as the extent of his knowledge. He there defined a straight line nearly as has been done here, viz. "A straight "line is that which cannot meet another straight line in more points "than one, otherwise they perfectly coincide, and are one and the "same." Dr. Reid also contends, that this must have been Euclid's own definition; because, in the first proposition of the eleventh Book, that author argues, "that two straight lines cannot have a common seg-"ment, for this reason, that a straight line does not meet a straight "line in more points than one, otherwise they coincide." Whether this amounts to a proof of the definition above having been actually Euclid's, I will not take upon me to decide: but it is certainly a proof that the writings of that geometer ought long since to have suggested this definition to his commentators; and it reminds me, that I might have learned from these writings what I have acknowledged above to be derived from a remoter source.

There is another characteristic, and obvious property of straight lines, by which I have often thought that they might be very conveniently defined, viz. that the position of the whole of a straight line is determined by the position of two of its points, in so much that, when two points of a straight line continue fixed, the line itself cannot change its position. It might therefore be said, that *a straight line is one in which, if the position of two points be determined, the position of the whole line is determined.* But this definition, though it amount in fact to the same thing with that already given, is rather more abstract, and not so easily made the foundation of reasoning. I therefore thought it best to lay it aside, and to adopt the definition given in the text.

V.

The definition of a plane is given from Dr. Simson, Euclid's being liable to the same objections with his definition of a straight line; for, he says, that a plane superficies is one which "lies evenly between "its extreme lines." The defects of this definition are completely removed in that which Dr. Simson has given. Another definition different from both might have been adopted, viz. That those superficies are called plane, which are such, that if three points of the one coincide with three points of the other, the whole of the one must coincide with the whole of the other. This definition, as it resembles that of a straight line, already given, might, perhaps, have been introduced with some advantage; but as the purposes of demonstration cannot be better answered than by that in the text, it has been thought best to make no farther alteration.

VI.

In Euclid, the general definition of a plane angle is placed before that of a rectilineal angle, and is meant to comprehend those angles which are formed by the meeting of the other lines than straight lines. A plane angle is said to be "the inclination of two lines to "one another which meet together, but are not in the same direc- "tion." This definition is omitted here, because that the angles formed by the meeting of curve lines, though they may become the subject of geometrical investigation, certainly do not belong to the Elements; for the angles that must first be considered are those made by the intersection of straight lines with one another. The angles formed by the contact or intersection of a straight line and a circle, or of two circles, or two curves of any kind with one another, could produce nothing but perplexity to beginners, and cannot possibly be understood till the properties of rectilineal angles have been fully explained. On this ground, I am of opinion, that in an elementary treatise, it may fairly be omitted. Whatever is not useful, should, in explaining the elements of a science, be kept out of sight altogether; for, if it does not assist the progress of the understanding, it will certainly retard it.

AXIOMS.

Among the Axioms there have been made only two alterations. The 10th Axiom in Euclid is, that "two straight lines cannot inclose "a space;" which, having become a corollary to our definition of a straight line, ceases of course to be ranked with self-evident propositions. It is therefore removed from among the Axioms, and that which was before the 11th is accounted the 10th.

The 12th Axiom of Euclid is, that "if a straight line meets two "straight lines, so as to make the two interior angles on the same "side of it taken together less than two right angles, these straight "lines being continually produced, shall at length meet upon that "side on which are the angles which are less than two right angles." Instead of this proposition, which, though true, is by no means self-evident; another that appeared more obvious, and better entitled to be accounted an Axiom, has been introduced, viz. "that two straight "lines, which intersect one another, cannot be both parallel to the "same straight line." On this subject, however, a fuller explanation is necessary, for which see the note on the 29th Prop.

PROP. IV. and VIII. B. I.

The fourth and eighth propositions of the first book are the foundation of all that follows with respect to the comparison of triangles. They are demonstrated by what is called the method of supraposition, that is, by laying the one triangle upon the other, and proving that they must coincide. To this some objections have been made, as if it were ungeometrical to suppose one figure to be removed from its place and applied to another figure. "The laying," says Mr. Thomas Simson in his Elements, "of one figure upon another, whatever evi-"dence it may afford, is a *mechanical* consideration, and depends on "no postulate." It is not clear what Mr. Simson meant here by the word *mechanical:* but he probably intended only to say, that the method of supraposition involves the idea of motion, which belongs rather to mechanics than geometry; for I think it is impossible that such a Geometer as he was could mean to assert, that the evidence derived from this method is like that which arises from the use of instruments, and of the same kind with what is furnished by experience and observation. The demonstrations of the fourth and eighth, as they are given by Euclid, are as certainly a process of pure reasoning, depending solely on the idea of equality, as established in the

8th Axiom, as any thing in geometry. But, if still the removal of the triangle from its place be considered as creating a difficulty, and as inelegant, because it involves an idea, that of motion, not essential to geometry, this defect may be entirely remedied, provided that, to Euclid's three postulates, we be allowed to add the following, viz. *That if there be two equal straight lines, and if any figure whatsoever be constituted on the one, a figure every way equal to it may be constituted on the other.* Thus if AB and DE be two equal straight lines, and ABC a triangle on the base AB, a triangle DEF every way equal to ABC may be supposed to be constituted on DE as a base. By this it is not meant to assert that the method of describing the triangle DEF is actually known, but merely that the triangle DEF may be conceived to exist in all respects equal to the triangle ABC. Now, there is no truth whatsoever that is better entitled than this to be ranked among the Postulates or Axioms of geometry; for the straight lines AB and DE being every way equal, there can be nothing belonging to the one that may not also belong to the other.

On the strength of this postulate the fourth Proposition is thus demonstrated.

If ABC, DEF be two triangles, such that the two sides AB and AC of the one are equal to the two ED, DF of the other, and the angle BAC, contained by the sides AB, AC of the one, equal to the angle EDF, contained by the sides ED, DF of the other; the triangles ABC and EDF are every way equal.

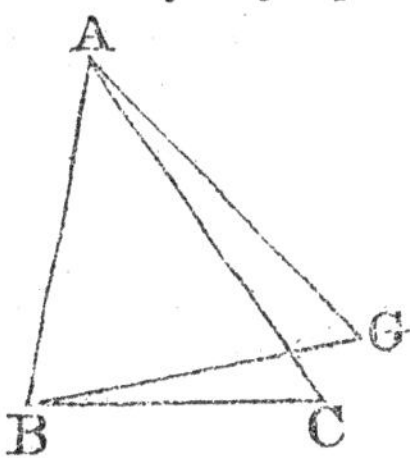

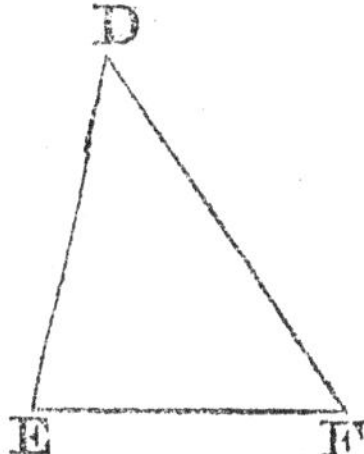

On AB let a triangle be constituted every way equal to the triangle DEF; then if this triangle coincide with the triangle ABC, it is evident that the proposition is true, for it is equal to DEF by hypothesis, and to ABC, because it coincides with it; wherefore ABC, DEF are equal to one another. But if it does not coincide with ABC, let it have the position ABG; and first suppose G not to fall on AC; then the angle BAG is not equal to the angle BAC. But the angle BAG is equal to the angle EDF, therefore EDF and ABC are not equal, and they are also equal by hypothesis, which is impossible. Therefore the point G must fall upon AC; now, if it fall upon AC but not at C, then AG is not equal to AC; but AG is equal to DF, therefore DF and AC are not equal, and they are also equal by supposition, which is impossible. Therefore G must coincide with C, and the triangle AGB with the triangle ACB. But AGB is every way equal to DEF, therefore, ACB and DEF are also every way equal. Q. E. D.

By help of the same postulate, the 5th may also be very easily demostrated.

Let ABC be an isosceles triangle, in which AB, AC are the equal sides; the angle ABC, ACB opposite to these sides are also equal.

Draw the straight line EF equal to BC, and suppose that on EF the triangle DEF is constituted every way equal to the triangle ABC, that is, having DE equal to AB, DF to AC, the angle EDF to the angle BAC, the angle ACB to the angle DFE, &c.

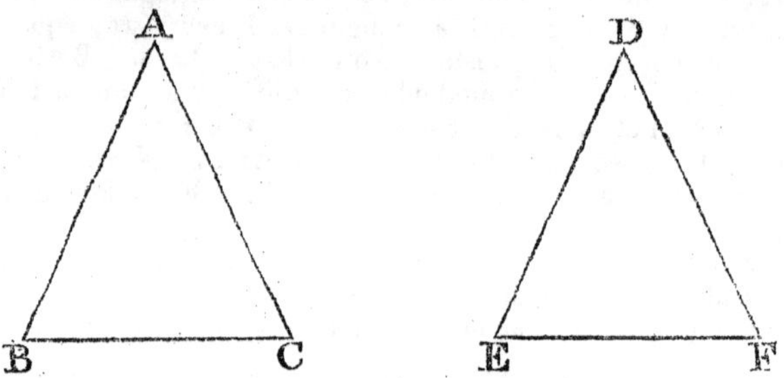

Then because DE is equal to AB, and AB is equal to AC, DE is equal to AC; and for the same reason, DF is equal to AB. And because DF is equal to AB, DE to AC, and the angle FDE to the angle BAC, the angle ABC is equal to the angle DFE, (4. 1.). But the angle ACB is also, by hypothesis, equal to the angle DFE; therefore the angles ABC, ACB are equal to one another. Q. E. D.

Thus also, the 8th proposition may be demonstrated independently of the 7th.

Let ABC, DEF be two triangles, of which the sides AB, AC are equal to the sides DE, DF each to each, and also the base BC to the base EF; the angle BAC is to equal the angle EDF.

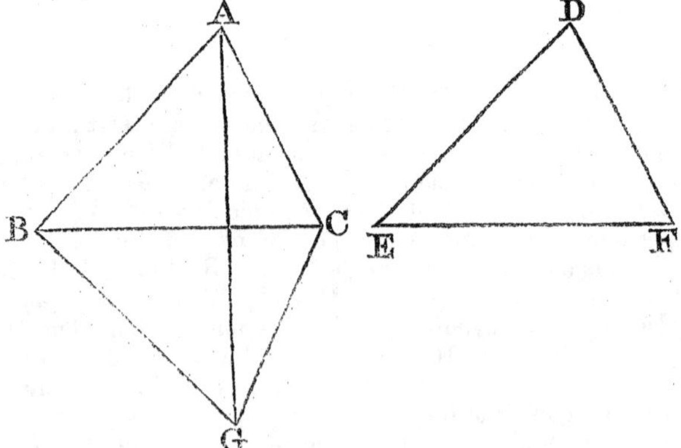

On BC, which is equal to EF, and on the side of it opposite to the triangle ABC, let a triangle BGC be constituted every way equal to the triangle DEF, that is, having GB equal to DE, GC to DF, the angle BGC to the angle EDF, &c.: join AG.

Because GB and AB are each equal, by hypothesis, to DE, AB and GB are equal to one another, and the triangle ABG is isosceles. Wherefore also (5. 1.) the angle BAG is equal to the angle BGA. In the same way, it is shown that AC is equal to GC, and the angle CAG to the angle CGA. Therefore adding equals to equals, the two angles BAG, CAG together are equal to the two angles BGA, CGA together, that is, the whole angle BAC to the whole BGC. But the angle BGC is, by hypothesis, equal to the angle EDF, therefore also the angle BAC is equal to the angle EDF. Q. E. D.

Such demonstrations, it must, however, be acknowledged, trespass against a rule which Euclid has uniformly adhered to throughout the Elements, except where he was forced by necessity to depart from it; This rule is, that nothing is ever supposed to be done, the manner of doing which has not been already taught, so that the construction is derived either directly from the three postulates laid down in the beginning, or from problems already reduced to those postulates. Now, this rule is not essential to geometrical demonstration, where, for the purpose of discovering the properties of figures, we are certainly at liberty to suppose any figure to be constructed, or any line to be drawn, the existence of which does not involve an impossibility. The only use, therefore, of Euclid's rule is to guard against the introduction of impossible hypotheses, or the taking for granted that a thing may exist which in fact implies a contradiction; from such suppositions, false conclusions might, no doubt, be deduced, and the rule is therefore useful in as much as it answers the purpose of excluding them. But the foregoing postulatum could never lead to suppose the actual existence of any thing that is impossible; for it only assumes the existence of a figure equal and similar to one already existing, but in a different part of space from it, or having one of its sides in an assigned position. As there is no impossibility in the existence of one of these figures, it is evident that there can be none in the existence of the other.

PROP. VII.

Dr. Simson has very properly changed the enunciation of this proposition, which, as it stands in the original, is considerably embarrassed and obscure. His enunciation, with very little variation, is retained here.

PROP. XXI.

It is essential to the truth of this proposition, that the straight lines drawn to the point within the triangle be drawn from the two extremities of the base; for, if they be drawn from other points of the base, their sum may exceed the sum of the sides of the triangle in any ratio that is less than that of two to one. This is demonstrated by Pappus

Alexandrinus in the 3d Book of his *Mathematical Collections*, but the demonstration is of a kind that does not belong to this place. If it be required simply to show, that in certain cases the sum of the two lines drawn to the point within the triangle may exceed the sum of the sides of the triangle, the demonstration is easy, and is given nearly as follows by Pappus, and also by Proclus, in the 4th Book of his Commentary on Euclid.

Let ABC be a triangle, having the angle at A a right angle: let D be any point in AB; join CD, then CD will be greater than AC, because in the triangle ACD the angle CAD is greater than the angle ADC. From DC cut off DE equal to AC; bisect CE in F, and join BF; BF and FD are greater than BC and CA.

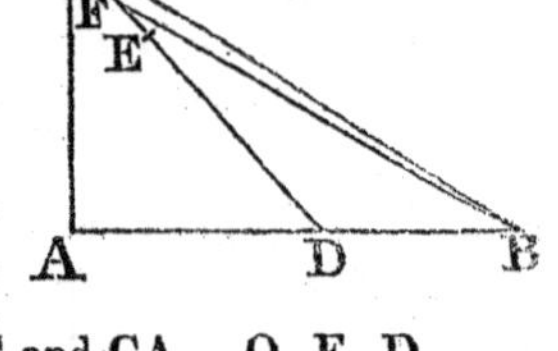

Because CF is equal to FE, CF and FB are equal to EF and FB, but CF and FB are greater than BC, therefore EF and FB are greater than BC. To EF and FB add EB, and to BC add AC, which is equal to ED by construction, and BF and FD will be greater than BC and CA. Q. E. D.

It is evident, that if the angle BAC be obtuse, the same reasoning may be applied.

This proposition is a sufficient vindication of Euclid for having demonstrated the 21st proposition, which some affect to consider as self-evident; for it proves, that the circumstance on which the truth of that proposition depends is not obvious, nor that which at first sight it is supposed to be, viz. that of the one triangle being included within the other. For this reason I cannot agree with M. Clairaut, that Euclid demonstrated this proposition only to avoid the cavils of the Sophists. But I must, at the same time, observe, that what the French Geometer has said on the subject has certainly been misunderstood, and, in one respect, unjustly censured by Dr. Simson. The exact translation of his words is as follows: "If Euclid has taken the trou-"ble to demonstrate, that a triangle included within another has the "sum of its sides less than the sum of the sides of the triangle in "which it is included. we are not to be surprised. That geometer "had to do with those obstinate Sophists, who made a point of refus-"ing their assent to the most evident truths," &c. (Elemens de Geometrie par M. Clairaut. Pref.)

Dr. Simson supposes M. Clairaut to mean, by the proposition which he enunciates here, that when one triangle is included in another, the sum of the two sides of the included triangle is necessarily less than the sum of the two sides of the triangle in which it is included, whether they be on the same base or not. Now this is not only not Euclid's proposition, as Dr. Simson remarks, but it is not true, and is directly contrary to what has just been demonstrated from Proclus. But the fact seems to be, that M. Clairaut's meaning is entirely different, and that he intends to speak not of two of the sides of a triangle, but of all the three; so that his proposition is, "that when one "triangle is included within another, the sum of all the three sides of "the included triangle is less than the sum of all the three sides of the "other," and this is without doubt true, though I think by no means self-evident. It must be acknowledged also, that it is not exactly Euclid's proposition, which, however, it comprehends under it, and is the general theorem, of which the other is only a particular case. Therefore, though M. Clairaut may be blamed for maintaining that to be an Axiom which requires demonstration, yet he is not to be accused of mistaking a false proposition for a true one.

PROP. XXII.

Thomas Simpson in his Elements has objected to Euclid's demonstration of this proposition, because it contains no proof, that the two circles made use of in the construction of the Problem must cut one another; and Dr. Simson, on the other hand, always unwilling to acknowledge the smallest blemish in the works of Euclid, contends, that the demonstration is perfect. The truth, however, certainly is, that the demonstration admits of some improvement; for the limitation that is made in the enunciation of any Problem ought always to be shown to be necessarily connected with the construction of it, and this is what Euclid has neglected to do in the present instance. The defect may easily be supplied, and Dr. Simson himself has done it in effect in his note on this proposition, though he denies it to be necessary.

Because that of the three straight lines DF, FG, GH, any two are greater than the third, by hypothesis, FD is less than FG and GH, that is, than FH, and therefore the circle described from the centre F, with the distance FD must meet the line FE between F and H; and, for

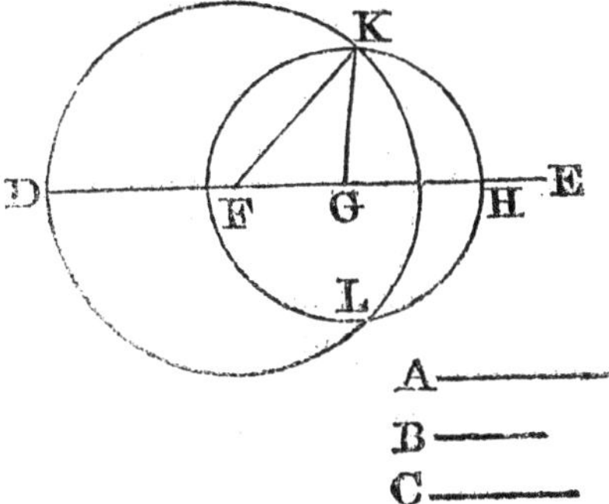

the like reason, the circle described from the centre G at the distance GH, must meet DG between D and G, and therefore, the one of these circles cannot be wholly within the other. Neither can the one be wholly without the other, because DF and GH are greater than FG; the two circles must therefore intersect one another.

PROP. XXVII. and XXVIII.

Euclid has been guilty of a slight inaccuracy in the enunciations of these propositions, by omitting the condition, that the two straight lines on which the third line falls, making the alternate angles, &c. equal, must be in the same plane, without which they cannot be parallel, as is evident from the definition of parallel lines. The only editor, I believe, who has remarked this omission, is M. de Foix Duc de Candalle, in his translation of the Elements published in 1566. How it has escaped the notice of subsequent commentators is not easily explained, unless because they thought it of little importance to correct an error by which nobody was likely to be misled.

PROP. XXIX.

The subject of parallel lines is one of the most difficult in the Elements of Geometry. It has accordingly been treated of in a great variety of different ways, of which, perhaps, there is none that can be said to have given entire satisfaction. The difficulty consists in *converting* the 27th and 28th of Euclid, or in demonstrating, that parallel straight lines, or such as do not meet one another, when they meet a third line, make the alternate angles with it equal, or, which comes to the same, are equally inclined to it, and make the exterior angle equal to the interior and opposite. In order to demonstrate this proposition, Euclid assumed it as an Axiom, that "if a straight line meet two straight "lines, so as to make the interior angles on the same side of it less "than two right angles, these straight lines being continually produc- "ed, will at length meet on the side on which the angles are that are "less than two right angles." This proposition, however, is not self-evident, and ought the less to be received without proof, that, as Proclus has observed, the converse of it is a proposition that confessedly requires to be demonstrated. For the converse of it is, that two straight lines which meet one another make the interior angles, with any third line, less than two right angles; or, in other words, that the two interior angles of any triangle are less than two right angles, which is the 17th of the First Book of the Elements: and it should seem, that a proposition can never rightly be taken for an Axiom, of which the converse requires a demonstration.

The methods by which Geometers have attempted to remove this blemish from the elements are of three kinds. 1. By a new definition of parallel lines. 2. By introducing a new Axiom concerning parallel lines, more obvious than Euclid's. 3. By reasoning merely from the definition of parallels, and the properties of lines already demonstrated, without the assumption of any new Axiom.

1. One of the definitions that has been substituted for Euclid's is, that straight lines are parallel, which preserve always the same distance from one another, by the word distance being understood, a per-

pendicular drawn to one of the lines from any point whatever in the other. If these perpendiculars be every where of the same length, the straight lines are called parallel. This is the definition given by Wolfius, by Boscovich, and by Thomas Simpson, in the first edition of his Elements. It is however a faulty definition, for it conceals an Axiom in it, and takes for granted a property of straight lines, that ought either to be laid down as self-evident, or demonstrated, if possible, as a Theorem. Thus, if from the three points A, B, and C of the straight line AC, perpendiculars AD, BE, CF be drawn all equal to one another, it is implied in the definition, that the points D, E and F are in the same straight line, which, though it be true, it was not the business of the definition to inform us of. Two perpendiculars, as AD and CF, are alone sufficient to determine the position of the straight line DF, and therefore the definition ought to be, "that two straight lines are parallel, when there are "two points in the one, from which the perpendiculars drawn to the "other are equal, and on the same side of it."

This is the definition of parallels which M. D'Alembert seems to prefer to all others; but he acknowledges, and very justly, that it still remains a matter of difficulty to demonstrate, that all the perpendiculars drawn from the one of these lines to the other are equal. (*Encyclopedie, Art. Parallele.*)

Another definition that has been given of parallels is, that they are lines which make equal angles with a third line, toward the same parts, or such as make the exterior angle equal to the interior and opposite. Varignon, Bezout, and several other mathematicians, have adopted this definition, which, it must be acknowledged, is a perfectly good

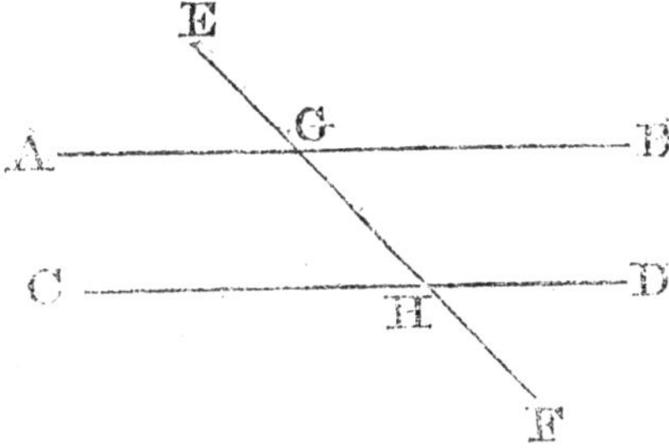

one, if it be understood by it, that the two lines called parallel, are such as make equal angles with a *certain* third line, but not with *any* line that falls upon them. It remains, therefore, to be demonstrated, That if AB and CD make equal angles with GH, they will do so also with any other line whatsoever. The definition, therefore, must be thus understood, That parallel lines are such as make equal angles.

with a *certain* third line, or, more simply, lines which are perpendicular to a given line. It must then be proved, 1. That straight lines which are equally inclined to a *certain* line or perpendicular to a *certain* line, must be equally inclined to all the other lines that fall upon them; and also, 2. That two straight lines which do not meet when produced, must make equal angles with any third line that meets them.

The demonstration of the first of these propositions is not at all facilitated by the new definition, unless it be previously shown, that all the angles of a triangle are equal to two right angles.

The second proposition would hardly be necessary if the new definition were employed; for when it is required to draw a line that shall not meet a given line, this is done by drawing a line that shall have the same inclination to a third line that the first or given line has. It is known that lines so drawn cannot meet. It would no doubt be an advantage to have a definition that is not founded on a condition purely negative.

2. As to the Mathematicians who have rejected Euclid's Axiom, and introduced another in its place, it is not necessary that much should be said. Clavius is one of the first in this class; the Axiom he assumes is, "That a line of which the points are all equidistant from a "certain straight line in the same plane with it, is itself a staight line." This proposition he does not, however, assume altogether, as he gives a kind of metaphysical proof of it, by which he endeavours to connect it with Euclid's definition of a straight line, with which proof at the same time he seems not very well satisfied. His reasoning, after this proposition is granted (though it ought not to be granted as an Axiom), is logical and conclusive, but is prolix and operose, so as to leave a strong suspicion that the road pursued is by no means the shortest possible.

The method pursued by Simson, in his Notes in the First Book of Euclid, is not very different from that of Clavius. He assumes this Axiom, "That a straight line cannot first come nearer to another "straight line, and then go farther from it without meeting it." (Notes, &c. English Edition.) By coming nearer is understood, conformably to a previous definition, the diminution of the perpendiculars drawn from the one line to the other. This Axiom is more readily assented to than that of Clavius, from which, however, it is not very different: but it is not very happily expressed, as the idea not merely of motion, but of time, seems to be involved in the notion of *first* coming nearer, and *then* going farther off. Even if this inaccuracy is passed over, the reasoning of Simson, like that of Clavius, is prolix, and evidently a circuitous method of coming at the truth.

Thomas Simpson, in the second edition of his Elements, has presented this Axiom in a simpler form. "If two points in a straight "line are posited at unequal distances from another straight line "in the same plane, those two lines being indefinitely produced on the "side of the least distance will meet one another."

By help of this Axiom it is easy to prove, that if two straight lines

AB, CD are parallel, the perpendiculars to the one, terminated by the other, are all equal, and are also perpendicular to both the parallels. That they are equal is evident, otherwise the lines would meet by the Axiom. That they are perpendicular to both, is demonstrated thus:

If AC and BD, which are perpendicular to AB, and equal to one another, be not also perpendicular to CD, from C let CE be drawn at right angles to BD. Then, because AB and CE are both perpendicular to BD, they are parallel, and therefore the perpendiculars AC and BE are equal. But AC is equal to BD, (by hypothesis,) therefore BE and BD are equal, which is impossible; BD is therefore at right angles to CD.

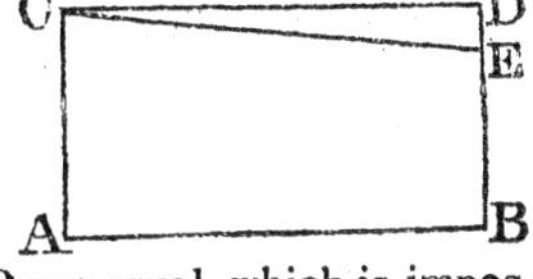

Hence the proposition, that "if a straight line fall on two parallel "lines, it makes the alternate angles equal," is easily derived. Let

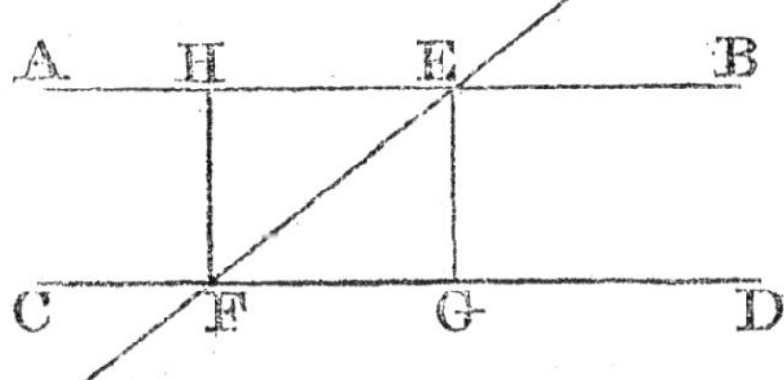

FH and GE be perpendicular to CD, then they will be parallel to one another, and also at right angles to AB, and therefore FG and HE are equal to one another, by the last proposition. Wherefore in the triangles EFG, EFH, the sides HE and EF are equal to the sides GF and FE, each to each, and also the third side HF to the third side EG, therefore the angle HEF is equal to the angle EFG, and they are alternate angles. Q. E. D.

This method of treating the doctrine of parallel lines is extremely plain and concise, and is perhaps as good as any that can be followed, when a new Axiom is assumed. In the text above, I have, however, followed a different method, employing as an Axiom, "That two "straight lines, which cut one another, cannot be both parallel to the "same straight line." This Axiom has been assumed by others, particularly by Ludlam, in his very useful little tract, entitled *Rudiments of Mathematics*.

It is a proposition readily enough admitted as self-evident, and leads to the demonstration of Euclid's 29th Proposition, even with more brevity than Simson's.

3. All the methods above enumerated leave the mind somewhat dissatisfied, as we naturally expect to discover the properties of parallel lines, as we do those of other geometric quantities, by comparing the definition of those lines, with the properties of straight lines already

known. The most ancient writer who appears to have attempted to do this is Ptolemy the astronomer, who wrote a treatise expressly on the subject of Parallel Lines. Proclus has preserved some account of this work in the Fourth Book of his commentaries: and it is curious to observe in it an argument founded on the principle which is known to the moderns by the name of the *sufficient reason.*

To prove, that if two parallel straight lines, AB and CD be cut by a third line EF, in G and H, the two interior angles AGH, CHG will

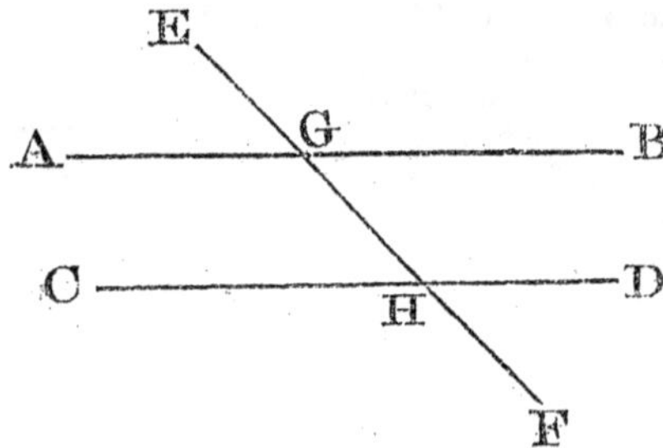

be equal to two right angles, Ptolemy reasons thus: If the angles AGH, CHG be not equal to two right angles, let them, if possible, be greater than two right angles: then, because the lines AG and CH are not more parallel than the lines BG and DH, the angles BGH, DHG are also greater than two right angles. Therefore, the four angles AGH, CHG, BGH, DHG are greater than four right angles; and they are also equal to four right angles, which is absurd. In the same manner it is shown, that the angles AGH, CHG cannot be less than two right angles. Therefore they are equal to two right angles.

But this reasoning is certainly inconclusive. For why are we to suppose that the interior angles which the parallels make with the line cutting them, are either in every case greater than two right angles, or in every case less than two right angles? For any thing that we are yet supposed to know, they may be sometimes greater than two right angles, and sometimes less, and therefore we are not entitled to conclude, because the angles AGH, CHG are greater than two right angles, that therefore the angles BGH, DHG are also necessarily greater than two right angles. It may safely be asserted, therefore, that Ptolemy has not succeeded in his attempt to demonstrate the properties of parallel lines without the assistance of a new Axiom.

Another attempt to demonstrate the same proposition without the assistance of a new Axiom has been made by a modern geometer, Franceschini, Professor of Mathematics in the University of Bologna, in an essay, which he entitles, *La Teoria delle parallele rigorosamente dimonstrata,* printed in his *Opuscoli Mathematici,* at Bassano in 1787.

The difficulty is there reduced to a proposition nearly the same with this, That if BE make an acute angle with BD, and if DE be perpendicular to BD at any point, BE and DE, if produced will meet. To demonstrate this, it is supposed, that BO, BC are two parts taken in BE, of which BC is greater than BO, and that the perpendiculars ON, CL are drawn to BD; then shall BL be greater than BN. For, if not, that is, if the perpendicular CL falls either at N, or between B and N, as at F; in the first of these cases the angle CNB is equal to the angle ONB, because they are both right angles, which is impossible; and, in the second, the two angles CFN, CNF of the triangle CNF, exceed two right angles. Therefore, adds our author, since, as BC increases, BL also increases, and since BC may be increased without limit, so BL may become greater than any given line, and therefore may be greater than BD; wherefore, since the perpendiculars to BD from points beyond D meet BC, the perpendicular from D necessarily meets it. Q. E. D.

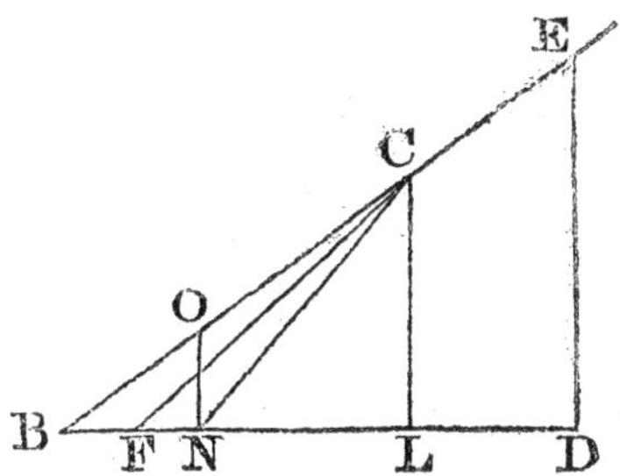

Now it will be found, on examination, that this reasoning is no more conclusive than the preceding. For, unless it be proved, that whatever multiple BC is of BO, the same is BL of BN, the indefinite increase of BC does not necessarily imply the indefinite increase of BL, or that BL may be made to exceed BD. On the contrary, BL may always increase, and yet may do so in such a manner as never to exceed BD: In order that the demonstration should be conclusive, it would be necessary to show, that when BC increases by a part equal to BO, BL increases always by a part equal to BN; but to do this will be found to require the knowledge of those very properties of parallel lines that we are seeking to demonstrate.

Legendre, in his Elements of Geometry, a work entitled to the highest praise, for elegance and accuracy, has delivered the doctrine of parallel lines without any new Axiom. He has done this in two different ways, one in the text, and the other in the notes. In the former he has endeavoured to prove, independently of the doctrine of parallel lines, that all the angles of a triangle are equal to two right angles; from which proposition, when it is once established, it is not difficult to deduce every thing with respect to parallels. But, though his demonstration of the property of triangles just mentioned is quite logical and conclusive, yet it has the fault of being long and indirect, proving first, that the three angles of a triangle cannot be greater than two right angles, next, that they cannot be less, and doing both by reasonings abundantly subtle, and not of a kind readily apprehended by those who are only beginning to study the Mathematics.

The demonstration which he has given in the notes is extremely ingenious, and proceeds on this very simple and undeniable Axiom, that we cannot compare an angle and a line, as to magnitude, or cannot have an equation of any sort between them. This truth is involved in the distinction between homogeneous and heterogeneous quantities, (Euc. v. def. 4.), which has long been received in Geometry, but led only to negative consequences, till it fell into the hands of Legendre. The proposition which he deduces from it is, that if two angles of one triangle be equal to two angles of another, the third angles of these triangles are also equal. For, it is evident, that when two angles of a triangle are given, and also the side between them, the third angle is thereby determined; so that if A and B be any two angles of a triangle, P the side interjacent, and C the third angle, C is determined, as to its magnitude, by A, B and P; and, besides these, there is no other quantity whatever which can affect the magnitude of C. This is pl in, because if A, B and P are given, the triangle can be constructed, all the triangles in which A, B and P are the same, being equal to one another.

But of the quantities by which C is determined, P cannot be one; for if it were, then C must be a *function* of the quantities A, B, P; that is to say, the value of C can be expressed by some combination of the quantities A, B and P. An equation, therefore, may exist between the quantities A, B, C, and P; a d consequently the value of P is equal to some combination, that is, to some function of the quantities A, B and C; but this is impossible, P being a line, and A, B, C being angles; so that no function of the first of these quantities can be equal to any function of the other three. The angle C must therefore be determined by the angles A and B alone, without any regard to the magnitude of P, the side interjacent. Hence in all triangles that have two angles in one equal to two in another, each to each, the third angles are also equal.

Now, this being demonstrated, it is easy to prove that the three angles of any triangle are equal to two right angles.

Let ABC be a triangle right angled at A, draw AD perpendicular to BC. The triangles ABD, ABC have the angles BAC, BDA right angles, and the angle B common to both; therefore by what has just been proved, their third angles BAD, BCA are also equal. In the same way t is shown, that CAD is equal to CBA; therefore the two angles BAD, CAD are equal to the two BCA, CBA; but BAD +CAD is equal to a right angle, therefore the angles BCA, CBA are together equal to a right angle, and consequently the three angles of the right angled triangle ABC are equal to two right angles.

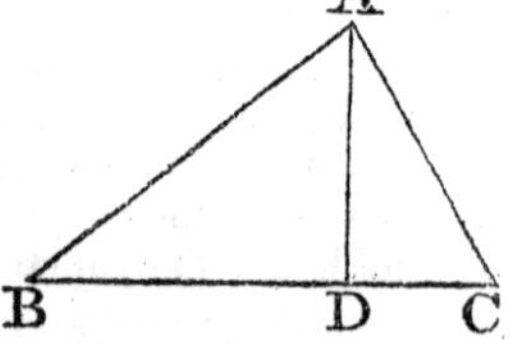

And since it is proved that the oblique angles of every right angled triangle are equal to two right angles, and since every triangle may be divided into two right angled triangles, the four oblique angles of which are equal to the three angles of the triangle, therefore the three angles of every triangle are equal to two right angles. Q. E. D.

Though this method of treating the subject is strictly demonstrative, yet, as the reasoning in the first of the two preceding demonstrations is not perhaps sufficiently simply to be apprehended by those just entering on mathematical studies, I shall submit to the reader another method, not liable to the same objection, which I know, from experience, to be of use in explaining the Elements. It proceeds, like that of the French Geometer, by demonstrating, in the first place, that the angles of any triangle are together equal to two right angles, and deducing from thence, that two lines, which make with a third line the interior angles. less than two right angles, must meet if produced. The reasoning used to demonstrate the first of these propositions may be objected to by some as involving the idea of motion, and the transference of a line from one place to another This, however, is no more than Euclid has done himself on some occasions; and when it furnishes so short a road to the truth as in the present instance, and does not impair the evidence of the conclusion, it seems to be in no respect inconsistent with the utmost rigour of demonstration. It is of importance in explaining the Elements of Science, to connect truths by the shortest chain possible; and till that is done, we can never consider them as being placed in their *natural order*. The reasoning in the first of the following propositions is so simple, that it seems hardly susceptible of abbreviation, and it has the advantage of connecting immediately two truths so much alike, that one might conclude, even from the bare enunciations, that they are but different cases of the same general theorem, viz. That all the angles about a point, and all the exterior angles of any rectilineal figure, are constantly of the same magnitude, and equal to four right angles

DEFINITION.

If, while one extremity of a straight line remains fixed at A, the line itself turns about that point from the position AB to the position AC, it is said to describe the angle BAC contained by the line AB and AC.

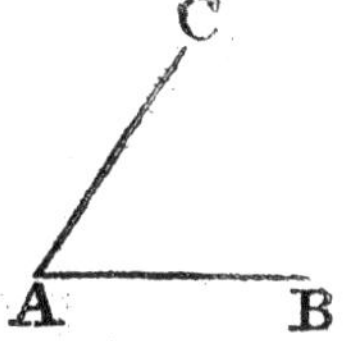

Cor. If a line turn about a point from the position AC till it come into the position AB again, it describes angles which are together equal to four right angles. This is evident from the second Cor. to the 15th.

PROP. I.

All the exterior angles of any rectilineal figure are together equal to four right angles.

1. Let the rectilineal figure be the triangle ABC, of which the exterior angles are DCA, FAB, GBC; these angles are together equal to four right angles.

Let the line CD, placed in the direction of BC produced, turn about the point C till it coincide with CE, a part of the side CA, and have described the exterior angle DCE or DCA. Let it then be carried along the line CA, till it be in the position AF, that is, in the direction of CA produced, and the point A remaining fixed, let it turn about A till it describe the angle FAB, and coincide with a part of the line AB. Let it next be carried along AB till it come into the position BG, and by turning about B, let it describe the angle GBC, so as to coincide with a part of BC. Lastly, Let it be carried along BC till it coincide with CD, its first position. Then, because the line CD has turned about one of its extremities till it has come into the position CD again, it has by the corollary to the above definition described angles which are together equal to four right angles; but the angles which it has described are the three exterior angles of the triangle ABC, therefore the exterior angles of the triangle ABC are equal to four right angles.

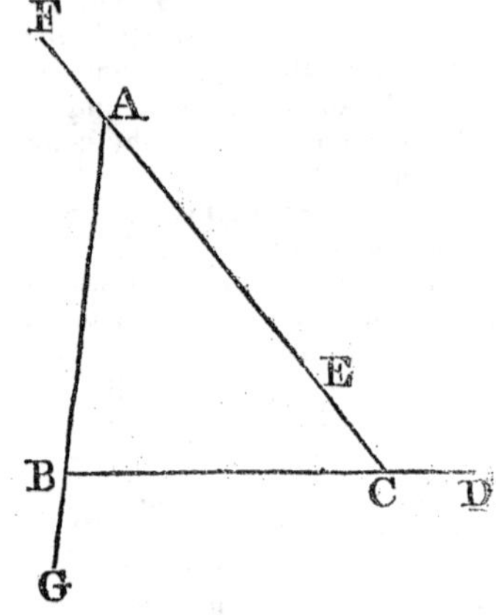

2. If the rectilineal figure have any number of sides, the proposition is demonstrated just as in the case of a triangle. Therefore all the exterior angles of any rectilineal figure are together equal to four right angles. Q. E. D.

Cor. 1. Hence, all the interior angles of any triangle are equal to two right angles. For all the angles of the triangle, both exterior and interior, are equal to six right angles, and the exterior being equal to four right angles, the interior are equal to two right angles.

Cor. 2. An exterior angle of any triangle is equal to the two interior and opposite, or the angle DCA is equal to the angles CAB, ABC. For the angles CAB, ABC, BCA are equal to two right angles; and the angles ACD, ACB are also (13. 1.) equal to two right angles; therefore the three angles CAB, ABC, BCA are equal to the two ACD, ACB; and taking ACB from both, the angle ACD is equal to the two angles CAB, ABC.

Cor. 3. The interior angles of any rectilineal figure are equal to twice as many right angles as the figure has sides, wanting four. For all the angles exterior and interior are equal to twice as many right angles as the figure has sides; but the exterior are equal to four right angles; therefore the interior are equal to twice as many right angles as the figure has sides, wanting four.

PROP. II.

Two straight lines, which make with a third line the interior angles on the same side of it less than two right angles, will meet on that side, if produced far enough.

Let the straight lines AB, CD, make with AC the two angles BAC, DCA less than two right angles; AB and CD will meet if produced toward B and D.

In AB take AF=AC; join CF, produce BA to H, and through C draw CE, making the angle ACE equal to the angle CAH.

Because AC is equal to AF, the angles AFC, ACF are also equal (5. 1.); but the exterior angle HAC is equal to the two interior and opposite angles ACF, AFC, and therefore it is double of either of them, as of ACF. Now ACE is equal to HAC by construction, therefore ACE is double of ACF, and is bisected by the line CF. In the same manner, if FG be taken equal to FC, and if CG be drawn, it may be shown that CG bisects the angle ACE, and so on continually. But if from a magnitude, as the angle ACE, there be taken its half, and from the remainder FCE its half FCG, and from the remainder GCE its half, &c. a remainder will at length be found less than the given angle DCE.*

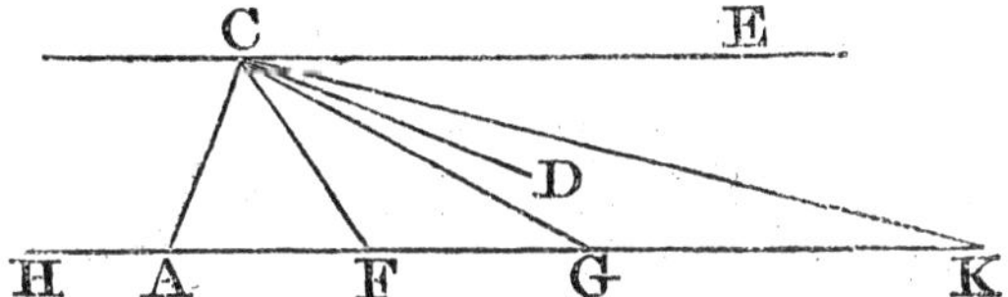

Let GCE be the angle, whose half ECK is less than DCE, then a straight line CK is found, which falls between CD and CE, but nevertheless meets the line AB in K. Therefore CD, if produced, must meet AB in a point between G and K. Therefore, &c. Q. E. D.

This demonstration is indirect; but this proposition, if the definition of parallels were changed, as suggested at p. 302, would not be necessary; and the proof, that lines equally inclined to any one line must be so to every line, would follow directly from the angles of a triangle being equal to two right angles. The doctrine of parallel lines would in this manner be freed from all difficulty.

PROP. III. 29. 1. Euclid.

If a straight line fall on two parallel straight lines, it makes the alternate angles equal to one another; the exterior equal to the interior

---

* Prop. 1. 1. Sup. The reference to this proposition involves nothing inconsistent with good reasoning, as the demonstration of it does not depend on any thing that has gone before, so that it may be introduced in any part of the Elements.

and opposite on the same side; and likewise the two interior angles, on the same side equal to two right angles.

Let the straight line EF fall on the parallel straight lines AB, CD; the alternate angles AGH, GHD are equal, the exterior angle EGB is equal to the interior and opposite GHD; and the two interior angles BGH, GHD are equal to two right angles.

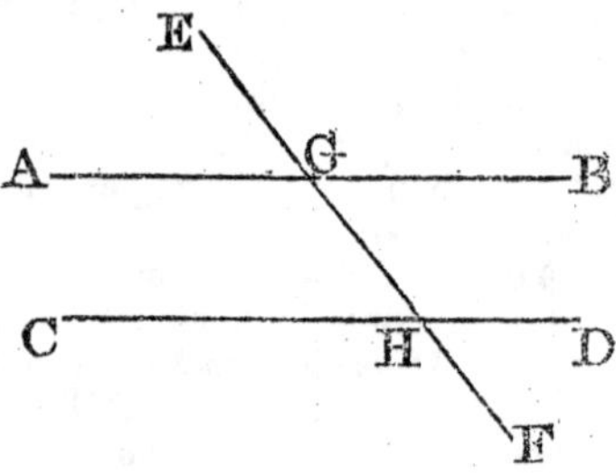

For if AGH be not equal to GHD, let it be greater, then adding BGH to both, the angles AGH, HGB are greater than the angles DHG, HGB. But AGH, HGB are equal to two right angles, (13.), therefore BGH, GHD are less than two right angles, and therefore the lines AB, CD will meet, by the last proposition, if produced toward B and D. But they do not meet, for they are parallel by hypothesis, and therefore the angles AGH, GHD are not unequal, that is, they are equal to one another.

Now the angle AGH is equal to EGB, because these are vertical, and it has also been shown to be equal to GHD, therefore EGB and GHD are equal. Lastly, to each of the equal angles EGB, GHD add the angle BGH, then the two EGB, BGH are equal to the two DHG, BGH. But EGB, BGH are equal to two right angles, (13. 1.), therefore BGH, GHD are also equal to two right angles. Therefore, &c. Q. E. D.

The following proposition is placed here, because it is more connected with the First Book than with any other. It is useful for explaining the nature of Hadley's sextant; and, though involved in the explanations usually given of that instrument, it has not, I believe, been hitherto considered as a distinct Geometric Proposition, though very well entitled to be so on account of it simplicity and elegance, as well as its utility.

THEOREM.

If an exterior angle of a triangle be bisected, and also one of the interior and opposite, the angle contained by the bisecting lines is equal to half the other interior and opposite angle of the triangle.

Let the exterior angle ACD of the triangle ABC be bisected by the straight line CE, and the interior and opposite ABC by the straight line BE, the angle BEC is equal to half the angle BAC.

The lines CE, BE will meet; for since the angle ACD is greater than ABC, the half of ACD is greater than the half of ABC, that is, ECD is greater than EBC; add ECB to both and, the two angles ECD, ECB are greater than EBC, ECB. But ECD, ECB are equal to two right angles; therefore ECB, EBC are less than two right angles, and therefore the lines CE, BE must meet on the same side of BC on which the triangle ABC is. Let them meet in E.

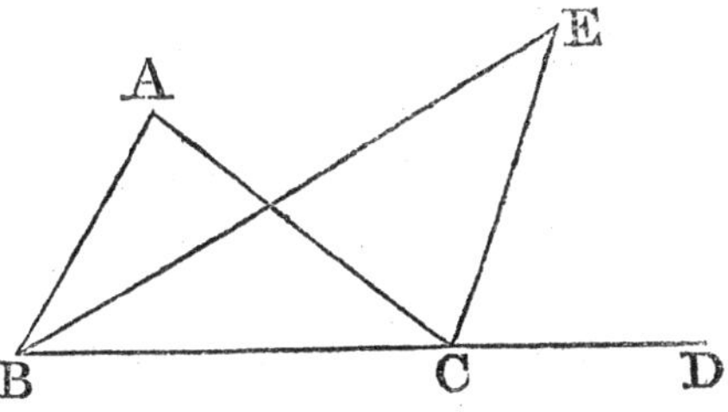

Because DCE is the exterior angle of the triangle BCE, it is equal to the two angles CBE, BEC, and therefore twice the angle DCE, that is, the angle DCA is equal to twice the angles CBE, and BEC. But twice the angles CBE is equal to the angle ABC, therefore the angle DAC is equal to the angle ABC, together with twice the angle BEC; and the same angle DCA being the exterior angle of the triangle ABC, is equal to the two angles ABC, CAB, wherefore the two angles ABC, CAB are equal to ABC and twice BEC. Therefore, taking away ABC from both, there remains the angle CAB equal to twice the angle BEC, or BEC equal to the half of BAC. Therefore, &c. Q. E. D.

BOOK II.

The Demonstrations of this Book are no otherwise changed than by introducing into them some characters similar to those of Algebra, which is always of great use where the reasoning turns on the addition or subtraction of rectangles. To Euclid's demonstrations, others are sometimes added, serving to deduce the propositions from the fourth, witnout the assistance of a diagram.

PROP. A and B.

These Theorems are added on account of their great use in geometry, and their close connection with the other propositions which are the subject of this Book. Prop. A is an extension of the 9th and 19th.

BOOK III.

DEFINITIONS.

The definition which Euclid makes the first of this Book is that of equal circles, which he defines to be "those of which the diameters "are equal." This is rejected from among the definitions, as being a Theorem, the truth of which is proved by supposing the circles applied to one another, so that their centres may coincide, for the whole of the one must then coincide with the whole of the other. The converse, viz. That circles which are equal have equal diameters, is proved in the same way.

The definition of the angle of a segment is also omitted, because it does not relate to a rectilineal angle, but to one understood to be contained between a straight line and a portion of the circumference of a circle. In like manner, no notice is taken in the 16th proposition of the angle comprehended between the semicircle and the diameter, which is said by Euclid to be greater than an acute rectilineal angle. The reason for these omissions has already been assigned in the notes on the fifth definition of the first Book.

PROP. XX.

It has been remarked of this demonstration, that it takes for granted, that if two magnitudes be double of two others, each of each, the sum or difference of the first two is double of the sum or difference of the other two, which are two cases of the 1st and 5th of the 5th Book. The justness of this remark cannot be denied; and though the cases of the Propositions here referred to are the simplest of any, yet the truth of them ought not in strictness to be assumed without proof. The proof is easily given. Let A and B, C and D be four magnitudes, such that $A=2C$, and $B=2D$; then $A+B=2(C+D)$. For since $A=C+C$, and $B=D+D$, adding equals to equals, $A+B=(C+D)+(C+D)=2(C+D)$. So also, if A be greater than B, and therefore C greater than D, since $A=C+C$, and $B=D+D$, taking equals from equals $A-B=(C-D)+(C-D)$, that is, $A-B=2(C-D)$.

BOOK V.

The subject of proportion has been treated so differently by those who have written on elementary geometry, and the method which Euclid has followed has been so often, and so inconsiderately censured, that in these notes it will not perhaps be more necessary to account for the changes that I have made, than for those that I have not made. The changes are but few, and relate to the language, not to the essence of the demonstrations; they will be explained after some of the definitions have been particularly considered.

DEF. III.

The definition of ratio given here has been greatly extolled by some authors; but whatever value it may have in the eyes of a metaphysician, it has but little in those of the geometer, because nothing concerning the properties of ratios, can be deduced from it. Dr. Barrow has very judiciously remarked concerning it, "that Euclid had probably "no other design in making this definition, than to give a general sum-"mary idea of ratio to beginners, by premising this metaphysical defi-"nition to the more accurate definitions of ratios that are equal to one "another, or one of which is greater or less than the other; I call "it a metaphysical, for it is not properly a mathematical definition, "since nothing in mathematics depends on it, or is deduced, nor, as I "judge, can be deduced, from it." (Barrow's Lectures, Lect. 3.). Dr. Simson thinks the definition has been added by some unskilful editor; but there is no ground for that supposition, other than what arises from the definition being of no use. We may, however, well enough imagine, that a certain idea of order and method induced Euclid to give some general definition of ratio before he used the term in the definition of equal ratios.

DEF. IV.

This definition is a little altered in the expression; Euclid has it, that "magnitudes are said to have a ratio to one another, when the "less can be multiplied so as to exceed the greater."

DEF. V.

One of the chief obstacles to the ready understanding of the 5th Book of Euclid, is the difficulty that most people find of reconciling the idea of proportion which they have already acquired, with the account of it that is given in this definition. Our first ideas of proportion, or of proportionality, are got by trying to compare together the magnitude of external bodies; and though they be at first abundantly vague and incorrect, they are usually rendered tolerably precise by the study of arithmetic; from which we learn to call four numbers proportionals, when they are such that the quotient which arises from dividing the first by the second, (according to the common rule for division), is the same with the quotient that arises from dividing the third by the fourth.

Now, as the operation of arithmetical division is applicable as readily to any two magnitudes of the same kind, as to two numbers, the notion of proportion thus obtained may be considered as perfectly general. For, in arithmetic, after finding how often the divisor is contained in the dividend, we multiply the remainder by 10, or 100, or 1000, or any power, as it is called, of 10, and proceed to inquire how oft the divisor is contained in this new dividend; and, if there be any remainder, we go on to multiply it by 10, 100, &c. as before, and to divide the product by the original divisor, and so on, the division sometimes terminating when no remainder is left, and sometimes going on *ad infinitum*, in consequence of a remainder being left at each operation. Now, this process may easily be imitated with any two magnitudes A and B, providing they be of the same kind, or such that the one can be multiplied so as to exceed the other. For, suppose that B is the least of the two ; take B out of A as oft as it can be found, and let the quotient be noted, and also the remainder, if there be any ; multiply this remainder by 10, or 100, &c. so as to exceed B, and let B be taken out of the quantity produced by this multiplication as oft as it can be found ; let the quotient be noted, and also the remainder, if there be any. Proceed with this remainder as before, and so on continually ; and it is evident, that we have an operation that is applicable to all magnitudes whatsoever, and that may be performed with respect to any two lines, any two plane figures, or any two solids, &c.

Now, when we have two magnitudes and two others, and find that the first divided by the second, according to this method, gives the very same series of quotients that the third does when divided by the fourth, we say of these magnitudes, as we did of the numbers above described, that the first is to the second as the third to the fourth. There are only two more circumstances necessary to be considered, in order to bring us precisely to Euclid's definition.

First, It is known from arithmetic, that the multiplication of the successive remainders each of them by 10, is equivalent to multiplying the quantity to be divided by the product of all those tens ; so that multiplying, for instance, the first remainder by 10, the second by 10, and the third by 10, is the same thing, with respect to the quotient, as if the quantity to be divided had been at first multiplied by 1000 ; and therefore, our standard of the proportionality of numbers may be expressed thus : If the first multiplied any number of times by 10, and then divided by the second, gives the same quotient as when the third is multiplied as often by 10, and then divided by the fourth, the four magnitudes are proportionals.

Again, it is evident, that there is no necessity in these multiplications for confining ourselves to 10, or the powers of 10. and that we do so, in arithmetic, only for the conveniency of the decimal notation ; we may therefore use any multipliers whatsoever, providing we use the same in both cases. Hence, we have this definition of proportionals, When there are four magnitudes, and any multiple whatsoever of the first, when divided by the second, gives the same quotient with the like multiple of the third, when divided by the fourth, the four magnitudes are proportionals, or the first has the same ratio to the second that the third has to the fourth.

We are now arrived very nearly at Euclid's definition ; for, let A, B, C, D be four proportionals, according to the definition just given, and m any number ; and let the multiple of A by m, that is mA, be divided by B ; and first, let the quotient be the number n exactly, then also, when mC is divided by D, the quotient will be n exactly. But, when mA divided by B gives n for the quotient, $m\text{A}=n\text{B}$ by the nature of division, so that when $m\text{A}=n\text{B}$, $m\text{C}=n\text{D}$, which is one of the conditions of Euclid's definition.

Again, when mA is divided by B, let the division not be exactly performed, but let n be a whole number less than the exact quotient, then $nB \angle mA$, or $mA \urcorner nB$; and, for the same reason, $mC \urcorner nD$, which is another of the conditions of Euclid's definition.

Lastly, when mA is divided by B, let n be a whole number greater than the exact quotient, then $mA \angle nB$, and because n is also greater than the quotient of mC divided by D, (which is the same with the other quotient), therefore $mC \angle nD$.

Therefore, uniting all these three conditions, we call A, B, C, D, proportionals, when they are such, that if $mA \urcorner nB$, $mC \urcorner nD$; if $mA = nB$, $mC = nD$; and if $mA \angle nB$, $mC \angle nD$, m and n being any numbers whatsoever Now, this is exactly the criterion of proportionality established by Euclid in the 5th definition, and is derived here by generalizing the common and most famliar idea of proportion.

It appears from this, that the condition of mA containing B, whether with or without a remainder, as often as mC contains D, with or without a remainder, and of this being the case whatever value be assigned to the number m, includes in it all the three conditions that are mentioned in Euclid's definition; and hence, that definition may be expressed a little more simply by saying, that *four magnitudes are proportionals, when any multiple of the first contains the second, (with or without remainder,) as oft as the same multiple of the third contains the fourth.* But, though this definition is certainly, in the expression, more simple than Euclid's, it is not, as will be found on trial, so easily applied to the purpose of demonstration. The three conditions which Euclid brings together in his definition, though they somewhat embarrass the expression of it, have the advantage of rendering the demonstrations more simple than they would otherwise be, by avoiding all discussion about the magnitude of the remainder left, after B is taken out of mA as oft as it can be found. All the attempts, indeed, that have been made to demonstrate the properties of proportionals rigorously, by means of other definitions than Euclid's, only serve to evince the excellence of the method followed by the Greek Geometer, and his singular address in the application of it.

The great objection to the other methods is, that if they are meant to be rigorous, they require two demonstrations to every proposition, one when the division of mA into parts equal to B can be exactly performed, the other when it cannot be exactly performed whatever value be assigned to m, or when A and B are what is called incommensurable; and this last case will generally be found to require an indirect demonstration, or a *reductio ad absurdum.*

M. D'Alembert, speaking of the doctrine of proportion, in a discourse that contains many excellent observations, but in which he has overlooked Euclid's manner of treating this subject entirely, has the following remark: " On ne peut démontrer que de cette manière, " (la réduction à absurde,) la plupart des propositions qui regardent " les incommensurables. L'idée de l'infini entre au moins implicite- " mens dans la notion de ces sortes de quantités; et comme nous n'a-

"vous qu'une idée negative de l'infini, on ne peut démontrer directé "ment, et *a priori*, tout ce qui concerne l'infini mathématique. (*Encyclopédie, mot Géométrie.*)

This remark sets in a strong and just light the difficulty of demonstrating the propositions that regard the proportion of incommensurable magnitudes, without having recourse to the *reductio ad absurdum*; but it is surprising, that M. D'Alembert, a geometer no less learned than profound, should have neglected to make mention of Euclid's method, the only one in which the difficulty he states is completely overcome. It is overcome by the introduction of the idea of indefinitude, (if I may be permitted to use the word), instead of the idea of infinity; for m and n, the multipliers employed, are supposed to be indefinite, or to admit of all possible values, and it is by the skilful use of this condition that the necessity of indirect demonstrations is avoided. In the whole of geometry, I know not that any happier invention is to be found; and it is worth remarking, that Euclid appears in another of his works to have availed himself of the idea of indefinitude with the same success, viz. in his books of Porisms, which have been restored by Dr. Simson, and in which the whole analysis turned on that idea, as I have shown at length in the Third Volume of the Transactions of the Royal Society of Edinburgh. The investigations of these propositions were founded entirely on the principle of certain magnitudes admitting of innumerable values; and the methods of reasoning concerning them seem to have been extremely similar to those employed in the fifth of the Elements. It is curious to remark this analogy between the different works of the same author; and to consider, that the skill, in the conduct of this very refined and ingenious artifice, acquired in treating the properties of proportionals, may have enabled Euclid to succeed so well in treating the still more difficult subject of Porisms.

Viewing in this light Euclid's manner of treating proportion, I had no desire to change any thing in the principle of his demonstrations. I have only sought to improve the language of them, by introducing a concise mode of expression, of the same nature with that which we use in arithmetic, and in algebra. Ordinary language conveys the ideas of the different operations supposed to be performed in these demonstrations so slowly, and breaks them down into so many parts, that they make not a sufficient impression on the understanding. This indeed will generally happen when the things treated of are not represented to the senses by Diagrams, as they cannot be when we reason concerning magnitude in general, as in this part of the Elements. Here we ought certainly to adopt the language of arithmetic or algebra, which, by its shortness, and the rapidity with which it places objects before us, makes up in the best manner possible for being merely a conventional language, and using symbols that have no resemblance to the things expressed by them. Such a language, therefore, I have endeavoured to introduce here; and I am convinced, that if it shall be found an improvement, it is the only one of which the fifth of Euclid will admit. In other respects I have followed Dr. Simson's edition, to the accuracy of which it would be difficult to make any addition.

In one thing I must observe, that the doctrine of proportion, as laid down here, is meant to be more general than in Euclid's Elements. It is intended to include the properties of proportional numbers as well as of all magnitudes. Euclid has not this design, for he has given a definition of proportional numbers in the seventh Book, very different from that of proportional magnitudes in the fifth; and it is not easy to justify the logic of this manner of proceeding; for we can never speak of two numbers and two magnitudes both having the same ratios, unless the word ratio have in both cases the same signification. All the propositions about proportionals here given are therefore understood to be applicable to numbers; and accordingly, in the eighth Book, the proposition that proves equiangular parallelograms to be in a ratio compounded of the ratios of the numbers proportional to their sides, is demonstrated by help of the propositions of the fifth Book.

On account of this, the word *quantity*, rather than *magnitude*, ought in strictness to have been used in the enunciation of these propositions, because we employ the word Quantity to denote not only things extended, to which alone we give the name of Magnitude, but also numbers. It will be sufficient, however, to remark, that all the propositions respecting the ratios of magnitudes relate equally to all things of which multiples can be taken, that is, to all that is usually expressed by the word Quantity in its most extended signification, taking care always to observe, that ratio takes place only among like quantities. (See Def. 4.)

DEF. X.

The definition of *compound ratio* was first given accurately by Dr. Simson; for, though Euclid used the term, he did so without defining it. I have placed this definition before those of *duplicate* and *triplicate ratio*, as it is in fact more general, and as the relation of all the three definitions is best seen when they are ranged in this order. It is then plain, that two equal ratios compound a ratio duplicate of either of them; three equal ratios, a ratio triplicate of either of them, &c.

It was justly observed by Dr. Simson, that the expression, *compound ratio*, is introduced merely to prevent circumlocution, and for the sake principally of enunciating those propositions with conciseness that are demonstrated by reasoning *ex æquo*, that is, by reasoning from the 22d or 23d of this Book. This will be evident to any one who considers carefully the Prop. F. of this, or the 23d of the 6th Book.

An objection which naturally occurs to the use of the term *compound ratio*, arises from its not being evident how the ratios described in the definition determine in any way the ratio which they are said to compound, since the magnitudes compounding them are assumed at pleasure. It may be of use for removing this difficulty, to state the matter as follows: if there be any number of ratios (among magnitudes of the same kind) such that the consequent of any of them is the antecedent of that which immediately follows, the first of the antecedents has to the last of the consequents a ratio which evidently depends on the intermediate ratios, because if they are determined, it is determined also ; and this dependence of one ratio on all the other ratios, is expressed by saying that it is compounded of them. Thus, if $\frac{A}{B}, \frac{B}{C}, \frac{C}{D}, \frac{D}{E}$ be any series of ratios, such as described above, the ratio $\frac{A}{E}$, or of A to E, is said to be compounded of the ratios $\frac{A}{B}, \frac{B}{C}$, &c. The ratio $\frac{A}{E}$ is evidently determined by the ratios $\frac{A}{B}, \frac{B}{C}$, &c. because if each of the latter is fixed and invariable, the former cannot change. The exact nature of this dependence, and how the one thing is determined by the other, it is not the business of the definition to explain, but merely to give a name to a relation which it may be of importance to consider more attentively.

BOOK VI.

DEFINITION II.

This definition is changed from that of *reciprocal figures*, which was of no use, to one that corresponds to the language used in the 14th and 15th propositions, and in other parts of geometry.

PROP. XXVII, XXVIII, XXIX.

As considerable liberty has been taken with these propositions, it is necessary that the reasons for doing so should be explained. In the first place, when the enunciations are translated literally from the Greek, they sound very harshly, and are, in fact, extremely obscure. The phrase of applying to a straight line, a parallelogram deficient, or exceeding by another parallelogram, is so elliptical, and so little analogous to ordinary language, that there could be no doubt of the propriety of at least changing the enunciations.

It next occurred, that the Problems themselves in the 28th and 29th propositions are proposed in a more general form than is necessary in an elementary work, and that, therefore, to take those cases of them that are the most useful, as they happen to be the most simple, must be the best way of accommodating them to the capacity of the learner. The problem which Euclid proposes in the 28th is, "To a given "straight line to apply a parallelogram equal to a given rectilineal "figure, and deficient by a parallelogram similar to a given parallelo-"gram;" which may be more intelligibly enunciated thus: "To cut "a given line, so that the parallelogram which has in it a given angle, "and is contained under one of the segments of the given line, and a "straight line which has a given ratio to the other segment, may be "equal to a given space;" instead of which problem I have substituted this other; "To divide a given straight line so that the rectangle "under its segments may be equal to a given space." In the actual solution of problems, the greater generality of the former proposition is an advantage more apparent than real, and is fully compensated by the simplicity of the latter, to which it is always easily reducible.

The same may be said of the 29th, which Euclid enunciates thus: "To a given straight line to apply a parallelogram equal to a given "rectilineal figure exceeding by a parallelogram similar to a given pa-"rallelogram." This might be proposed otherwise: "To produce a "given line, so that the parallelogram having in it a given angle, and "contained by the whole line produced, and a straight line that has a "given ratio to the part produced, may be equal to a given rectilineal "figure." Instead of this, is given the following problem, more simple, and, as was observed in the former instance, very little less general. "To produce a given straight line, so that the rectangle contain-"ed by the segments, between the extremities of the given line, and "the point to which it is produced, may be equal to a given space."

PROP. A, B, C, &c.

Nine propositions are added to this Book on account of their utility and their connection with this part of the Elements. The first four of them are in Dr. Simson's edition, and among these Prop. A is given immediately after the third, being, in fact, a second case of that proposition, and capable of being included with it, in one enunciation. Prop. D. is remarkable for being a theorem of Ptolemy the Astronomer, in his Μεγαλη Συνταξις, and the foundation of the construction of his trigonometrical tables. Prop. E is the simplest case of the former; it is also useful in trigonometry, and, under another form, was the 97th, or, in some editions, the 94th of Euclid's Data. The propositions F and G are very useful properties of the circle, and are taken from the *Loci Plani* of Apollonius. Prop. H is a very remarkable property of the triangle; and K is a proposition which, though it has been hitherto considered as belonging particularly to trigonometry, is so often of use in other parts of the mathematics, that it may be properly ranked among the elementary theorems of Geometry.

SUPPLEMENT.

BOOK I.

PROP. V. and VI, &c.

THE demonstrations of the 5th and 6th propositions require the method of exhaustions, that is to say, they prove a certain property to belong to the circle, because it belongs to the rectilineal figures inscribed in it, or described about it according to a certain law, in the case when those figures approach to the circles so nearly as not to fall short of it, or to exceed it by any assignable difference. This principle is general, and is the only one by which we can possibly compare curvilineal with rectilineal spaces, or the length of curve lines with the length of straight lines, whether we follow the methods of the ancient or of the modern geometers. It is therefore a great injustice to the latter methods to represent them as standing on a foundation less secure than the former: they stand in reality on the same, and the only difference is, that the application of the principle, common to them both, is more general and expeditious in the one case than in the other. This identity of principle, and affinity of the methods used in the elementary and the higher mathematics, it seems the more necessary to observe, that some learned mathematicians have appeared not to be sufficiently aware of it, and have even endeavoured to demonstrate the contrary. An instance of this is to be met with in the preface of the valuable edition of the works of Archimedes, lately printed at Oxford. In that preface, Torelli, the learned commentator, whose labours have done so much to elucidate the writings of the Greek Geometer, but who is so unwilling to acknowledge the merit of the modern analysis, undertakes to prove, that it is impossible, from the relation which the rectilineal figures inscribed in, and circumscribed about, a given curve, have to one another, to conclude any thing concerning the properties of the curvilineal space itself, except in certain circumstances which he has not precisely described. With this view be attempts to show, that if we are to reason from the relation which certain rectilineal figures belonging to the circle have to one another, notwithstanding that those figures may approach so near to the circular spaces within which they are inscribed, as not to differ from them by any assignable magnitude, we shall be led into error, and shall seem to prove, that the circle is to the square of its diameter exactly as 3 to 4. Now, as this is a conclusion which the discoveries of Archimedes himself prove so clearly to be false, Torelli argues, that the principle from which it is deduced must be false also; and in this he would no doubt be right, if his former conclusion had been fairly drawn. But the truth is, that a very gross paralogism is to be found in that part of his reasoning, where he makes a transition from the ratios of the small rectangles, inscribed in the circular spaces, to the ratios of the sums of those rectangles, or of the whole rectilineal figures. In doing this, he takes for granted a proposition, which, it is wonderful, that one who had studied geometry in the school of Archimedes, should for a moment have supposed to be true. The proposition is this: If A, B, C, D, E, F, be any number of magnitudes, and a, b, c, d, e, f, as many others; and if

A : B : : a : b,

C : D : : c : d,

E : F : : e : f, then the sum of A, C and E will be to the sum of B, D and F, as the sum of a, c and e, to the sum of b, d and f, or A + C + E : B + D + F : : a+c+e : b+d+f. Now, this proposition, which Torelli supposes to be perfectly general, is not true, except in two cases, viz. either first, when A : C : : a : c, and

A : E : : a : e; and consequently,

B : D : : b : d, and

B : F : : b : f; or, secondly, when all the ratios of A to B, C to D, E to F, &c. are equal to one another. To demonstrate this, let us suppose that there are four magnitudes, and four others,

thus A : B : : a : b, and

C : D : : c : d, then we cannot have A+C : B + D : : a+c : b+d, unless either, A : C : : a : c, and B : D : : b : d; or A : C : : b : d, and consequently a : b : : c : d.

Take a magnitude K, such that a : c : : A : K, and another L, such that b : d : : B : L; and suppose it true, that A+C : B+D : : a+c : b+d. Then, because by inversion; K : A : : c : a, and, by hypothesis, A : B : : a : b, and also B : L : : b : d, ex æquo, K : L : : c : d; and consequently, K : L : : C : D.

K,	A,	B,	L,
c,	a,	b,	d.

Again, because A : K : : a : c, by addition,

A+K : K : : a+c : c; and, for the same reason,

B+L : L : : b+d : d, or, by inversion,

L : B+L : : d : b+d. And, since it has been shown, that K : L : : c : d; therefore ex æquo,

A+K,	K,	L,	B+L,
a+c,	c,	d,	b+d.

A+K : B+L : : a+c : b+d; but by hypothesis,

A+C : B+D : : a+c : b+d, therefore

A+K : A+C : : B+L : B+D.

Now, first, let K and C be supposed equal, then it is evident, that L and D are also equal; and therefore, since by construction $a : c :: A : K$, we have also $a : c :: A : C$; and, for the same reason, $b : d :: B : D$, and these analogies from the first of the two conditions, of which one is affirmed above to be always essential to the truth of Torelli's proposition.

Next, if K be greater than C, then since

$A+K : A+C :: B+L : B+D$, by division,
$A+K : K-C :: B+L : L-D$. But, as was shown
$K : L :: C : D$, by conversion and alternation,
$K-C : K :: L-D : L$, therefore, ex æquo,
$A+K : K :: B+L : L$, and lastly, by division,
$A : K :: B : L$, or $A : B :: K : L$, that is,
$A : B :: C : D$.

Wherefore, in this case the ratio of A to B is equal to that of C to D, and consequently, the ratio of a to b equal to that of c to d. The same may be shown, if K is less than C; therefore in every case there are conditions necessary to the truth of Torelli's proposition, which he does not take into account, and which, as is easily shown, do not belong to the magnitudes to which he applies it.

In consequence of this, the conclusion which he meant to establish respecting the circle, falls entirely to the ground, and with it the general inference aimed against the modern analysis.

It will not, I hope, be imagined, that I have taken notice of these circumstances with any design to lessen the reputation of the learned Italian, who has in so many respects deserved well of the mathematical sciences, or to detract from the value of a posthumous work, which by its elegance and correctness, does so much honour to the English editors. But I would warn the student against that narrow spirit which seeks to insinuate itself even into the abstractions of geometry, and would persuade us, that elegance, nay truth itself, are possessed exclusively by the ancient methods of demonstration. The high tone in which Torelli censures the modern mathematics is imposing, as it is assumed by one who had studied the writings of Archimedes with uncommon diligence. His errors are on that account the more dangerous, and require to be the more carefully pointed out.

PROP. IX.

This enunciation is the same with that of the third of the *Dimensio Circuli* of Archimedes; but the demonstration is different, though it proceeds, like that of the Greek Geometer, by the continual bisection of the 6th part of the circumference.

The limits of the circumference are thus assigned; and the method of bringing it about, notwithstanding many quantities are neglected in the arithmetical operations, that the errors shall in one case be all on the side of defect, and in another all on the side of excess, (in which I have followed Archimedes,) deserves particularly to be observed, as affording a good introduction to the general methods of approximation.

BOOK II.

DEF. VIII. and PROP. XX.

SOLID angles, which are defined here in the same manner as in Euclid, are magnitudes of a very peculiar kind, and are particularly to be remarked for not admitting of that accurate comparison, one with another, which is common in the other subjects of geometrical investigation. It cannot, for example, be said of one solid angle, that it is the half, or the double of another solid angle; nor did any geometer ever think of proposing the problem of bisecting a given solid angle. In a word, no multiple or sub-multiple of such an angle can be taken, and we have no way of expounding, even in the simplest cases, the ratio which one of them bears to another.

In this respect, therefore, a solid angle differs from every other magnitude that is the subject of mathematical reasoning, all of which have this common property, that multiples and sub-multiples of them may be found. It is not our business here to inquire into the reason of this anomaly, but it is plain, that on account of it, our knowledge of the nature and the properties of such angles can never be very far extended, and that our reasonings concerning them must be chiefly confined to the relations of the plane angles, by which they are contained. One of the most remarkable of those relations is that which is demonstrated in the 21st of this Book, and which is, that all the plane angles which contain any solid angle must together be less than four right angles. This proposition is the 21st of the 11th of Euclid.

This proposition, however, is subject to a restriction in certain cases, which, I believe, was first observed by M. le Sage of Geneva, in a communication to the Academy of Sciences of Paris in 1756. When the section of the pyramid formed by the planes that contain the solid angle is a figure that has none of its angles exterior, such as a triangle, a parallelogram, &c. the truth of the proposition just enunciated cannot be questioned. But, when the aforesaid section is a figure like that which is annexed, viz. ABCD, having some angles such as BDC, exterior, or, as they are sometimes called, re-entering angles, the proposition is not necessarily true; and it is plain, that in such cases the demonstration which we have given, and which is the same with Euclid's, will no longer apply. Indeed, it were easy to show, that on bases of this kind, by multiplying the number of sides, solid angles may be formed, such that the plane

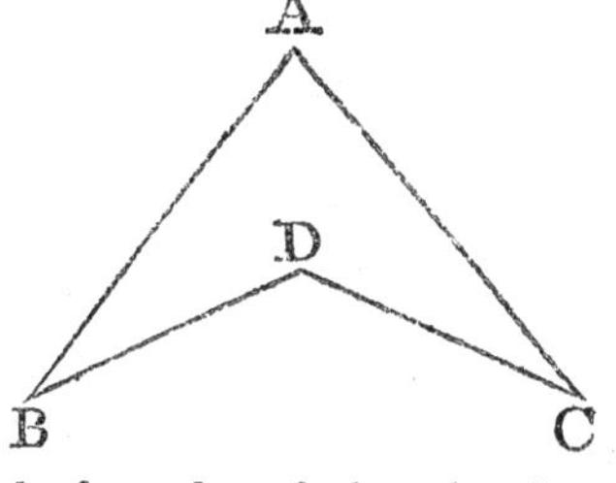

angles which contain them shall exceed four right angles by any quantity assigned. An illustration of this from the properties of the sphere is perhaps the simplest of all others. Suppose that on the surface of a hemisphere there is described a figure bounded by any number of arches of great circles making angles with one another, on opposite sides alternately, the plane angles at the centre of the sphere that stand on these arches may evidently exceed four right angles, and that too, by multiplying and extending the arches in any assigned ratio. Now, these plane angles contain a solid angle at the centre of the sphere, according to the definition of a solid angle.

We are to understand the proposition in the text, therefore, to be true only of those solid angles in which the inclination of the plane angles are all the same way, or all directed toward the interior of the figure. To distinguish this class of solid angles from that to which the proposition does not apply, it is perhaps best to make use of this criterion, that they are such, that when any two points whatsoever are taken in the planes that contain the solid angle, the straight line, joining those points, falls wholly within the solid angle: or thus, they are such, that a straight line cannot meet the planes which contain them in more than two points. It is thus, too, that I would distinguish a plane figure that has none of its angles exterior, by saying, that it is a rectilineal figure, such that a straight line cannot meet the boundary of it in more than two points.

We, therefore, distinguish solid angles into two species; one in which the bounding planes can be intersected by a straight line only in two points; and another where the bounding planes may be intersected by a straight line in more than two points: to the first of these the proposition in the text applies, to the second it does not.

Whether Euclid meant entirely to exclude the consideration of figures of the latter kind, in all that he has said of solids, and of solid angles, it is not now easy to determine: it is certain, that his definitions involve no such exclusion; and as the introduction of any limitation would considerably embarrass these definitions, and render them difficult to be understood by a beginner, I have left it out, reserving to this place a fuller explanation of the difficulty. I cannot conclude this note without remarking, with the historian of the Academy, that it is extremely singular, that not one of all those who had read or explained Euclid before M. le Sage, appears to have been sensible of this mistake. (*Memoires de l' Acad. des Sciences* 1756, *Hist. p.* 77.) A circumstance that renders this still more singular is, that another mistake of Euclid on the same subject, and perhaps of all other geometers, escaped M. le Sage also, and was first discovered by Dr. Simson, as will presently appear.

PROP. IV.

This very elegant demonstration is from Legendre, and is much easier than that of Euclid.

The demonstration given here of the 6th is also greatly simpler than that of Euclid. It has even an advantage that does not belong to Legendre's, that of requiring no particular construction or determination of any one of the lines, but reasoning from properties common to every part of them. This simplification, when it can be introduced, which, however, does not appear to be always possible, is, perhaps, the greatest improvement that can be made on an elementary demonstration.

PROP. XIX.

The problem contained in this proposition, of drawing a straight line perpendicular to two straight lines not in the same plane, is certainly to be accounted elementary, although not given in any book of elementary geometry that I know of before that of Legendre. The solution given here is more simple than his, or than any other that I have yet met with : it also leads more easily, if it be required, to a trigonometrical computation.

BOOK III.

DEF. II. and PROP. I.

These relate to similar and equal solids, a subject on which mistakes have prevailed not unlike to that which has just been mentioned. The equality of solids, it is natural to expect, must be proved like the equality of plane figures, by showing that they may be made to coincide, or to occupy the same space. But, though it be true that all solids which can be shown to coincide are equal and similar, yet it does not hold conversely, that all solids which are equal and similar can be made to coincide. Though this assertion may appear somewhat paradoxical, yet the proof of it is extremely simple.

Let ABC be an isosceles triangle, of which the equal sides are AB and AC ; from A draw AE perpendicular to the base BC, and BC will be bisected in E. From E draw ED perpendicular to the plane ABC, and from D, any point in it, draw DA, DB, DC to the three angles of the triangle ABC. The pyramid DABC is divided into two pyramids DABE, DACE, which, though their equality will not be disputed, cannot be so applied to one another as to coincide. For, though the triangles ABE, ACE are equal, BE being equal to CE, EA common to both, and the angles AEB, AEC equal, because they are right angles, yet if these

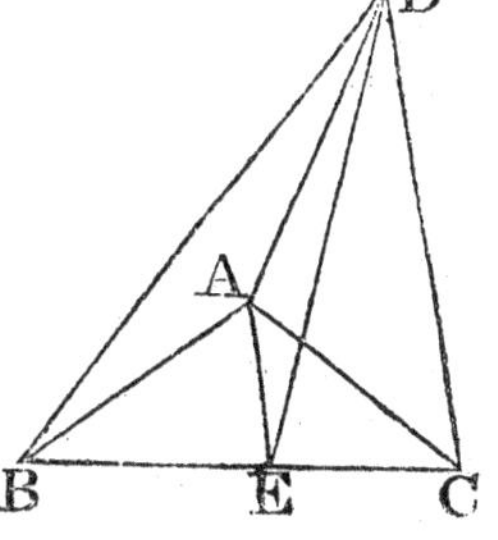

two triangles be applied to one another, so as to coincide, the solid DACE will nevertheless, as is evident, fall without the solid DABE, for the two solids will be on the opposite sides of the plane ABE. In the same way, though all the planes of the pyramid DABE may easily be shown to be equal to those of the pyramid DACE, each to each; yet will the pyramids themselves never coincide, though the equal planes be applied to one another, because they are on the opposite sides of those planes.

It may be said, then, on what ground do we conclude the pyramids to be equal? The answer is, because their construction is entirely the same, and the conditions that determine the magnitude of the one identical with those that determine the magnitude of the other. For the magnitude of the pyramid DABE is determined by the magnitude of the triangle ABE, the length of the line ED, and the position of ED, in respect of the plane ABE; three circumstances that are precisely the same in the two pyramids, so that there is nothing that can determine one of them to be greater than another.

This reasoning appears perfectly conclusive and satisfactory; and it seems also very certain, that there is no other principle equally simple, on which the relation of the solids DABE, DACE to one another can be determined. Neither is this a case that occurs rarely; it is one, that in the comparison of magnitudes having three dimensions, presents itself continually; for, though two plane figures that are equal and similar can always be made to coincide, yet, with regard to solids that are equal and similar, if they have not a certain similarity in their position, there will be found just as many cases in which they cannot, as in which they can coincide. Even figures described on surfaces, if they are not plane surfaces, may be equal and similar without the possibility of coinciding. Thus, in the figure described on the surface of a sphere, called a spherical triangle, if we suppose it to be isosceles, and a perpendicular to be drawn from the vertex on the base, it will not be doubted, that it is thus divided into two right angled spherical triangles equal and similar to one another, and which, nevertheless, cannot be so laid on one another as to agree. The same holds in innumerable other instances, and therefore it is evident, that a principle, more general and fundamental than that of the equality of coinciding figures, ought to be introduced into Geometry. What this principle is has also appeared very clearly in the course of these remarks; and it is indeed no other than the principle so celebrated in the philosophy of Leibnitz, under the name of THE SUFFICIENT REASON. For it was shown, that the pyramids DABE and DACE are concluded to be equal, because each of them is determined to be of a certain magnitude, rather than of any other, by conditions that are the same in both, so that there is no REASON for the one being greater than the other. This Axiom may be rendered general by saying, That things of which the magnitude is determined by conditions that are exactly the same, are equal to one another; or, it might be expressed thus; Two magnitudes A and B are equal, when there is no reason that A should exceed B, rather than that B should exceed A. Either of these will serve as the fundamental principle for comparing geometrical magnitudes of every kind; they will apply in those cases where the coincidence of magnitudes with one another has no place; and they will apply with great readiness to the cases in which a coincidence may take place, such as in the 4th, the 8th, or the 26th of the First Book of the Elements.

The only objection to this Axiom is, that it is somewhat of a metaphysical kind, and belongs to the doctrine of the *sufficient reason*, which is looked on with a suspicious eye by some philosophers. But this is no solid objection; for such reasoning may be applied with the greatest safety to those objects with the nature of which we are perfectly acquainted, and of which we have complete definitions, as in pure mathematics. In physical questions, the same principle cannot be applied with equal safety, because in such cases we have seldom a complete definition of the thing we reason about, or one that includes all its properties. Thus, when Archimedes proved the spherical figure of the earth, by reasoning on a principle of this sort, he was led to a false conclusion, because he knew nothing of the rotation of the earth on its axis, which places the particles of that body, though at equal distances from the centre, in circumstances very different from one another. But, concerning those things that are the creatures of the mind altogether, like the objects of mathematical investigation, there can be no danger of being misled by the principle of the sufficient reason, which at the same time furnishes us with the only single Axiom, by help of which we can compare together geometrical quantities, whether they be of one, of two, or of three dimensions.

Legendre in his Elements has made the same remark that has been just stated, that there are solids and other Geometric Magnitudes, which, though similar and equal, cannot be brought to coincide with one another, and he has distinguished them by the name of *Symmetrical* Magnitudes. He has also given a very satisfactory and ingenious demonstration of the equality of certain solids of that sort, though not so concise as the nature of a simple and elementary truth would seem to require, and consequently not such as to render the axiom proposed above altogether unnecessary.

But a circumstance for which I cannot very well account is, that Legendre, and after him Lacroix, ascribe to Simson the first mention of such solids as we are here considering. Now I must be permitted to say, that no remark to this purpose is to be found in any of the writings of Simson, which have come to my knowledge. He has indeed made an observation concerning the Geometry of Solids, which was both new and important, viz. that solids may have the condition which Euclid thought sufficient to determine their quality, and may nevertheless be unequal; whereas the observation made here is, that solids may be equal and similar, and may yet want the condition of being able to coincide with one another. These propositions are widely different; and how so accurate a writer as Legendre should have mistaken the one for the other, is not easy to be explained. It must be observed, that he does not seem in the least aware of the observation which Simson has really made. Perhaps having himself made the remark we now speak of, and on looking slightly into Simson, having found a limitation of the usual description of equal solids, he had without much inquiry, set it down as the same with his own notion; and so, with a great deal of candour, and some precipitation, he has ascribed to Simson a discovery which really belonged to himself. This at least seems to be the most probable solution of the difficulty.

I have entered into a fuller discussion of Legendre's mistake than I should otherwise have done, from having said in the first edition of these elements, in 1795, that I believed the non-coincidence of similar and equal solids in certain circumstances, was then made for the first time. This it is evident would have been a pretension as ridiculous as ill-founded, if the same observation had been made in a book like Simson's, which in this country was in every body's hands, and which I had myself professedly studied with attention. As I have not seen any edition of Legendre's Elements earlier than that published in 1802, I am ignorant whether he or I was the first in making the remark here referred to. That circumstance is, however, immaterial; for I am not interested about the originality of the remark, though very much interested to show that I had no intention of appropriating to myself a discovery made by another.

Another observation on the subject of those solids, which, with Legendre we shall call Symmetrical, has occurred to me, which I did not at first think of, viz. that Euclid himself certainly had these solids in view when he formed his definition (as he very improperly calls it) of equal and similar solids. He says that those solids *are equal and similar*, which are contained under the same number of equal and similar planes. But this is not true, as Dr. Simson has shown in a passage just about to be quoted, because two solids may easily be assigned, bounded by the same number of equal and similar planes, which are obviously unequal, the one being contained within the other. Simson observes, that Euclid needed only to have added, that the equal and similar planes must be similarly situated, to have made his description exact. Now, it is true, that this addition would have made it exact in one respect, but would have rendered it imperfect in another; for though all the solids having the conditions here enumerated, are equal and similar, many others are equal and similar which have not those conditions, that is, though bounded by the same equal number of similar planes, those planes are not similarly situated. The symmetrical solids have not their equal and similar planes similarly situated, but in an order and position directly contrary. Euclid, it is probable, was aware of this, and by seeking to render the description of equal and similar solids so general, as to comprehend solids of both kinds, has stript it of an essential condition, so that solids obviously unequal are included in it, and has also been led into a very illogical proceeding, that of defining the equality of solids, instead of proving it, as if he had been at liberty to fix a new idea to the word *equal* every time that he applied it to a new kind of magnitude. The nature of the difficulty he had to contend with, will perhaps be the more readily admitted as an apology for this error, when it is considered that Simson, who had studied the matter so carefully, as to set Euclid right in one particular was himself wrong in another, and has treated of equal and similar solids, so as to exclude the symmetrical altogether, to which indeed he seems not to have at all adverted.

I must, therefore, again repeat, that I do not think that this matter can be treated in a way quite simple and elementary, and at the same time general, without introducing the principle of the *sufficient reason* as stated above. It may then be demonstrated, that similar and equal solids are those contained by the same number of equal and similar planes, either with similar or contrary situations. If the word *contrary* is properly understood, this description seems to be quite general.

Simson's remark, that solids may be unequal, though contained by the same number of equal and similar planes, extends also to solid angles which may be unequal, though contained by the same number of equal plane angles These remarks he published in the first edition of his Euclid in 1756, the very same year that M. le Sage communicated to the Academy of Sciences the observation on the subject of solid angles, mentioned in a former note; and, it is singular, that these two geometers, without any communication with one another, should almost at the same time have made two discoveries very nearly connected, yet neither of them comprehending the whole truth, so that each is imperfect without the other.

Dr. Simson has shown the truth of his remark, by the following reasoning.

"Let there be any plane rectilineal figure as the triangle ABC, and from a point D within it, draw the straight line DE at right angles to the plane ABC; in DE take DE, DF equal to one another, upon the opposite sides of the plane, and let G be any point in EF; join DA, DB, DC; EA, EB, EC; FA, FB, FC; GA, GB, GC: Because the straight line EDF is at right angles to the plane ABC, it makes right angles with DA, DB, DC, which it meets in that plane; and in the triangles EDB, FDB, ED and DB are equal to FD and DB, each to each, and they contain right angles: therefore the base EB is equal to the base FB; in the same manner EA is equal to FA, and EC to FC: and in the triangles EBA, FBA, EB, BA are equal to FB, BA, and the base EA is equal to the base FA; wherefore the angle EBA is equal to the angle FBA, and the triangle EBA equal to the triangle FBA, and the other angles equal to the other angles; therefore these triangles are similar: In the same manner the triangle EBC is similar to the triangle FBC, and the triangle EAC to FAC; therefore there are two solid figures, each of which is contained by six triangles, one of

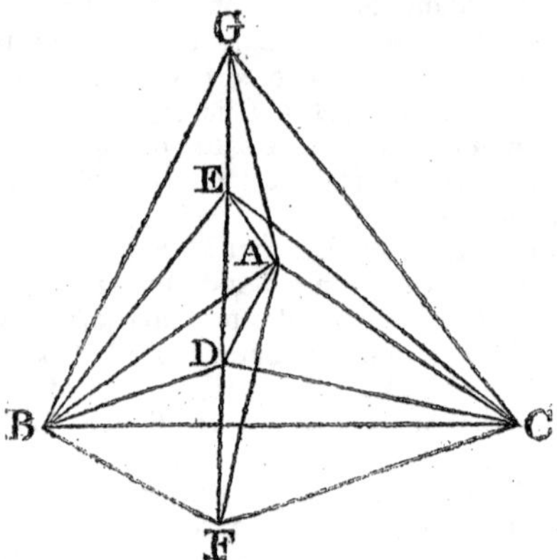

them by three triangles, the common vertex of which is the point G, and their bases the straight lines AB, BC, CA and by three other triangles the common vertex of which is the point E, and their bases the same lines AB, BC, CA. The other solid is contained by the same three triangles, the common vertex of which is G, and their bases AB, BC, CA; and by three other triangles, of which the common vertex is the point F, and their bases the same straight lines AB, BC, CA: Now, the three triangles GAB, GBC, GCA are common to both solids, and the three others EAB, EBC, ECA, of the first solid have been shown to be equal and similar to the three others FAB, FBC, FCA of the other solid, each to each; therefore, these two solids are contained by the same number of equal and similar planes: But that they are not equal is manifest, because the first of them is contained in the other: Therefore it is not universally true, that solids are equal which are contained by the same number of equal and similar planes."

COR. From this it appears, that two unequal solid angles may be contained by the same number of equal plane angles."

"For the solid angle at B, which is contained by the four plane angles EBA, EBC, GBA, GBC is not equal to the solid angle at the same point B, which is contained by the four plane angles FBA, FBC, GBA, GBC; for the last contains the other: And each of them is contained by four plane angles, which are equal to one another, each to each, or are the self-same, as has been proved: And indeed, there may be innumerable solid angles all unequal to one another, which are each of them contained by plane angles that are equal to one another, each to each: It is likewise manifest, that the before-mentioned solids are not similar, since their solid angles are not all equal."

PLANE TRIGONOMETRY.

DEFINITIONS, &c.

TRIGONOMETRY is defined in the text to be the application of Number to express the relations of the sides and angles of triangles. It depends, therefore, on the 47th of the first of Euclid, and on the 7th of the first of the Supplement, the two propositions which do most immediately connect together the sciences of Arithmetic and Geometry.

The sine of an angle is defined above in the usual way, viz. the perpendicular drawn from one extremity of the arch, which measures the angle on the radius passing through the other; but in strictness the sine is not the perpendicular itself, but the ratio of that perpendicular to the radius, for it is this ratio which remains constant, while the angle continues the same, though the radius vary. It might be convenient, therefore, to define the sine to be the quotient which arises from dividing the perpendicular just described by the radius of the circle.

So also, if one of the sides of a right angled triangle about the right angle be divided by the other, the quotient is the tangent of the angle opposite to the first-mentioned side, &c. But though this is certainly the rigorous way of conceiving the sines, tangents, &c. of angles, which are in reality not magnitudes, but the ratios of magnitudes; yet as this idea is a little more abstract than the common one, and would also involve some change in the language of trigonometry, at the same time that it would in the end lead to nothing that is not attained by making the radius equal to unity, I have adhered to the common method, though I have thought it right to point out that which should in strictness be pursued.

A proposition is left out in the Plane Trigonometry, which the astronomers make use of in order, when two sides of a triangle, and the angle contained by them, are given, to find the angles at the base, without making use of the sum or difference of the sides, which, in some cases, when only the Logarithms of the sides are given, cannot be conveniently found.

THEOREM.

If, as the greater of any two sides of a triangle to the less, so the radius to the tangent of a certain angle; then will the radius be to the tangent of the difference between that angle and half a right angle, as the tangent of half the sum of the angles, at the base of the triangle to the tangent of half their difference.

Let ABC be a triangle, the sides of which are BC and CA, and the base AB, and let BC be greater than CA. Let DC be drawn at right angles to BC, and equal to AC; join BD, and because (Prop. 1.) in the right angled triangle BCD, BC : CD :: R : tan CBD, CBD is the angle of which the tangent is to the radius as CD to BC, that is, as CA to BC, or as the least of the two sides of the triangle to the greatest.

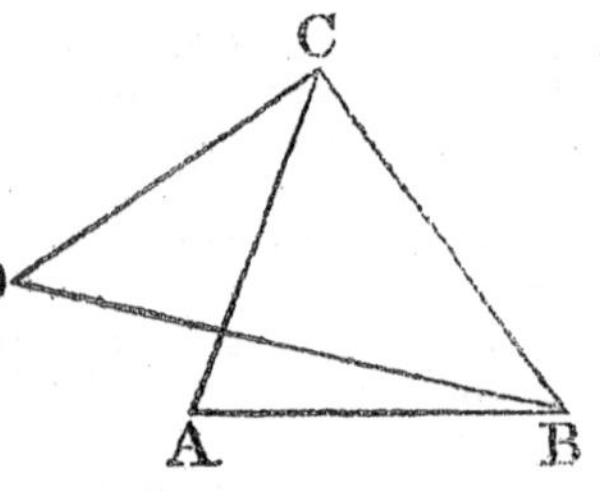

But BC+CD : BC−CD :: tan $\frac{1}{2}$ (CDB+CBD) : tan $\frac{1}{2}$ (CDB−CBD) (Prop. 5.) ;
and also, BC+CA : BC−CA :: tan $\frac{1}{2}$ (CAB+CBA) : tan $\frac{1}{2}$ (CAB−CBA). Therefore, since CD=CA,
tan $\frac{1}{2}$ (CDB+CBD) : tan $\frac{1}{2}$ (CDB—CBD) ::
tan $\frac{1}{2}$ (CAB+CBA) : tan $\frac{1}{2}$ (CAB−CBA). But because the angles CDB+CBD=90°, tan $\frac{1}{2}$ (CDB+CBD :
tan $\frac{1}{2}$ (CDB−CBD :: R : tan (45°−CBD), (2 Cor. Prop. 3.) ;
therefore. R : tan (45°−CBD) :: tan $\frac{1}{2}$ (CAB+CBA) :
tan $\frac{1}{2}$ (CAB−CBA); and CBD was already shown to be such an angle that BC : CA :: R : tan CBD. Therefore, &c. Q. E. D.

COR. If BC, CA, and the angle C are given to find the angles A and B; find an angle E such, that BC : CA :: R : tan E; then R : tan (45°−E) :: tan $\frac{1}{2}$ (A+B) : tan $\frac{1}{2}$ (A−B). Thus $\frac{1}{2}$ (A−B) is found, and $\frac{1}{2}$ (A+B) being given, A and B are each of them known, Lem. 2.

In reading the elements of Plane Trigonometry, it may be of use to observe, that the first five propositions contain all the rules absolutely necessary for solving the different cases of plane triangles. The learner, when he studies Trigonometry for the first time, may satisfy himself with these propositions, but should by no means neglect the others in a subsequent perusal.

PROP. VII. and VIII.

I have changed the demonstration which I gave of these propositions in the first edition, for two others considerably simpler and more concise, given me by Mr. JARDINE, teacher of the Mathematics in Edinburgh, formerly one of my pupils, to whose ingenuity and skill I am very glad to bear this public testimony.

SPHERICAL

TRIGONOMETRY.

PROP. V.

THE angles at the base of an isosceles spherical triangle are *symmetrical* magnitudes, not admitting of being laid on one another, nor of coinciding, notwithstanding their equality. It might be considered as a sufficient proof that they are equal, to observe that they are each determined to be of a certain magnitude rather than any other, by conditions which are precisely the same, so that there is no reason why one of them should be greater than another. For the sake of those to whom this reasoning may not prove satisfactory, the demonstration in the text is given, which is strictly geometrical.

For the demonstrations of the two propositions that are given in the end of the Appendix to the Spherical Trigonometry, see Elementa Sphæricorum, Theor. 66. apud Wolfii Opera Math. tom. iii.; Trigonometrie par Cagnoli, § 463; Trigonometrie Spherique par Mauduit, § 165.

FINIS.

www.ingramcontent.com/pod-product-compliance
Lightning Source LLC
LaVergne TN
LVHW010209110826
845151LV00002B/646

9781425534578